FRACTURE OF NON-METALLIC MATERIALS

ON MATERIALS, ENGINEERING AND MECHANICAL SCIENCE

A series devoted to the publication of courses and educational seminars given at the Joint Research Centre, Ispra Establishment, as part of its education and training program. Published for the Commission of the European Communities, Directorate-General Telecommunications, Information Industry and Innovation.

The publisher will accept continuation orders for this series which may be cancelled at any time and which provide for automatic billing and shipping of each title in the series upon publication. Please write for details.

FRACTURE OF NON-METALLIC MATERIALS

Proceeding of the 5th Advanced Seminar on Fracture Mechanics,
Joint Research Centre, Ispra, Italy,
14–18 October 1985
in collaboration with the
EUROPEAN GROUP ON FRACTURE

Edited by

KLAUS P. HERRMANN

Laboratorium für Technische Mechanik,
University of Paderborn, Paderborn, F.R.G.

and

LARS HANNES LARSSON

Commission of the European Communities,
Joint Research Centre, Ispra Establishment,
Ispra, Italy

D. REIDEL PUBLISHING COMPANY

A MEMBER OF THE KLUWER ACADEMIC PUBLISHERS GROUP

DORDRECHT / BOSTON / LANCASTER / TOKYO

Library of Congress Cataloging in Publication Data

Advanced Seminar on Fracture Mechanics (5th : 1985 : Joint Research Centre, Ispra, Italy)
   Fracture of non-metallic materials.

   (Ispra courses on materials, engineering and mechanical science)
   Includes index.
   1. Fracture   mechanics—Congresses.   2. Nonmetallic   materials—Fracture—Congresses.   I.   Herrmann, Klaus P.   II.   Larsson, Lars Hannes.   III.   European Group on Fracture.   IV.   Title.   V.   Series.
   TA409.A38   1985   620.1'126   86–29710
   ISBN-13: 978-94-010-8621-9   e-ISBN-13: 978-94-009-4784-9
   DOI: 10.1007/978-94-009-4784-9

Commission of the European Communities   Joint Research Centre Ispra (Varese), Italy

Publication arrangements by
Commission of the European Communities
Directorate-General Telecommunications, Information Industry and Innovation, Luxembourg

EUR 9834
© 1987 ECSC, EEC, EAEC, Brussels and Luxembourg
Softcover reprint of the hardcover 1st edition 1987

LEGAL NOTICE
Neither the Commission of the European Communities nor any person acting on behalf of the Commission is responsible for the use which might be made of the following information.

Published by D. Reidel Publishing Company
P.O. Box 17, 3300 AA Dordrecht, Holland

Sold and distributed in the U.S.A. and Canada
by Kluwer Academic Publishers,
101 Philip Drive, Assinippi Park, Norwell, MA 02061, U.S.A.

In all other countries, sold and distributed
by Kluwer Academic Publishers Group,
P.O. Box 322, 3300 AH Dordrecht, Holland

CONTENTS

FOREWORD

This book contains the lectures of the 5th Advanced Seminar on Fracture Mechanics (ASFM 5) held at the Joint Research Centre, Ispra, on 14-18 October 1985. The series of the ASFMs is one of the two main regularly scheduled international events sponsored by the European Group on Fracture (EGF), alternating with the European Conferences of Fracture (ECFs). Whereas ECFs are held in a different place on even years (the last, ECF6, was in Amsterdam in June 1986), ASFMs are hosted at the JRC-Ispra on odd years. This establishment belonging to the Commission of the European Communities performs research work of common interest to the EC Member countries. One of the activities of the JRC-Ispra is the organization of Ispra-Courses aiming at the transfer of knowledge and the strengthening of exchanges and ties between European scientific workers.

ASFMs are designed to give an advanced level treatment in selected areas of fracture mechanics. Previous ASFMs had been devoted to elastic-plastic fracture mechanics and to subcritical crack growth due to fatigue, stress corrosion and creep. In the early stages of preparation of ASFM5 it was decided to concentrate on a new theme, the fracture phenomena and fracture mechanics of non-metallic materials.

Whereas fracture mechanics started with the study of glass by Griffith, its later developments centered predominantly on metallic alloys. However, in recent years non-metallic materials have found increasing uses and correspondingly efforts have been made to develop testing and prediction methods for these materials.

As for the previous ASFMs, the EGF nominated for ASFM5 an Advisory Board with the following members :

Prof. P. Boch	Ecole Nationale Supérieure de la Céramique Industrielle, Limoges, France
Dr. E. Sommer	Fraunhofer Institut für Werkstoffmechanik Freiburg, FRG
Prof. J.G. Williams	Imperial College of Science and Technology London, U.K.
Prof. F.H. Wittmann	Ecole Polytechnique Fédérale de Lausanne Switzerland

The Advisory Board played an essential role by drafting and optimizing the programme and choosing the lecturers.

The main theme of the seminar is the fracture of non-metallic materials possessing a heterogeneous microscopical structure and containing material interfaces or inherent material defects like microcracks or voids. Thereby several important classes of materials: ceramics and glasses, composites and polymers as well as concrete are considered as compound systems with respect to their general characteristics and the associated fracture problems which they pose under mechanical and/or thermal loading. Because of the heterogeneous structure governing the mechanical behaviour of the above mentioned non-metallic materials the application of fracture mechanics developed for metals on those materials has to be performed very carefully. Therefore, one has to distinguish between two different approaches, namely classical fracture mechanics applied to an effective material on a macromechanical level and the approach to be used on a micromechanical level.

The lectures presented at ASFM5 have been revised and updated to constitute the chapters of this book. The four first lectures form an introductory part in which concepts of general interest for the rest of the book are discussed. Sommer explains the basic findings concerning brittle fracture processes. The two following lectures by Bui, and Bui & Stolz, deal with new approaches to fracture mechanics which study the damage of the material. Schmitt discusses the possibilities of numerical methods in fracture mechanics. Some of the examples presented in the three last lectures of the introductory part concern metals. These are well suited to illustrate the methodologies of ductile fracture (e.g. the R-curve approach) which are beginning to be applied to some non-metallic materials.

Ceramics are treated in the four following lectures. Because of the low fracture toughness of ceramics the critical microcracks in ceramic articles are very small: two orders of magnitude smaller than in metallic parts. Therefore, the prevention and detection of defects is more difficult in ceramics than in metals. Further, microcracking can appear due to expansion mismatch at grain boundaries as well as due to phase transitions. Because of the strength scattering of ceramics the usual description is in terms of Weibull's statistics which relates the probability of rupture to ultimate strength. Thereby one of the main aims of ceramists is the increase of Weibull's modulus for ceramic materials. The important phenomenon of static fatigue is a chemically activated phenomenon of slow crack growth. The latter is investigated via crack velocity versus stress intensity factor curves. A probabilistic treatment of the lifetime of ceramics can be given by the Strength-Probability-Time (S.P.T.) diagram which links data from

Weibull's diagram and from v = f(K) curves. Proof testing is a technique which guarantees a minimum lifetime for the survivors of a batch of specimens. In this section on ceramics Boch first presents the main definitions of ceramic materials and processes and reviews the brittle behaviour of these materials. Davidge shows how fracture mechanics and damage concepts can be used to predict the engineering performance of ceramics. Thereafter Boch discusses the behaviour of ceramics in situations where components are exposed to thermal shocks. Finally, the lecture by Claussen concerns the new and fast developing technology of toughening of ceramic materials by utilizing the phase transformation of ZrO_2 particles.

The first lecture in this book, by Sommer, already used examples on glass to illustrate the features of brittle fracture and introduced the phenomenon of slow crack growth at $K < K_{IC}$. An understanding of the mechanisms controlling subcritical crack growth allows the prediction of lifetimes of stressed structures and an improvement of the mechanical properties of glasses. The lecture of Fantozzi and Orange discusses in detail the behaviour of cracks in glasses and presents the corresponding testing techniques. Herrman's lecture concerns studies of thermal crack growth in self-stressed glassy compounds by analytical solutions, finite elements and experiments.

Wood is a renewable raw material which consumes only a small amount of energy for its transformation into structures. It has an interesting ratio of strength to density and allows the construction of elegant glue-laminated structures. In the lecture by Jodin and Pluvinage it is shown how fracture mechanics can be applied to this very anisotropic material. Due to the high variability of wood and its natural defects (knots) probabilistic methods are a useful tool.

For the characterization of toughness of polymers and adhesives exhibiting a viscoelastic behaviour time-dependent phenomena are of a certain significance. Therefore the usual Griffith type of energy balance as well as the definition of an energy release rate concept present some problems. In those cases, however, where polymers show only a slightly visco-elastic material behaviour, the classical linear-elastic and elastic-plastic fracture mechanics can be applied for the definition of toughness. But some precaution is necessary because thermoplastic resins undergo two basically different plastic deformation processes, namely shear yielding and normal yielding (crazing). The latter can be described by the Dugdale model of elastic-plastic fracture mechanics. Thereby the COD-value is very useful as a basis for predicting also time dependent behaviour. In this book three lectures are devoted to polymers. The use of fracture mechanics to characterize the toughness of polymers and adhesives is reviewed by Williams. Doell presents the powerful optical methods used for fracture testing of brittle polymers which allow the measurement of crack speed, the size and shape of the craze zone as well as the stress intensity factor. Kausch and Stalder discuss the effects of structural features on the fracture behaviour of polymers.

The last family of materials treated in this book is concrete. Wittmann introduces three levels (micro-, meso-, and macro-levels) needed for understanding the effect of structural features on crack formation and failure of concrete and discusses in detail the micro-level properties of hardened cement paste. Roelfstra's lecture is devoted to the treatment of the meso-level by "numerical concrete" in which the structure of concrete and crack behaviour are simulated by finite elements modelling the aggregates with a random geometry. This promising approach may be of interest for other composite materials. Cedolin discusses the macro-level: modelling of the non-linear effects arising due to the fracture process zone and applicability of R-curve concepts through the use of a size effect law.

The families of materials treated in the above-mentioned sections of the book have many common features. For example, multi-phase glassy compounds, wood and concrete are all compound materials (fiber-reinforced composites and laminates were presented at ASFM5, but are not included in this book). Although the approaches used for testing and analysing various materials are often quite different the Editors have the feeling that some of the methodologies developed for one specific material could be advantageously applied to other materials. This has been already mentioned for numerical concrete.

The reader will find in the lectures not only a general presentation of the methods but also some practical applications of non-metallic materials, like hip-joints, grinding wheels and wooden pallets. Some keywords and concepts are collected in the Index.

The camera-ready text technique was used in this book due to its advantages in cost and time. Hopefully the reader will excuse any unavoidable lack of homogeneity in the presentation of the chapters.

The Editors would like to express their thanks to the EGF President H.C. van Elst, and Secretary A. Bakker, for their support during the preparation of ASFM5. The competent work of the Advisory Board whose members are listed above was highly appreciated. Mr. R. Misenta, Manager of the Ispra Education and Training Service, and his Ispra Courses team took care of the practical arrangements for ASFM5. We are most indebted to the lecturers and authors of the chapters of this book for their contributions and their efforts during the editing process.

K. Herrmann   L.H. Larsson

FUNDAMENTALS OF BRITTLE FRACTURE

E. Sommer
Fraunhofer-Institut für Werkstoffmechanik
Wöhlerstrasse 11
7800 Freiburg
Federal Republik of Germany

ABSTRACT. The basic findings explaining brittle fracture processes will
be reviewed – starting with the necessity resulting from the difference
in ultimate and technical strength to introduce flaw or crack concepts.
The stress-strain field surrounding such flaws or cracks will be discussed. For the assessment of fracture initiation and propagation criteria will be provided, i.e. a short introduction into fracture mechanics
will be given. The criterion of maximum principal stress is considered
as the guiding criterion when brittle behaviour prevails. Characteristic
features of the fracture appearance as a consequence of kinematics of
the fracture process itself or a superposition of different fracture
modes will be analyzed.
    These considerations will be illustrated by examples showing the
behaviour of glass samples since glass turned out to be an ideal model
material for brittle fracture. Nevertheless, throughout the lecture
general features of brittle fracture are discussed.

1. INTRODUCTION

Typical features of brittle fracture are well known from every day life
experience and can be demonstrated by the first example in Fig. 1:
- the fracture process occurs without considerable macroscopic plastic
  deformation; i.e. generally without a prewarning
- the fracture process takes place in a rapid manner; i.e. there is only
  a small chance of influencing the fracture process after initiation
- the fracture process often completely destroys the structural member.
    In the early years of fracture research [2, 4] glass – the material
which has been chosen for the first and most of the following examples –
has received wide attention for studying principal fracture phenomena.
Due to its linear elastic behaviour up to fracture for a wide range of
temperatures, due to its structural homogeneity glass can be regarded as
an almost ideal model material for demonstrating typical features of
brittle fracture. The phenomenon "brittle fracture", however, is not restricted to the material glass at all. Dependent on the temperature range,
the rate of loading, the local stress state and the material properties,

1

K. P. Herrmann and L. H. Larsson (eds.), Fracture of Non-Metallic Materials, 1–19.
© *1987 by ECSC, EEC, EAEC, Brussels and Luxembourg.*

Figure 1. Broken window [1]

brittle fracture may occur as well in other materials more widely used than glasses.

Extreme parameter conditions may cause brittle fracture even in metals - known for their ductile behaviour. A ductile fracture process is in general less dangerous since a large amount of plastic work is dissipated in a volume around the actual crack tip. Such processes can be characterized by the methods of elastic-plastic fracture mechanics (EPFM). If plastic deformation is negligible the methods of EPFM reduce to those of linear elastic fracture mechanics (LEFM) which are extremely helpful for the characterization of brittle fracture.

These fracture concepts also find application to non-metallic materials. Nevertheless it has to be proved individually, whether the basic assumptions required for a valid use of EPFM or LEFM are met by the material to be investigated. Complex structural conditions as in polymers and ceramics which are of increasing practical importance may hinder a straight forward application and ask for individual corrections.

2. PRINCIPAL CONSIDERATIONS

2.1. Brittle Behaviour

Brittleness is not an absolute property of a material, but depends on external parameters as temperature, loading rate, stress-strain state and environment. In the context of this paper a fracture process is considered as brittle if the integral stress-strain behaviour of a specimen or structural member remains elastic during the complete process. This definition allows non linear and plastic deformation on a microscale for instance in the surrounding of a crack tip - and includes implicitly a geometry dependent scaling factor. It does not distinguish between the initiation and propagation phase of fracture. Therefore the analysis of individual experiments has to take into account whether the field of parameters influencing the property "brittleness" affects a global or only a local region.

Although a plastic deformation on a microscale has not been excluded it is generally assumed that fracture in brittle materials is always initiated by tensile stresses and propagates perpendicularly to the maximum principal stress of the local field [2, 3]. This hypothesis is in very good agreement with experimental observations, therefore, there is no need to introduce other failure criteria.

2.2. Ultimate and Technical Strength

It results from the application of the maximum principal stress criterion that in an unnotched specimen the onset of fracture should occur when the maximum principal stress overcomes the molecular strength. But even for an ideal material as glass which is homogeneous, isotropic and extremely brittle in a wide range of the parameters temperature and loading rate this assumption leads to discrepancies which have to be explained in more detail.

2.2.1. Ultimate Strength. The most simple model which characterizes the molecular strength is based on the "load-displacement" situation of two atoms (Fig. 2):

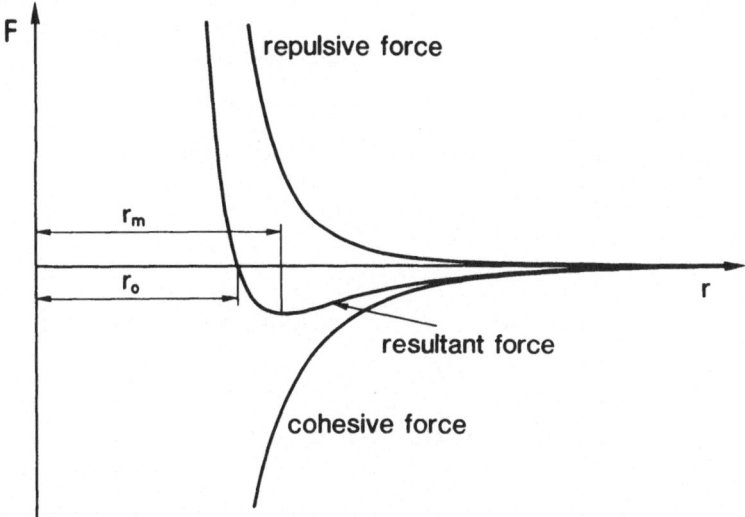

Figure 2. Interatomic forces versus the distance of separation

The force F between the atoms is a function of their distance r. Depending whether the distance is small or great repulsive ($F_r = \gamma r^{-n}$) or cohesive forces ($F_1 = -\beta r^{-m}$) are acting. The total force - distance of separation curve results from a superposition of the two curves

$$F = \gamma r^{-n} - \beta r^{-m}.$$

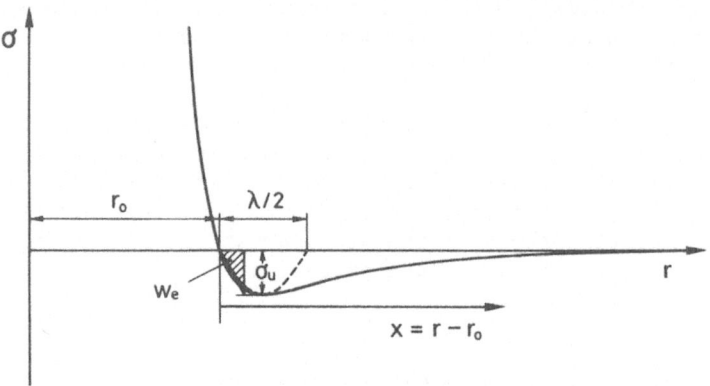

Figure 3. Sine-curve approximation for the separation process

Only if the long range cohesive force is dominant, i.e. if m < n, binding of the atoms will occur. A stable distance r_0 is given for F = 0. When the atoms shall be separated F_{max} has to be compensated by external forces at a corresponding elongation of $r_{max} - r_0$.

Further simplifying assumptions on the structure of a material, its threedimensional periodicity, and the type of bonds to be broken are incorporated if this model is used to predict the ultimate strength σ_u. Generally the important part of the force-displacement curve is approximated by a sine curve (Fig. 3)

$$\sigma = \sigma_u \sin\frac{2\pi x}{\lambda} \qquad (2.1)$$

where

σ = stress between the two surfaces to be separated
σ_u = ultimate strength
x = $r - r_0$ = elongation or displacement
λ = fitting parameter $(x/(\lambda/4) = 1)$
ε = x/r_0 = strain.

Then the work done per unit area to separate the material is:

$$\int_0^{\lambda/2} \sigma_u \cdot \sin\frac{2\pi x}{\lambda} \ dx = \sigma_u \cdot \frac{\lambda}{\pi} \qquad (2.2)$$

The parameter λ can be expressed by the modulus of elasticity E and the distance r_0 by calculating the initial slope

$$\frac{d\sigma}{dx} = \frac{d\sigma}{r_0 \ d\varepsilon} = \frac{E}{r_0} \qquad \text{respectively,} \qquad (2.3)$$

$$\frac{d\sigma}{dx} = \sigma_u \frac{2\pi}{\lambda} \cos \frac{2\pi x}{\lambda} \qquad \xrightarrow[x \to o]{} \qquad \sigma_u \frac{2\pi}{\lambda} \qquad (2.4)$$

Solving for λ by equating (2.3) and (2.4)

$$\lambda = \frac{2\pi r_o \sigma_u}{E} \qquad (2.5)$$

and inserting in equation (2.2) results for the total work done

$$W_t = 2 \frac{\sigma_u^2}{E/r_o} \qquad (2.6)$$

This expression is by a factor of 4 higher than that which would result from the elastic contribution only (Fig. 3, hatched area)

$$W_e = \frac{1}{2} \frac{\sigma_u^2}{E/r_o} \qquad (2.7)$$

If it is assumed that the total work done to separate the material is equal to the surface energy 2S necessary to form the two new fracture surfaces from 2.6 results

$$\sigma_u = (S E/r_o)^{1/2} \qquad (2.8)$$

According to Kerkhof [2] for instance for silica glass the following data can be inserted:

$S \approx 3 \cdot 10^{-1}$ N/m
$E \approx 7 \cdot 10^4$ MPa
$r_o \approx 2 \cdot 10^{-10}$ m

and will lead to an ultimate strength of approximately

$$\sigma_u \approx 10^4 \text{ MPa} \qquad (2.9)$$

The ultimate strain ε_u is given for the peak value of σ in Fig. 3:

$$\varepsilon_u = (\lambda/4)/r_o \qquad (2.10)$$

With equation (2.5) and the data used before follows

$$\varepsilon_u = \frac{\pi}{2} \frac{\sigma_u}{E} \approx 15 \%.$$

Both the technical strength and the integral strain at fracture deviate by approximately two orders of magnitude from the obtained results. Griffith [4, 5] tried to explain this discrepancy by assuming preexisting flaws in the material.

Although this consideration has been carried out for the special
material glass it can be generalized. As a matter of fact this result is
one of the main driving forces to further develop the methods of fracture mechanics.

2.2.2. Technical Strength. The technical strength usually is determined
from tensile or bending tests. Load at a constant rate is applied to unnotched specimens under controlled environmental conditions (temperature
and surrounding medium) until fracture occurs. Since the material is
supposed to be brittle the registered stress-strain curve will be linear
up to the initiation of fracture. Compared to the loading phase the
fracture process of the brittle specimen needs an extremely short time
which in general cannot be registered by the recording system. Therefore, the resulting stress at the onset of fracture represents the technical strength σ_t. - Of course, this consideration has to be modified
if short time loading has to be analyzed. -
    In spite of these almost ideal experimental conditions for a deter-
mination of the technical strength the outcome is rather unsatisfactory.
The obtained data vary for a given material with the size of the specimens used and their surface properties; and - in addition - show a pronounced scatter. Since the technical strength turns out to be two orders
of magnitude lower than the ultimate strength the Griffith hypothesis of
preexisting flaws helps to explain these discrepancies.
    For the same type of glass discussed before for instance the data
are
$\sigma_t \approx 70$ MPa   technical strength
$\varepsilon_t \simeq 1 \%$ technical strain at fracture.

2.3. Effect of Flaws and Cracks

In order to prove the Griffith [4, 5] assumption that preexisting flaws
or cracks are mainly responsible for the initiation of fracture in
brittle materials a short information on their behaviour in stress
fields will be provided. For the sake of simplicity in a first step only
two dimensional flaws and cracks subjected to uniaxial or biaxial tensile load perpendicular to the main axes are considered.

2.3.1. Stress Distribution.
Flaw: A flaw modelled by an elliptical hole in a plate [6] (Fig. 4)
shows the following maximum stress σ_m at the end of the major axis a if
subjected
to uniaxial tensile load σ

$$\sigma_m = (1 + 2 (a/\rho)^{1/2})\sigma \qquad (2.11)$$

where
a,b = major, minor axis of the ellipse
$\rho = b^2/a$ = radius of curvature at the end of the major axis
to the biaxial tensile load $\sigma_1 = \sigma_2 = \sigma$

$$\sigma_m = 2 (a/\rho)^{1/2} \sigma \qquad (2.12)$$

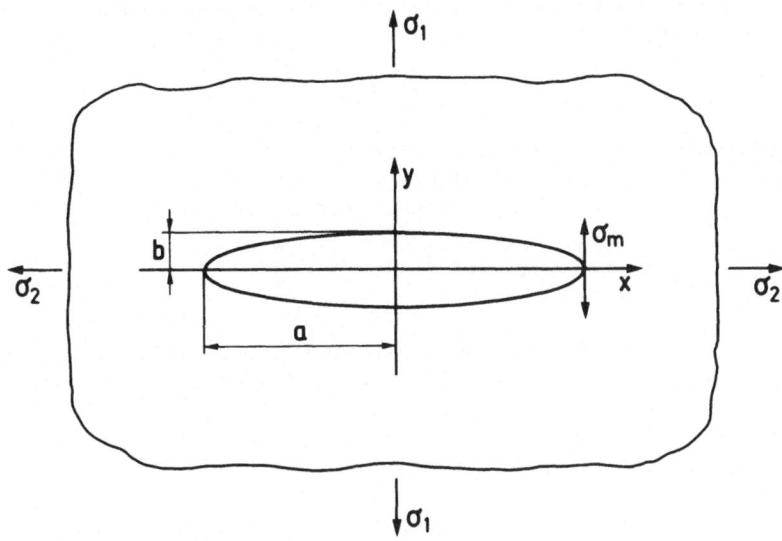

Figure 4. Elliptical hole in a plate subjected to tension

This model allows to explain even in a quantitative manner the increase of the technical strength due to blunting of surface cracks. This can be achieved e.g. for glasses by surface etching with certain concentrations of hydrofluoric acid.

Sharp crack:

If $\rho \to o$ the well known crack model applies and the near tip stress field is [7, 8, 9, 10]:

$$\sigma_{yy} = \frac{K_1}{(2\pi r)^{1/2}} \quad \cos \frac{\phi}{2} \; [1 + \sin \frac{\phi}{2} \sin \frac{3\phi}{2}]$$

(2.13)

$$\sigma_{xx} = \frac{K_1}{(2\pi r)^{1/2}} \quad \cos \frac{\phi}{2} \; [1 - \sin \frac{\phi}{2} \sin \frac{3\phi}{2}]$$

$$\sigma_{zz} = 0 \qquad\qquad\qquad\qquad\qquad \text{for plane stress}$$

$$\sigma_{zz} = \nu(\sigma_{xx} + \sigma_{yy}) \qquad\qquad\qquad \text{for plane strain}$$

$$\tau_{xy} = \frac{K_1}{(2\pi r)^{1/2}} \quad \sin \frac{\phi}{2} \; \cos \frac{\phi}{2} \cos \frac{3\phi}{2}$$

where
r, ϕ = in plane coordinates at the crack tip
K_I = stress intensity factor
ν = Poisson's ratio

For a central crack in an infinite plate under uniaxial load the stress intensity factor is:

$$K_I = \sigma(\pi\, a)^{1/2} \tag{2.14}$$

This model requires a physical interpretation of the stress singularity and leads to the additional discrepancy [2]: for a constant distance r the maximal values of the principal stress

$$\sigma_{1,2} = \frac{K_I}{(2\pi r)^{1/2}}\, \cos\frac{\phi}{2}\,(1 \pm \sin\frac{\phi}{2}) \tag{2.15}$$

result at $\phi = \pm\,60°$ and are 30 % higher than the value obtained at $\phi = 0°$. That means if crack initiation and propagation is assisted by the formation of secondary cracks in the process zone at the primary crack tip this is more likely to occur at the angles $\pm\,60°$ than on the ligament at $\phi = 0°$.

2.3.2. Energy Balance. According to Irwin [11] the energy available for crack extension is related to the stress intensity factor:

$$G_I = K^2/E \qquad\qquad \text{for plane stress}$$
$$G_I = (K^2/E)(1-\nu^2) \qquad \text{for plane strain} \tag{2.16}$$

G_I = strain energy release rate

E = modulus of Elasticity.

If G_I would have to exceed only the surface energy 2S needed to form the new fracture surfaces the energy balance is

$$G_I = 2S \tag{2.17}$$

Accordingly the critical stress required for failure σ_c can be calculated in the case of an externally loaded central crack of the total length 2a

$$\sigma_c = (2S\, E/\pi a)^{1/2} \qquad\qquad \text{for plane stress}$$
$$\sigma_c = (2S\, E/\pi a(1-\nu^2))^{1/2} \qquad \text{for plane strain.} \tag{2.18}$$

This consideration has led Griffith [4, 5] to the conclusion by relating the surface energy and the technical strength of glass that the structure of glass must contain small cracks. Using the data for glass introduced in the previous sections and setting $\sigma_c = \sigma_t$

$$a_c = (\frac{2S\, E}{\sigma_t^2\,\pi(1-\nu^2)}) \simeq 3\cdot10^{-6}\ \text{m.}$$

In reality the assumption (2.17) is as oversimplified as the assumption of preexisting intrinsic cracks. For the same glass the energy to initiate fracture G_{Ic} is about one order of magnitude higher than the surface energy 2S, because more energy is consumed for instance by neglected dissipative processes.

2.4. Influence of Time and Environment

The lifetime of a component or specimen consisting of brittle materials is strongly dependent upon the parameters time and environment as inert or aggressive media and temperature. Since the contribution to lifetime is negligible as soon as a crack runs fast the end of life can be defined by the onset of unstable crack growth, i.e.

$$G_I \geq G_{Ic} \quad \text{and} \quad dG_I/da > dG_{Ic}/da$$

The characteristic phenomena to be expected in a tensile test will be deduced from (2.14) and (2.16). According to these equations the strain energy release rate $G_I \propto \sigma^2 a$ has to exceed the crack resistance $G_{Ic} \propto \sigma_c^2 a_c$ where σ_c = critical stress, a_c = critical crack length at the point of instability. Thus the lifetime of a component is determined by the decay of the crack resistance with time. This results either from a reduction in strength or an increase of crack length or an interaction of both. In general both effects are hardly to be separated. Only under idealized circumstances, but mainly for the sake of simplicity the following consideration which assumes independency can be justified.

A variation of the strength only can be observed when the loading rate, the temperature or the surrounding medium is changed. As an example which cannot be generalized for other materials the investigations of Mould [12] on soda-lime glass will be cited. The effect of temperature and loading rate - in this case expressed as time to failure - on the relative strength, i.e. the observed strength related to the strength in liquid Nitrogen is shown in Fig. 5. At very low temperatures and small loading rates the strength is assumed to be almost independent of temperature and time since physical and chemical processes cannot be activated (Range A). With increasing temperature and time to failure a decay of strength is found (Range B). However, for further increasing temperature the opposite tendency can be observed (Range C) till finally failure occurs by viscous flow (Zone D).

In general a careful explanation has to be based on microstructural, strongly material related effects which are beyond the scope of this review article (see e.g. [13, 14]). In some cases even simple mechanical models are of analytical help; as mentioned in 2.3.1. for instance the model for crack tip blunting due to environmental influences allows to predict the increase in strength.

An increase in crack length prior to catastrophic failure can be mainly attributed to the effect of subcritical crack growth. The determining factors again can be found on a microstructural level; they are manifold and extremely material related. Nevertheless, for the purpose of technical use a helpful empirical correlation can be established by means of fracture mechanics since a unique function between the fracture

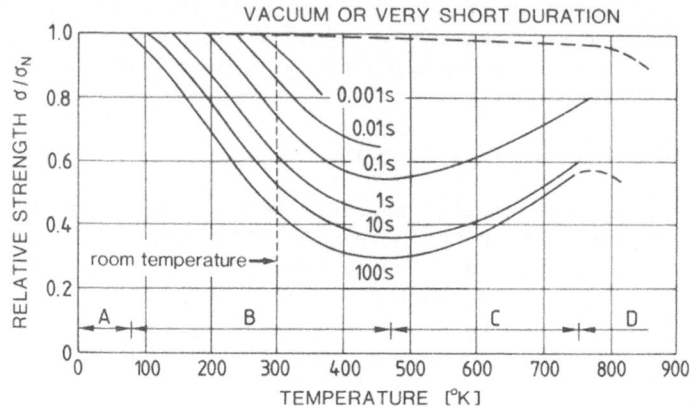

Figure 5. Relative strength of soda-lime glass as a function of
temperature (after [3, 12])

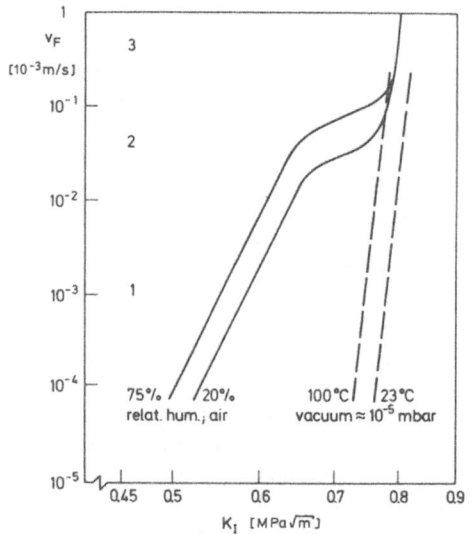

Figure 6. Fracture velocity v_F as a function of stress intensity
factor K_I for glass investigated at different temperatures
and humidities [15, 16]

velocity v_F and the stress intensity factor K_I in dependence of the environment and temperature can be observed. The typical behaviour shown
for glasses in Fig. 6 has also been found for a broad variety of materials and environments [15, 16, 17]. According to these findings three
characteristic ranges can be distinguished.

Range 1: The resulting fracture velocity is strongly influenced by the
corrosive environment – which means water vapour for glasses.
Range 2: In this transition region the transport velocity of the corro-

sive environment to the crack tip determines mainly the effect of corrosion on the crack propagation rate. There is only a slow increase of
the fracture velocity.
Range 3: The effect of the corrosive environment is negligible; the
curves meet those investigated under high vacuum conditions.
    As a good approximation the correlation between fracture velocity
and stress intensity factor can be described by

$$v_F = A K_I^n \tag{2.20}$$

where
A,n = constants experimentally to be determined.
    The range of validity of this equation is, of course, limited. The
minimum lifetime, i.e. the most dangerous situation of a component can
be obtained from this equation by integration.
    Although the fast fracture process under the aspect of lifetime
consideration has been neglected, it should be mentioned that a lot of
work has been devoted to this interesting field (e.g. [2, 18]).

2.5. Fracture Appearance and Superposition of Fracture Modes

The fracture appearance and morphology of fracture surfaces are strongly
related to
- the type of material investigated
- the geometry of the specimen or component and of the fracture initiating flaw or crack
- the type of load application
- the environmental conditions as temperature, inert, or aggressive
  media
- the crack velocity as a result of the parameters mentioned before.
    Only for restricted or almost ideal parameter constellations a
unique correlation between guiding parameters and fracture appearance
can be established. A homogeneous isotropic material as glass investigated in the form of simple specimen geometries as rods or plates, sta-
tically loaded in a dry environment at room temperature fulfills these
requirements almost perfectly. Typical features of the normal fracture
process and those resulting from the superposition of the different
fracture modes will be discussed. Mode I stands for normal, mode II for
inplane shear, mode III for antiplane shear loading.

2.5.1. Mode I fracture. If an unnotched rod of glass is fractured,
under mode I condition three characteristic features can be observed on
the fracture surface:
- a very smooth part without markings - the fracture mirror,
- a part with increasing roughness and
- a part where the crack branched off.
The size of the mirror depends on the amount of energy available for
the fracture process. The experimental finding

$$a_m \, \sigma_c^2 = const. \tag{2.21}$$

where

a_m = radius of the fracture mirror

σ_c = load at failure

can be explained with (2.18). Roughness and branching occur when the crack has reached the terminal velocity. Figure 7 shows two fracture surfaces of this kind.

2.5.2. Superposition of fracture modes I and II. According to the failure criterion of "maximum principal stress" a crack is supposed to propagate in a plane perpendicular to the axis of the principal stress at the crack tip. A rotation of this axis can be induced by superposing inplane shear (mode II) to normal tensile (mode I) loading. The inclination of this axis depends on the contribution of each stress field and results from the transformation of the local stress tensor to principal axes [20]. The deviation of the crack path for small angles $\phi_o \ll \pi/2$ is given by the ratio

$$\tan \phi_o = -\ 2K_{II}/K_I \qquad (2.22)$$

where

K_{II} = stress intensity factor for mode II. Figure 8 shows the principle of superposition.

If the inplane shear is for instance caused by a transverse ultrasonic wave the crack path periodically changes its direction and becomes modulated in a sinusoidal manner [1]. In Fig. 9 is demonstrated that due to the change of sign of the shear loading at certain time steps t_i the inclination of the crack path is changed as well. The velocity of the crack v_F can be obtained from

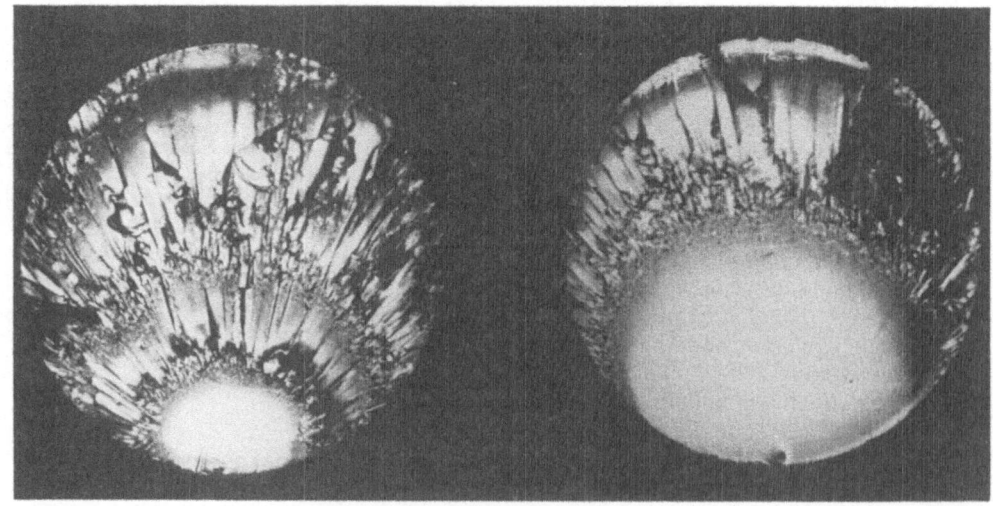

Figure 7. Fracture surfaces of glass rods; mode I condition; rod diameter 9.4 10^{-3}m terminal velocity $v_T \approx 1500$ m/s [19]

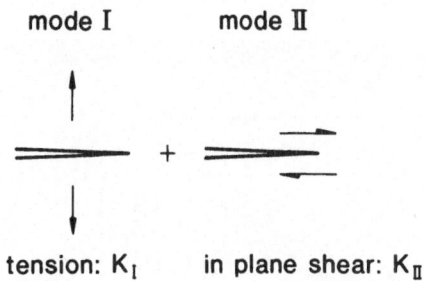

mode I   mode II

tension: K_I in plane shear: K_{II}

path of fracture:

$$\tan \varphi_0 \approx - \frac{2K_{II}}{K_I}$$

if $\varphi_0 \ll \pi / 2$

Figure 8. Superposition of fracture modes I and II

$$v_F = f \cdot \lambda_F \qquad (2.23)$$

where
f   = frequency of ultrasonic shear wave
λ_F = wave length of sinusoidal fracture modulation.
    This method developed by Kerkhof provides a precise experimental technique to determine the crack velocity in the range of 10^{-1} to $2 \ 10^3$ m/s. The photographs in Figs. 10 and 11 give evidence that the experimental findings perfectly agree with these considerations.

Example: $K_{II}(f)$ = shear wave,
            frequency f

K_I; $K_{II}(t)$ at time-steps t_i:

$K_{II}(t_1)$ $K_{II}(t_2)$ $K_{II}(t_3)$

path of fracture: sinusoidal modulation

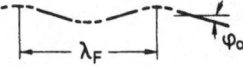

crack velocity (after Kerkhof): $v_F = f \cdot \lambda_F$

Figure 9. Modulation of fracture surfaces

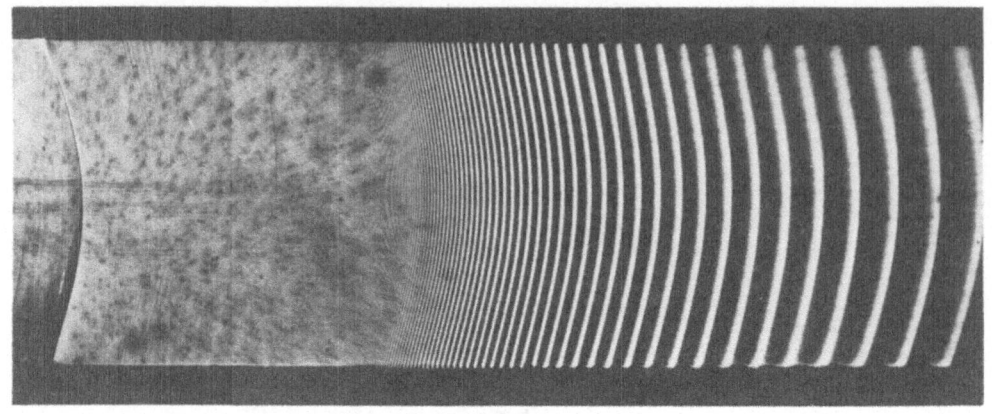

Figure 10. Sinusoidal modulation of a glass plate;
increasing crack speed [21]

2.5.3. Superposition of fracture modes I and III. The superposition
of tensile and anti-plane shear loadings, corresponding to fracture
mode I and III, causes again characteristic markings on the fracture
surfaces of brittle, homogeneous materials such as glass. Smekal [22]
called these marking "lances" because of the similarity of their shape

Figure 11. Interference pattern of a modulated fracture surface

to that of a medieval lance. The "lances" are separating lines between partial fracture surfaces.

If normal and anti-plane shear loadings are superimposed the orientation of the fracture surface should adjust to the resulting axis of principal stresses. However, a continuous adjustment of the fracture surface along the entire crack front is not possible. Thus, the crack breaks into partial fronts. Figures 12 and 13 demonstrate the principle of this process.

Experimentally lances can be produced on fracture surfaces of circular glass rods loaded in tension by the superposition of a small amount of torsion. In an undisturbed rod this torsion linearly increases from the center. If the rod contains a symmetrically located disc crack the stress field at the contour of this crack is determined by the axial stress and the hoop stress according to the mode I loading and the tangential shear stress according to mode III. If such a crack propagates, lances in radial direction will separate partial fracture surfaces – perpendicularly oriented to the axis of the maximum principal stress. The inclination of the partial fracture surfaces can be obtained by transforming the local stress field [23] to principal axes. As explained in Fig. 13 the angle of inclination α is related to the ratio of the stress intensity factors K_{III}, K_I for mode III and I and therefore linearly increases with the crack diameter if the ratio of the externally applied shear to tensile stress τ/σ and the rod diameter R are kept constant:

$$\tan \alpha = \frac{K_{III}}{K_I(1-2\nu)} = \frac{\pi a}{4R(1-2\nu)} \frac{\tau}{\sigma} \qquad (2.24)$$

(For $\alpha \ll \pi/2$ and $a/R \ll 1$).

Thus, the derivative with respect to the crack radius is a constant which mainly depends on the ratio of the applied torsion to tension.

$$\frac{d\tan \alpha}{da} \approx \frac{\pi}{4R(1-2\nu)} \frac{\tau}{\sigma} \qquad (2.25)$$

Experimental findings are in very good agreement with these principle considerations although obtained in a slightly different manner. Instead of the axial tensile stress fluid pressure is applied to the lateral surface of the round glass rod whose ends are free from pressure. The crack develops from the surface of the rod in a tongue-shaped manner. Figure 14 shows the resulting characteristic fracture markings [24, 25], Figure 15 the formed partial fracture surfaces. Since this picture is taken in an interference microscope the inclination can be determined from the interference fringe system. As long as the ratio a/R remains relatively small the derivative dtanα/da experimentally determined agrees very well to that calculated according to (2.25).

The experimental results prove that the inclination of fracture surfaces in the case of brittle fracture is strongly related to the local stress field. This means that a stress field can be reconstructed to a certain degree from the fracture appearance.

Example: disc-crack in a rod under tension

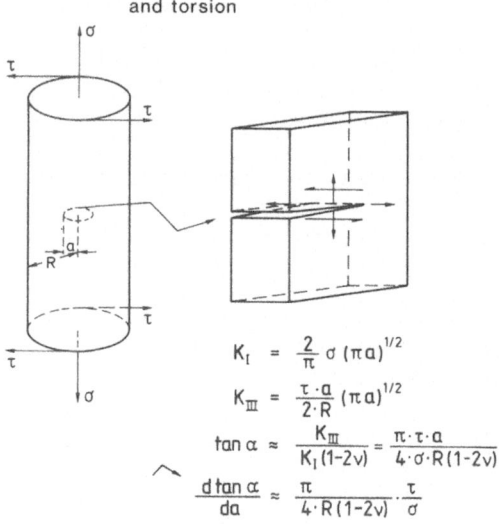

and torsion

$$K_I = \frac{2}{\pi}\,\sigma\,(\pi a)^{1/2}$$

$$K_{III} = \frac{\tau \cdot a}{2 \cdot R}\,(\pi a)^{1/2}$$

$$\tan \alpha \approx \frac{K_{III}}{K_I\,(1-2\nu)} = \frac{\pi \cdot \tau \cdot a}{4 \cdot \sigma \cdot R\,(1-2\nu)}$$

$$\frac{d \tan \alpha}{da} \approx \frac{\pi}{4 \cdot R\,(1-2\nu)} \cdot \frac{\tau}{\sigma}$$

Figure 12. Superposition of fracture modes I and III; K_I, K_{III} = stress intensity factors, α = angle of inclination of partial fracture surfaces

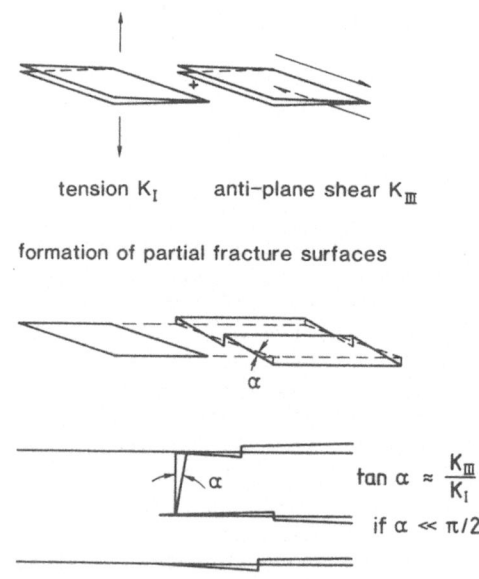

mode I   mode III

tension K_I anti-plane shear K_{III}

formation of partial fracture surfaces

$$\tan \alpha \approx \frac{K_{III}}{K_I}$$

if $\alpha \ll \pi/2$

Figure 13. Inclination of partial fracture surfaces, τ, σ = applied torsion and tension, α = angle of inclination

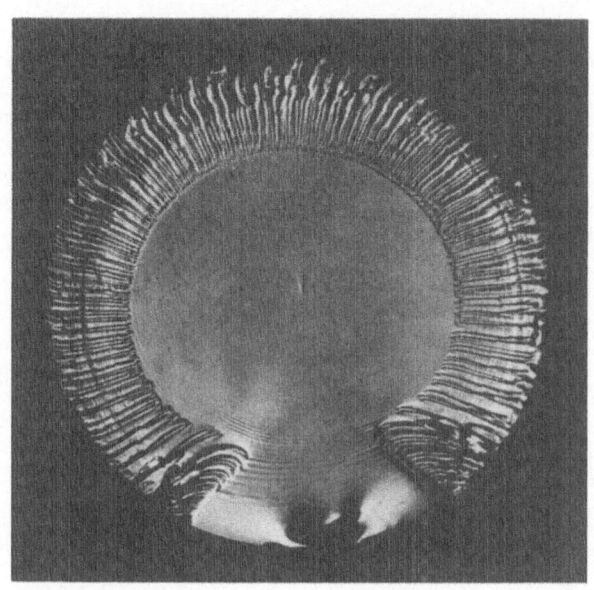

Figure 14. Fracture surface of a glass rod subjected to superposed fracture modes I and III [24];
$\tau/\sigma = 5.8 \ 10^2$; $R = 4.9 \ 10^{-3}$ m

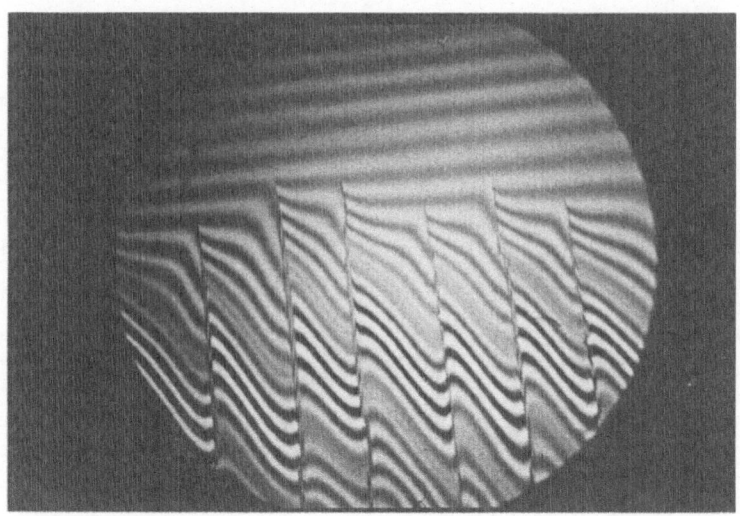

Figure 15. Inclination of partial fracture surfaces; interference pattern [24]

3. CONCLUSIONS

The concepts for describing brittle fracture processes have been developed to a very refined standard. Nevertheless, all these methods are based on restricting assumptions as necessary for the theoretical approach on the grounds of continuum mechanics or for the modelling of the properties of real materials. It belongs therefore to the responsibility of a user to validate the applicability of such concepts to his individual problem.

REFERENCES

[1]   Kerkhof, F.: Vortrag, Glastechn. Tagung 1974, Bad Neuenahr

[2]   Kerkhof, F.: Bruchvorgänge in Gläsern, Verlag der Deutsch. Glastechn. Gesellschaft, Frankfurt 1970

[3]   Liebowitz, H. (ed.): Fracture, Vol. VII, Fracture of Nonmetals and Composites, Academic Press, New York, London, 1972

[4]   Griffith, A.A.: 'The Phenomena of Rupture and Flow in Solids'. Phil. Trans. Roy. Soc. London A 221, 1920, 163-198

[5]   Griffith, A.A.: 'Theory of Rupture'. Proc. 1st Internt. Congress Appl. Mech., Delft, 1924, 55-63

[6]   Inglis, C.E.: 'Stresses in a Plate due to the Presence of Cracks and Sharp Corners'. Trans. Instn. Naval Archit. 55, 1913, 219-230

[7]   Wieghardt, K.: 'Über das Spalten und Zerreißen elastischer Körper', Z. f. Math. u. Phys., 55, 1907, 60-103

[8]   Sneddon, I.N.: The 'The Distribution of Stress in the Neighbourhood of a Crack in an Elastic Solid'. Proceedings of the Physical Society of London, 187, 1946, 229-260

[9]   Irwin, G.R.: 'Analysis of Stress and Strains Near the End of a Crack Traversing a Plate'. J. Applied Mechanics , 24, 1957, 361-364

[10] Williams, M.L.: 'On the Stress Distribution at the Base of a Stationary Crack'. J. Applied Mechanics, 24, 1957, 109-114

[11] Irwin, G.R.: 'Fracture', in: Flügge, S. (Hrsg.): Handbuch der Physik, Bd. 6, Berlin-Göttingen-Heidelberg, Springer-Verlag 1958, 551-590

[12] Mould, R.E.: 'Strength and Static Fatigue of Abraded Glass under Controlled Ambient Conditions', IV. Effect of Surrounding Medium, Am.Soc., 44, 1961, 481-491

[13]   Jayatilaka, Ayal de S.: 'Fracture of Engineering Brittle Mate-
        rials', Applied Science Publishers, Ltd., London, 1979

[14]   Evans, A.G.: 'Fracture in Ceramic Materials', Noyes Publications,
        Park Ridge, N.Y., U.S.A., 1984

[15]   Wiederhorn, S.M.: 'Influence of Water Vapor on Crack Propagation
        in Soda-lime Glass', J.Amer.Ceram.Soc., 50, 1967, 407-414

[16]   Richter, H.: 'Unterkritische Rißausbreitung in keramischen Werk-
        stoffen', Bericht d. Deutsch. Keram. Ges., 54, 1977, (12),
        405-409

[17]   Kleer, G.: Richter, H., Willmann, G. Heider, W., Popp, G.:
        'Bruchmechanische Charakterisierung des Ermüdungsverhaltens von
        SiSiC', Z.Werkstofftechnik, 16, 1985, 94-101

[18]   Kalthoff, J.F.: 'The Shadow Optical Method of Caustics', Ch. IX,
        Handbook on Experimental Mechanics, A.S. Kobayashi, ed.,
        Prentic Hall, Englewood Cliffs, New Jersey, to appear in 1986

[19]   Sommer, E.; 'Bruchmechanische Bewertung von Oberflächenrissen',
        Werkstoff-Forschung und Technik 1, B. Ilschner, ed., Springer-
        Verlag, Berlin-Heidelberg-New York-Tokyo, 1984

[20]   Erdogan, F. and Sih, G.C.: 'On the Crack Extension under Plane
        Loading and Transverse Shear', Trans.Amer.Soc.Mech.Engrs., 85,
        1963, 519-527

[21]   Schinker, M., Döll, W. und Weidmann, G.W.: 'Umgebungseffekte beim
        Bruch von dynamisch beanspruchten Glasplatten', Glastechnische
        Berichte, 50, 1977, 237-238

[22]   Smekal, A.G.: 'Zum Bruchvorgang bei sprödem Stoffverhalten unter
        ein- und mehrachsigen Beanspruchungen', Österr. Ing. Archiv, 7,
        1953, 49-70

[23]   Sneddon, I.N., Lowengrub, M.: Crack Problems in the Classical
        Theory of Elasticity, J. Wiley and Sons, Inc., New York-London-
        Sydney-Toronto, 1969

[24]   Sommer E.: 'Formation of Fracture "Lances" in Glass', Engineering
        Fracture Mechanics, 1, 1969, 539-546

[25]   Sommer, E.: 'Experimental Fracture Analysis', Murray Lecture,
        Las Vegas, SEM Spring Conference, June 1985

RECENT DEVELOPMENTS IN FRACTURE MECHANICS

H.D.Bui
Electricité de France, Dept. of MMN/IMA, Clamart
and Ecole Polytechnique, 91128 Palaiseau, France

ABSTRACT. The lecture deals with new approaches to fracture mechanics which take account of damage of the material. The main assumption is that a stress-strain law is only assumed for tensile strain less than a critical value ε_R beyond which the stress σ vanishes identically and irreversibly. It is shown with this model that a process zone of some characteristic size may exist. The dissipation rates by plastic deformation and by fracture, due to the release of stress, are introduced in the balance equation of energy. The dissipation analysis allows us to investigate the theoretical R-curves. Brittle and ductile fracture are discussed. The transition from damage fracture to classical fracture mechanics is shown by the limit $\varepsilon_R \to \infty$. Quite consistently, the dissipation rate by ductile fracture does not vanish.

1. INTRODUCTION

Fracture mechanics are known to provide useful means for engineering design and assessing integrity of structures. Most progresses in the past are concerned with mathematical analyses of crack tip singularity, computational methods for stress intensity factors, experimental approaches to fracture, microscopic investigations, etc. These topics are often regarded as independent aspects of fracture. In fact, there is obviously an interaction between these topics. There are also physical and mechanical dependence between macroscopic and microscopic levels. For example, in the highly strained zone near the crack tip, due to the nucleation and growth of voids, the conventional macroscopic behaviour of the material is no longer described by usual classical plasticity. One has to take account of damage for studying the process zone. Hence there is a need of a damage theory of the materials (See the next chapter, Bui and Stolz).
In recent works on local approaches to fracture, one focuses the discussion on the detailed analysis of the process zone, whose shape and size depend on the material property of the highly strained zone. The mathematical problem of determining the process zone turns out to be a cou-

21

K. P. Herrmann and L. H. Larsson (eds.), Fracture of Non-Metallic Materials, 21–32.

pled problem. In one hand, we have to consider damage theories for the process zone. In the other hand the boundary conditions are not prescribed on the contour of the process zone, but only on that of the entire body. We shall consider a particular boundary condition in the case where damage process zone and plastic zone are small compared to the size of the body. The elastic field near the tip , outside the process zone, is controlled by one parameter, for example K_I in mode I, or the J integral. In classical fracture mechanics, these values are considered as crack tip parameters which satisfy some criterion for example $K_I=K_{Ic}$ or $J=J_c$. When the latter criterion is used for ductile fracture, it is observed from experimental data that $J_c(\Delta a)$ depends on the crack growth Δa. The so-called R-curve appears to be a characteristic of the structure simply because of the dependence on the crack length. Therefore there is some inconsistency in the fracture mechanics of ductile materials.

In the present lecture, we review some recent works on local approaches to fracture. Then, using a very simple model of damage, we show how the problem of determining the process zone can be solved in some particular case. The local approach to fracture is based on a damage criterion which is assumed once for all, for any point of the solid. The constitutive equations, and the damage criterion are sufficient for the prediction of fracture and the analysis of the process zone evolution.
The main interest of local approaches is that stable or unstable propagation, brittle or ductile fracture, can be analysed by the same model, with only the change on constitutive equations and on damage criterions. The unstable propagation corresponds to the steady state solution under constant load $\dot{J}=0$. For stable propagation, the local approach leads to new investigations of the theoretical R-curves.

2. BALANCE EQUATION OF ENERGY

Due to the propagation of a crack or any other defects, there is a dissipation rate in the body because of irreveribility of these processes. It is useful to analyse the dissipation rates due to plasticity and fracture. The total dissipation rate $\dot{D}$, if we neglect thermal effects,

$$\dot{D} = \dot{D}_p + \dot{D}_f$$

consists of two terms due to plastic deformation and fracture respectively. Theoretical considerations [1] and experimental data [2], [3] , [4] show that only a small amount (10%) of the plastic rate is responsible for work-hardening of materials, while the remaining part (90%) is dissipated through internal slip friction. Therefore, the part of energy stored elastically can be neglected so that the plastic dissipation rate is given by

$$\dot{D}_p = \int_\Omega \sigma_{ij} \dot{\varepsilon}^p_{ij} \; dV \tag{1}$$

where σ is the stress and $\dot{\varepsilon}^p$ is the plastic strain rate, Ω is the plastic zone.

The fracture rate $\dot{D}_f$ does not have a unique interpretation. Let us consider one possible mechanism as illustrated by the example of a line elastic spring in pure tension which breaks down at the stress σ. The elastic energy stored $1/2\sigma^2/E$ in the spring is released into vibrations and then is dissipated by damping processes. Hence, fracture leads also to a dissipation rate because of irreversibility. We will see below how our model of damage leads to a simple relation between the fracture dissipation rate and the characteristic size of the process zone.

Let us recall the balance equation of energy :

$$P_{ext} = \dot{W}_{elast} + \dot{K} + \dot{D} \tag{2}$$

where P_{ext} is the power of external loads, $\dot{K}$ is the kinetic energy rate $\dot{D}$ is the dissipation rate and

$$\dot{W}_{elast} = \frac{d}{dt}\int_V W(\varepsilon - \varepsilon^P)\, dV$$

is the elastic strain energy rate of the body V. The equation (2) can be rewritten in the familiar form $G = \dot{D}/\dot{a}$ where

$$G = (P_{ext} - \dot{W}_{elast} - \dot{K})/\dot{a}$$

is the energy release rate. Two extreme situations are well known in classical Fracture Mechanics
i) Griffith's theory of brittle fracture

$$\dot{D}_p = 0 \quad , \quad \dot{D}_f = 2\gamma\dot{a}$$

ii) Ductile fracture

$$\dot{D}_p \neq 0 \quad , \quad \dot{D}_f = 0 \; .$$

In the above, $\dot{a}$ denotes the crack velocity, 2γ the Griffith constant. The last equality $\dot{D}_f = 0$ (Ductile fracture) means that, quite inconsistently, the separation of material at the crack tip does not require energy. This paradoxical result was established for a perfectly plastic material. We will show later that the fracture rate still vanishes during crack extension, irrespective of how ductile the material, when the unloading is linear elastic , because of the singular strain at the crack tip. This suggests us that, in order to obtain a non vanishing D_f, the strain must be bounded. Such a model implies a finite process zone and a damage criterion, as is shown in section 4.

Let us remark that another possible way to obtain a non vanishing fracture rate consists of the use of classical solutions to the crack problems and of the introduction of a criterion applied at some distance Δ to the crack tip. Kfouri and Miller [5] suggest the use of the crack separation work G^Δ over the distance Δ . These approaches cannot give information about the details of the process zone. The coupling between the damage process zone and the mechanical field outside this zone is simply ignored. Perhaps the uncoupled approach may be considered as the first approximation to the solution. However a better understanding of the ductile fracture requires the solution to the coupled problem.

3. THEORIES OF R-CURVE

In small scale yielding, the process zone size is small compared to that of the body. Let the stress field around the process zone be governed in the symmetrical mode I by the K_I factor or the J integral

$$J = \frac{(1-\nu^2)}{E} K_I^2 \qquad \text{(in plane strain)} \qquad (3)$$

As is indicated above, we shall consider J as the load parameter. That is the J-integral is defined here by its classical expression, but with the use of a contour Γ in the elastic zone. Therefore, the stress working density $W(\varepsilon)$ or the elastic strain energy density $W(\varepsilon-\varepsilon^P)$ are the same in the elastic zone. It is worth noticing that J is only path independent in the elastic zone, and it loses the physical meaning in the following cases :
i) Non linear elasticity. The J-integral is defined by the stress working density $W(\varepsilon)$. The path independency of J results from the one to one stress-strain relation $\sigma=\partial W(\varepsilon)/\partial\varepsilon$ which cannot accommodate the linear elastic unloading of the real material.
ii) Plasticity. The J-integral is defined by the elastic strain energy density [6], [11], ie $W=W(\varepsilon-\varepsilon^P)$. The value of J depends on the contour Γ in the plastic zone. The only case where J has a simple interpretation is the case where Γ is taken in the elastic zone and the mechanical fields are steady state solutions of crack propagation. It has been proved in [11] that, in this case, we have the following result

$$J\dot{a}=\dot{a}\int_{\Gamma} \{W(\varepsilon-\varepsilon^P)n_1 - \sigma_{ij}n_j u_{i,1}\}ds = \dot{D}_p + \dot{D}_f \qquad (4)$$

In crack model $\dot{D}_f=0$. Hence J is related to the plastic dissipation rate of the body. This is the fundamental reason why in ductile fracture one fails in searching a criterion of fracture which characterizes only the crack tip, without considering the plastic deformatoon of its neighbourhood. Moreover, the growth of the plastic zone during crack extension may explain why the J-integral increases with the crack ewtension Δa. Therefore the Eq. (4) is important for the R-curve analysis.

■Dugdale-Barenblatt model

The simplest model of the R-curve is given by a localized line plastic zone of length R in the front of each crack tip

$$R = a_0\{1/\cos(\pi\sigma/2\sigma_0) - 1\} \qquad (5)$$

where σ is the remote applied stress, σ_0 is the yield stress, and a_0 is the half crack length. If we introduce an effective crack $a=a_0 + R$

then from Eq.(5) we derive the R-curve as the applied load σ versus the effective crack length a or equivalently $\sigma(a-a_o)$. The normalized load σ/σ_o versus $(a-a_o)/a_o$ is plotted in Fig.1. This example shows that the R-curve results from the extension of the plastic zone.

■Cherepanov's model

Let us illustrate the applicability of Eq.(4) to the Cherepanov model of the R-curve [7]. According to [7], the plastic work increment dD_p is assumed to be proportional to the plastic area increment $d(\pi R^2)$, where R is the radius of the plastic zone the size of which is of order $R \approx K_I^2$. Hence

$$dD_p = AK_I^3 dK_I \tag{6}$$

The fracture energy dD_f due to the separation of material is proportional to the crack growth da and K_I^2

$$dD_f = BK_I^2 da \tag{7}$$

Finally, Cherepanov extended the Griffith criterion to the total dissipation J by assuming $dJ=2\gamma da$, where γ is a constant. Using Eq.(4), (6) and (7) we obtain from the relation $dJ=dD_p+dD_f$ the differential equation of the R-curve. The integration of the latter equation yields

$$(a-a_o)/a_o = \frac{AK_c^4}{8\gamma a_o} [-2K_I^2/K_c^2 - 2 \ln(1-K_I^2/K_c^2)] \tag{8}$$

where $K_c^2 = 2\gamma/B$ is the critical load. The normalized load K_I/K_c versus the crack growth is plotted in Fig.1 for some constant $\lambda=AK_c^4/8\gamma a_o$. Fig.1 shows that both models give similar result except for small value of the crack extension.

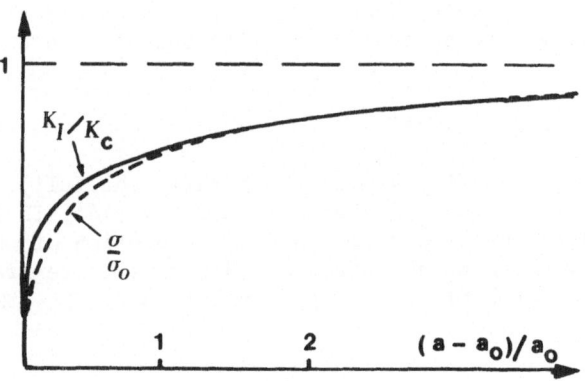

Figure 1. Theoretical R-curves.

4. LOCAL APPROACHES TO BRITTLE AND DUCTILE FRACTURE

Local models of fracture using various theories of damage are available in the recent literature. Let us discuss three kinds of models based on the growth of cavities, the continuous damage theory and the sudden damage theory respectively.

■Growth of cavities

In d'Escatha and Devaux [8] and Rousselier [9], damage is simulated by the growth of cavities. The damage criterion is the critical radius of cavity whose rate of evolution is given by the equation [10]

$$\dot{R}/R = 0.283 \dot{\varepsilon}_{eq} \exp(\frac{3}{2} \frac{\sigma^m}{\sigma_o})$$ (9)

where $\dot{\varepsilon}_{eq}$ is the equivalent strain rate, σ^m is the mean stress. Using a finite element method, assuming the presence of a cavity in each mesh of some characteristic size, these authors determined the response of the body under prescribed loading. When the critical radius is reached in the cell in front of the crack tip, the stress in the damaged element is released. Hence the crack growth is simulated by step increments.

■Continuous damage theories

There are continuous damage models which can be incorporated into crack analyses [12], [13], [14] (See the detailed discussion of these theories in the next chapter, Bui and Stolz). For the sake of simplicity we limit our discussion to scalar damage parameter ω. In Fig. 2. we have plotted the stress-strain curve of a material and the assumed evolution of damage which grows rapidly from 0 to 1. The classical behaviour corresponds to $\omega=0$ (undamaged state) while in the range $0<\omega<1$ we have a partially damaged state. The full damage corresponds to $\omega=1$. The corresponding elastic (I), plastic (II), partially damaged (III) and fully damaged (IV) zones are shown in Fing. 3. The crack itself may be considered as the zone (IV). The boundary between the zones (III) and (IV) is characterized by two conditions

$$\sigma_{ij} n_j = 0 \qquad \text{and} \qquad \omega = 1$$ (10)

Recently, Billardon and Moret-Bailly solved numerically the fully coupled equations of damage evolution and elasto-plasticity [14]. Their approximate analysis led to the conclusion that the evolution of the crack can be predicted by the coupled equations. However, there are some numerical difficulties due probably to a rapid evolution of the damage parameter.

■Sudden damage

The simplest theory of damage consists of assuming that ω takes only two possible values: either $\omega=0$ (undamaged state) or $\omega=1$ (full damage). The criterion $\omega=1$ is reached when the tensile strain satisfies $\varepsilon \geqslant \varepsilon_R$.

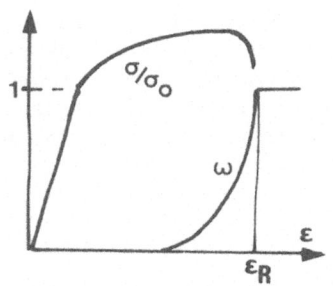

Figure 2. Stress-strain curve
Damage evolution

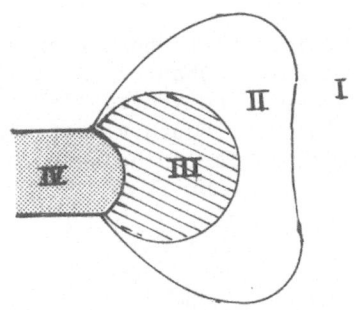

Figure 3. Damaged and plastic zones

The function $\omega(\varepsilon)$ would be represented by a step function at the critical strain, beyond which the stress tensor vanishes identically. The details near the process zone are simpler than those of Fig. 3. Since the
zone (III) is absent, the separation of particles or damage occurs only
along the boundary of the process zone (IV).
This model was first investigated in [11] for the steady state propagation of the damaged zone. The unsteady state case was studied in [20].
Let us summarize the equations.
i) Elasto-plastic zone: Classical equations (equilibrium, compatibility,
usual elasto-plastic constitutive equations, boundary conditions on the
outer contour)
ii) Damaged zone : $\sigma_{ij}=0$
iii) On the damage front: $\sigma_{ij}n_j=0$, $\varepsilon_{tt}=\varepsilon_R$, $\phi > 0$
where t is the tangent to the damage front, ϕ is the normal velocity
of the damage front. The conditions (10) in the moving boundary give
the relations between the stress rate $\dot{\sigma}$, the strain rate $\dot{\varepsilon}$ and the velocity ϕ[20].
iv) On the remaining part of the boundary of the process zone : $\sigma_{ij}n_j=0$,
$\varepsilon_{tt}<\varepsilon_R$ and $\phi=0$.
Finally the equations and conditions are similar to those of plasticity
with the introduction of a new field ϕ, Fig.4b.

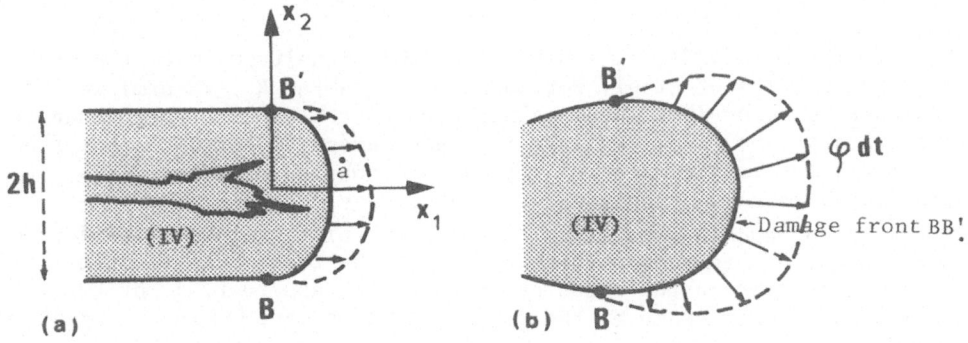

Figure 4a,b. Steady state and unsteady state propagation of damage

5. SOME IMPLICATIONS OF DAMAGE FRACTURE

We consider here the sudden damage theory and discuss its implications in Fracture Mechanics. As is illustrated by the rupture of a line elastic spring, the stored elastic energy $W(\varepsilon-\varepsilon^P)$ along the damage front BB' is released through vibrations and finally it is dissipated through heat conduction. It can be seen in Fig. 4b that the dissipation rate by rupture is given by

$$\dot{D}_f = \int_{BB'} W(\varepsilon-\varepsilon^P) \, \phi \, ds \tag{11}$$

where ds is the arc length. In the steady state case, Fig. 4a, the front BB' moves by pure translation with the velocity $\dot{a}$ ($\phi=\dot{a}n_1$), Eq.(11) can be rewritten as

$$\dot{D}_f = \dot{a} \int_{BB'} W(\varepsilon-\varepsilon^P) dx_2 = 2\dot{a} \, W_R h \tag{12}_a$$

where W_R is the constant value of the elastic energy at rupture, 2h is the width of the process zone. In plane strain :

$$\dot{D}_f = (1-\nu^2)\sigma_R^2 h\dot{a}/E \qquad (E : \text{Young Modulus}) \tag{12}_b$$

where σ_R is the maximum fracture stress in tension. We remark that Eq. (11) to (12)$_b$ are valid for brittle fracture, but also for ductile fracture.

■ The transition from damage fracture to Fracture Mechanics

Let the maximum stress be infinite. More precisely we assume that σ_R tends to infinity and that h tends to zero in such a way that $h\sigma_R^2$ remains constant. Therefore, the process zone becomes a flat crack h=0. The constant is simply related to the Griffith constant or the toughness of the material. Now instead of considering this limit process, let us assume finite values of σ_R and h.
Identifying $\dot{D}_f/\dot{a}$ to $(1-\nu^2)K_{Ic}^2/E$ leads to the formula

$$2h = 2K_{Ic}^2/\sigma_R^2 \tag{13}$$

For steel (σ_R=1000MPa, K_{Ic}=10MPa$\sqrt{m}$) we obtain h=100µm, i.e. the order of some grain sizes of steel. For concrete (σ_R=4MPa, K_{Ic}=2MPa$\sqrt{m}$) we obtain a higher value h=.25m. These values give us the order of magnitude of the process zone size. Perhaps, a sudden damage criterion is not realistic for non-metallic material for which a continuous damage theory [12] [13] may be more appropriate than the present model.

The model of sudden damage, due to its simplicity, allows us to derive exact solutions in some particular cases. The reader can refer to [11] for more details on the solution in mode III, and to [15] for the solution in mode I. For example, the damage front in mode III is a cusped cycloid.

■The paradox $D_f=0$ reexamined

Let us consider the J-integral as defined in Eq. (4), but with the path Γ on the boundary BB' of the process zone, Fig. 4a. Along BB' the stress vector vanishes $\sigma.n=0$. Hence only the first term Wn_1 contributes to the J-integral. Since the elastic strain energy density is constant along BB' we recover the equation (12) which means that the fracture energy rate is expressible by J along BB'. This dissipation rate does not vanish in damage models because of finite value of the process zone width. It is interesting to examine how the paradox $\dot{D}_f=0$ of ductile fracture mechanics can be reobtained by the limit process $\varepsilon_R \to \infty$.
For a perfectly plastic material, it is clear from Eq. (12) that the fracture energy rate tends to zero as $h \to 0$.
In fact, we can show that the latter result holds true for a power law $\sigma = C\varepsilon^n$ $(n < 1)$, for monotonic loading and for <u>linear</u> elastic unloading. The elastic strain energy density is proportional to σ^2, hence

$$W_R = A\ \varepsilon_R^{2n} \tag{14}$$

for material points in the critical state. To calculate the fracture energy rate by the equation $(12)_a$, we need an estimation of h. For this purpose, the Neuber formula for a notch [19] can be used. The product $\sigma\varepsilon$ does not depend on the material property and is equal to that calculated elastically. In the elastic notch, this product is of order $1/h$ (h being the radius of the notch). Hence, we obtain

$$h = B\ \varepsilon_R^{-(n+1)} \tag{15}$$

From Eq. (12), (14) and (15) we can see that

$$\dot{D}_f = 2\dot{a}\ AB\ \varepsilon_R^{n-1} \tag{16}$$

The limit of the above expression as ε_R tends to infinity is zero because of $n < 1$.

■Plastic zones

An exact solution to the coupled problem of damage and perfect plasticity has been found for the mode III, without propagation of the damage zone [11]. The active plastic zone is bounded by two cycloids (cusped

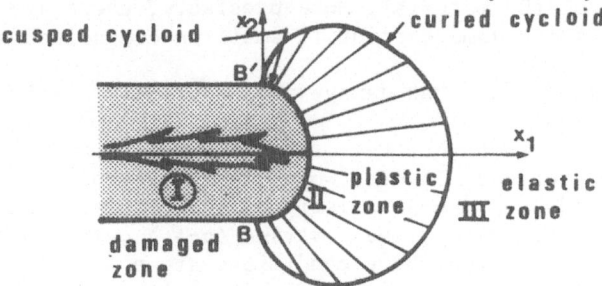

Figure 5. Plastic zone and damaged zone in mode III

and curled cycloids) Fig. 5. The characteristics are straigth lines normal to the boundary of the damaged zone, along which the shear stress
σ_{3n} is constant (n is the normal to the characteristic line).
The remote elastic stress field is governed by the K_{III} factor. This is
why we have considered J, for a contour in the elastic zone, as a load
parameter. The present solution allows us to calculate the plastic dissipation rate as follows

$$\dot{D}_p = K_{III}\dot{K}_{III}/\pi\mu\tau_0^2 - 2K_{III}\dot{K}_{III}h/\pi\mu \tag{17}$$

where μ is the shear modulus, τ_0 is the critical shear stress.
In [11], we generalized the result (17) to the case of propagation by
assuming that the wake plastic zone does not change significantly
the above solution in the active plastic zone. The unloading behind the
active plastic' zone may change the shape and the size of plastic and
damaged zones. It does not change the manner in which the dissipation rate depends on K and $\dot{K}$, as it can be shown by arguments based on the dimensions of the governing parameters. Therefore, for the mode I, the dissipation by plasticity (damage models)may be given by a similar formula

$$\dot{D}_p = AK_I^3\dot{K}_I - BK_I\dot{K}_Ih(K_I) \tag{18}$$

where A and B are some constants of the material, and h may be a function
of the load parameter in case of unsteady state propagation.
Various models of the R-curve have been discussed in [11] using the
balance equation of energy (4), and Eq. (3), $(12)_b$, (18). We obtain

$$\frac{(1-\nu^2)K_I^2}{E} = AK_I^3\dot{K}_I - BK_I\dot{K}_Ih(K_I) + \frac{(1-\nu^2)\sigma_R^2\dot{a}h(K_I)/E}{E} \tag{19}$$

The function $h(K_I)$ is yet an unknown which can only be determined by
solving the coupled equations between damage and fracture. However, by
assuming very simple forms of the function h, one can analyse the R-curve using the differential equation (19), [11].

6. BENDING SPECIMEN OF BRITTLE MATERIAL

We now discuss the use of a damage fracture model for the interpretation
of some experiments on brittle fracture. A valid test for K_{Ic} of brittle
material like concrete, requires a large size specimen compared to the
heterogeneities of the material. More presisely, there are two conditions
i) the size h of the damage front is small in comparison with the spe-
    cimen size
ii) at the critical load, the damage propagation is nearly in a steady
    state condition.
The last condition means that the critical state is not influenced by the
geometry of the specimen. In oter words, the critical value of the load
can be interpreted as a characteristic of the process zone only. In
terms of the R-curve theory, the J versus $(a-a_o)$ curve is horizontal at
the critical value J_c. The above conditions are not fulfilled in the
three point bend specimen. A finite element solution has been provided
by Terrien et al.[16]. It is shown that the evolution of the damage zone

is quite complex, Fig. 6. Initially, the damage zone is bounded by a circular arc. After a propagation in the direction of the loading point,
the damage front bifurcates symmetrically. The calculated load-displacement curve under displacement control is shown in Fig. 7. We can observe
the characteristic peak behaviour and the decrease of the stiffness.
This example illustrates the case where there is no criterion in terms
of K_I, J etc. However, the global response can be determined by solving
the coupled problem of damage and elasticity. The solution to this non
-linear problem can only be derived by step increments.

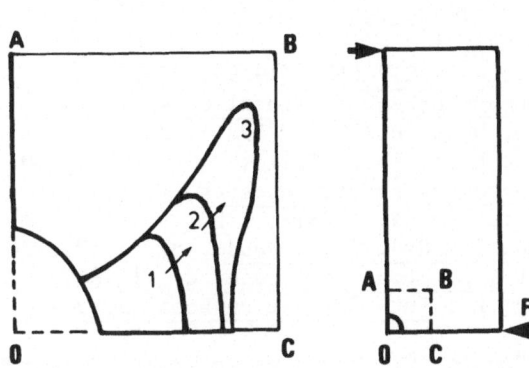

Figure 6. Damaged zone in three point
bend specimen of elastic-brittle boby.

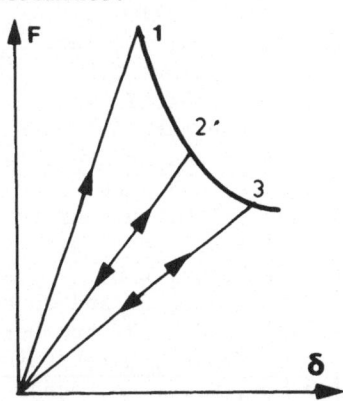

Figure 7. Calculated load
displacement curve under
displacement control.

7. CONCLUSIONS

By introducing a damage criterion in the stress-strain law, it is possible to describe fracture in brittle and ductile materials in a consistent manner.
1. Fracture is described by the propagation of a damage process zone.
2. Stable or unstable fracture can be established by means of the balance equation of energy.
3. Following an analysis of the dissipation rate, by fracture and by
plastic deformation, the R-curve theory can be derived. Generally, the
R-curve is not a characteristic of the material, but also of the geometry of the specimen. It is rather a structural characteristic curve,
particularly if the size of the damage front is large.
4. In case of small damage, both theories, damage fracture and classical
Fracture Mechanics, lead to the same result for brittle fracture. The
transition from damage fracture to Fracture Mechanics is obtained by a
limit process $\varepsilon_R \to \infty$.
5. In ductile fracture, damage fracture does not lead to a paradoxical
result concerning the fracture energy rate ($\dot{D}_f \neq 0$)

In the present lecture, we mainly discussed models of brittle and ductile
fracture. Modelling creep fracture can also be made by means of the same
idea of damage process zone, as already shown in [17], [18].

REFERENCES

[1] Mandel, J., Cahiers du Groupe Français de Rhéologie, 1, n°1,
    (1965), 1.
[2] Bui, H.D., Cahiers du Groupe Français de Rhéologie, 1, n° 1,
    (1965), 15.
[3] Taylor, G.I. and Quinney, H., Trans. Roy. Soc. London, (1931), 230.
[4] Nakada, Y., Phil. Mag., 11, n° 110, (1965), 1.
[5] Kfouri, A.P. and Miller, K.J., The Inst. of Mech. Engineers, 190,
    (1976), 571.
[6] Anderson, H., Int. J. Fract., 9 (1973), 231.
[7] Cherepanov, G.P., PMTF, 9, (1968), 1.
[8] d'Escatha, Y. and Devaux, J.C., ASTM STP 668, (1979), 229.
[9] Rousselier, G., IUTAM Symposium on Three-Dimensional Constitutive
    Relations and Ductile Fracture, North-Holland, (1981), 331.
[10]Rice, J.R. and Tracey, D.M., J. Mech. Phys. Solids, 17, (1969), 20.
[11]Bui, H.D. and Ehrlacher, A., Advances Fract. Research, 2,
    D. François et al. (Ed.) Pergamon Press, (1981), 533.
[12]Kachanov, L.N.,   Akad. Naud. SSR Ord. Tekh. Naud., 8, (1958), 26
[13]Lemaitre, J. and Chaboche, J.L., J. Méca. Appliquée, 2, (1978), 317
[14]Billardon, R. and Moret-Bailly, L., Int. Seminar on Local Approach
    of Fract., Moret/Loing (1986).
[15]Bui, H.D., Ehrlacher, A. and Renard, C., Advances in Fract. Research
    2, Valluri et al.(Ed.), Pergamon Press, (1984), 1081.
[16]Terrien, M., Bui, H.D., Acker, P., Rossi, P. Mazars, J. Legendre, D.
    François, D. and Redjel, B., Rupture du béton hydraulique en trac-
    tion, Report DGRST n°81 PO 713, Ecole Polytechnique, (1982).
[17]Bui, H.D., Dangvan,K. and de Langre, E. Proc. 2nd Int. Conf. on
    Creep & Fract. of Engng. Mat. and Struct., Swansea, April (1984)
[18]Bui, H.D., Dangvan,K. and de Langre, E. Advances in Fract. Research
    Valluri et al. (Ed.), Pergamon Press, (1984), 2207.
[19]Neuber, H., J. Appl. Mech., 83 (1961), 544.
[20]Bui, H.D., Dangvan, K. and Stolz, C., Comptes Rendus Acad. Sci.Paris
    292, (1981), 863.

DAMAGE THEORIES FOR BRITTLE AND DUCTILE MATERIALS

H. D. Bui and C. Stolz*
Electricité de France, Dept. of MMN/IMA, Clamart and
*Ecole Polytechnique, 91128 Palaiseau, France

ABSTRACT. The background of continuum damage theory is presented in the
framework of thermodynamics of irreversible processes. The assumption on
intrinsic dissipation yields the constitutive equations of the material.
Conjugate state variables are interpreted as "generalized forces".
Scalar and tensorial damage theories are discussed. Constitutive equations for elastic-plastic-fracturing solids are reviewed. The relationship between microscopic and macroscopic scales is examined in order to
understand some features of damage.

1. INTRODUCTION

Damage in continuum mechanics is induced by the initiation and growth
of microcracks and microcavities. It is an essentially discontinuous phenomenon at microscopic scale. The continuous damage theory deals with a
macroscopic approach to the constitutive equations of materials, in
which some damage parameters are considered as internal state variables.
    Kachanov [1] was the first to introduce a continuous variable ω
related to the scalar density of defects. He introduced the concept of
effective stress $\sigma_{ef}=\sigma/(1-\omega)$. This has been the starting point of damage theories developed for the analyses of fatigue in metals (Lemaitre [2]
Chaboche [3]), creep (Leckie and Hayhurst [4]), creep-fatigue interaction
(Lemaitre and Chaboche [5]).
    A different approach was proposed by Dougill [6], Bazant [7],
Dragon and Mroz [8] who applied the plasticity formalism to the description of inelastic behaviour of progressively fracturing solids. These
authors introduced the concepts of fracturing stress and fracturing
strain.
    The governing functions and parameters introduced in these theories
have still to be identified and determined further by experiments.
Because of the complex nature of damage and also the impossibility of
measuring directly the damage parameters, there little hope about
a simple theory of damage which should predict the behaviour of materials
with sufficient accuracy.

K. P. Herrmann and L. H. Larsson (eds.), Fracture of Non-Metallic Materials, 33–46.

There is a need for models which simulate the overall response of the inhomogeneous material at the microscale where developments of cracks, cavities and damaged zones are likely the main irreversible mechanisms. Such models exist in the theory of metal plasticity. At different micro-levels, the irreversibility may be induced by dislocation glides, shear slip in crystal, heterogeneous plastic deformation in polycrystalline aggregates.

Micromechanics approach to damage theory was the subject of some works, for instance Zaitsev [9], Zaitsev and Wittmann [10], Andrieux [11] [21], Marigo [12], Bui et al. [13], [14]. Such an approach presents the advantage of giving not only a comprehensive understanding of damage mechanism at the macroscale but also the guideline for a phenomenologi-cal approach to constitutive equations. The application of this analysis to concrete will be further discussed in this seminar by Wittmann and the corresponding numerical techniques will be the object of the lecture by Roelfstra.

2. THERMODYNAMICS OF IRREVERSIBLE PROCESSES

2.1. Generalized forces

Let us restrict ourself to isothermal and quasistatic loading. We neglect thermal dissipation and focus our attention on intrinsic dissipation . The thermodynamic description of a solid requires the knowledge of
i) the free energy $W(\varepsilon,\alpha)$, function of the strain ε and additional inter-nal variables α for characterizing its reversible behaviour
ii) and the intrinsic dissipation rate $\dot{D}$ for studying its irreversible behaviour, Germain et al. [15].

For pure elastic materials, the free energy $W(\varepsilon)$ is function of the strain only and the dissipation rate is identically zero. For inelastic materials, the internal variables are the set $\alpha = (\varepsilon^P, p, \omega)$ where ε^P is the plastic strain tensor, p is the cumulated plastic strain and ω is the damage parameter.

The terminology of plastic strain applies rather to metals than to non-metallic materials. For rocks or concrete it is more suitable to use the term "irreversible strain". True plastic deformation in rocks, as the result of shear slip in crystals, occurs only under very high hydro-static pressure. The irreversibility in non-metallic materials may be caused by various mechanisms such as friction in crack surfaces, propa-gation of microcracks, and as is shown further the "release of initial stress" by fracture at microlevel.

In most works, it is generally assumed that W depends on the total strain ε and the plastic strain ε^P in the following form $W(\varepsilon-\varepsilon^P, p, \omega)$ with an explicit dependence on the elastic strain $\varepsilon^e=\varepsilon-\varepsilon^P$. The conjugate variables corresponding to the strain ε and the set of internal variables α are called respectively the stress σ and the set of generalized forces A

$$\sigma = \frac{\partial W}{\partial \varepsilon} \qquad (1)$$

$$A = - \frac{\partial W}{\partial \alpha} \tag{2}$$

Let us denote the generalized forces corresponding to p and ω by R and Y respectively. Because of the assumed form of W, it is seen that the generalized force corresponding to ε^P is σ . Hence $A=(\sigma, R, Y)$. Generalized forces and fluxes yield the dissipation rate defined by the scalar product

$$\dot{D} = A. \dot{\alpha} \tag{3}$$

Equation (3) can also be written as

$$\dot{D} = \sigma_{ij} \dot{\varepsilon}^P_{ij} + R\dot{p} + Y.\dot{\omega} \tag{4}$$

2.2 Potential of dissipation

To obtain a complete description of the dissipation, we have to assume the existence of a potential $\psi(\varepsilon, \alpha, \dot{\alpha})$ which ensures that

$$A = \partial\psi/\partial\dot{\alpha} \tag{5}$$

Generally, the forces A belong to some convex of the A-space. This property is the extension of the "yield criterion" already known in the theory of plasticity. For example, in Rousselier [16], the convex C is

$$C : J(\sigma) - R + Y g(\sigma_{ii}) \leqslant 0 \tag{6}$$

where $J(\sigma)$ is the Von Misès stress deviator invariant $J=(s_{ij}s_{ij})^{1/2}$, $s_{ij} = \sigma_{ij} -\frac{1}{3}\sigma_{hh}\delta_{ii}$. Here, R can be interpreted as the radius of the yield surface. Rousselier considered a scalar damage parameter which is related to the mass density and found that the rate ω is proportional to the exponential of the mean stress.

In Lemaitre's model [2], there is no convex in the stress-space, but only a damage criterion in the strain-space. By experiments on concrete, Lemaitre and Mazars [17] have determined the function $\omega(\varepsilon)$ in the following form

$$\varepsilon \leqslant \varepsilon_o : \omega = 0$$

$$\varepsilon > \varepsilon_o : \omega = 1 - a\exp(-b(\varepsilon-\varepsilon_o)) - \varepsilon_o(1-a)/\varepsilon \tag{7}$$

where ε_o, a, b are some constants.

2.3 Evolution law

Let us consider the dual form of Eq.(5). Introduce the dual potential $\psi^*(\varepsilon, \alpha, A)$ obtained by a partial Legendre transform of the function $\psi(\varepsilon, \alpha, \dot{\alpha})$ with respect to the internal rate $\dot{\alpha}$

$$\psi^*(\varepsilon, \alpha, A) = \text{Sup} \{A.\dot{\alpha} - \psi(\varepsilon, \alpha, \dot{\alpha})\} \tag{8}$$

By differentiating the dual potential with respect to the generalized forces, one obtains the evolution law

$$\dot{\varepsilon} = \frac{\partial \psi^*}{\partial A} \tag{9}$$

Eq.(9) is nothing but the dual and equivalent form of Eq.(5). It is recognized that the evolution law for its first component $\dot{\varepsilon}^p$ is the well-known normality rule of the theory of plasticity.

2.4. Examples of dissipation in Fracture and Damage theory

So far we have considered the thermodynamic description of a homogeneous macroscopic element. The formalism presented here applies also to a system of solid, such as a cracked body. Following the analysis of [18], the crack length a can be considered as a state variable of the system and its dual variable is the "energy release rate" G. Hence the rate of energy dissipated by crack propagation is

$$\dot{D} = G\dot{a} \tag{10}$$

During crack propagation $G=G_c$, while at arrest $\dot{a}=0$ G satisfies the Griffith-Irwin criterion (the convex C)

$$G \leqslant G_c$$

As another example of dissipation, let us consider the propagation of a damaged zone in an elastic-brittle material, as already discussed in the preceeding lecture by Bui. The dissipation rate by damage fracture is given by

$$\dot{D} = \int_\Gamma W_R \phi(s)ds \tag{11}$$

where W_R is the elastic energy density at the critical strain ε_R. According to Eq.(11), the local generalized force is the elastic strain energy density prior to rupture. The local criterion of damage can be written as $W(\varepsilon) \leqslant W_R$.

3. CONTINUOUS DAMAGE MECHANICS

3.1. Scalar damage theory

The simplest model of continuous damage is that of Kachanov [1] developed further by Rabotnov [20], Lemaitre [2], Chaboche [3].
By considering the overall section area S and the effective section S_{ef} these authors introduced a scalar damage parameter ω

$$\omega = (S- S_{ef})/S \tag{12}$$

An effective stress σ_{ef} can be introduced and defined by the conservation of total force $\sigma_{ef}S_{ef}=\sigma S$. The effective stress can also be expressed in the familiar form

$$\sigma_{ef} = \sigma/(1-\omega) \tag{13}$$

The main assumption in Damage theory is that " the strain behaviour of the damaged material is represented by constitutive equation $\epsilon=g(\sigma)$ of the virgin material in the potential of which the applied stress is replaced by the effective stress".

If the elastic strain energy density of the virgin material is written as $\frac{1}{2}\epsilon^e.L_o.\epsilon^e$, the corresponding density of the damaged material is given by

$$W_d = \frac{1}{2}\epsilon^e.L_o.\epsilon^e(1-\omega) \tag{14}$$

Hence the elastic moduli tensor L_o is changed by a scalar factor. Following Lemaitre [22], applying the formalism of Section 2 to Eq.(14) one obtains the results

$$Y = - \partial W_d/\partial\omega = \frac{1}{2}\epsilon^e.L_o.\epsilon^e \tag{15}$$

$$Yd\omega = 1/2(\sigma_{ij}d\epsilon_{ij} - \epsilon_{ij}d\sigma_{ij}) \tag{16}$$

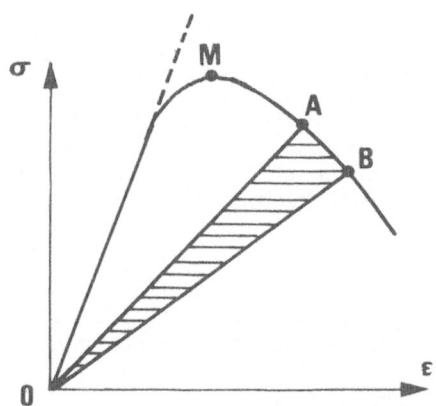

Figure 1. Elastic-brittle material : Hardening curve OM, Softening curve MB. Dissipation $Yd\omega$=(OABO).

The evolution of damage results in a decrease of the stiffness moduli, Fig.1. It might happen that the resistance of the solid first increases when ω is small (hardening behaviour) and then decreases when ω is large (softening benaviour) [13]. Generally, it is observed a light difference between the elastic response and the hardening curve. The unstable point M lies very close to the elastic line, Terrien [23]. Fig.1 shows the response of an elastic-brittle material under a cycle. The paths OA and OB are respectively elastic loading and unloading. Only along the path AB can the rate of damage be positive. The energy dissipated during the cycle OABO is equal to its area. This result is consistent with Eq. (16). It should be noted, Lemaitre and Chaboche [27], that Eq. (16) has some similarity with the well-known Irwin formula for the energy release rate in linear Fracture Mechanics.

The analysis of dissipation rate provides an additional and useful means for the identification of internal variables and generalized forces. The energy dissipated by heat can be measured by the temperature rise of the body. Modern techniques by use of the infra-red thermography with a very high accuracy ($.01°C$) have been applied to concrete and rocks by Luong [26], [34].

3.2 Anisotropic damage

Some works take account of anisotropic effects. Following Murakami and Ohno [24], the normal vectors to the surfaces S and S_{ef} might be changed by damage processes. A symmetrical second-rank tensor $\omega=(\omega_{ij})$ was introduced in the expression of the effective stress tensor

$$\sigma_{ef} = \tfrac{1}{2}(\sigma.P + P.\sigma) \tag{17}$$

where $P=(I-\omega)^{-1}$ and I is the second-rank unit tensor. A different approach was proposed by Chaboche who considered a fourth-rank tensor Ω [25]. The damage tensor is related to the applied and effective stresses by

$$(I-\Omega).\sigma_{ef} = \sigma \tag{18}$$

where I is the unit fourth-rank tensor. The equation (18) is somewhat of a generalization of Eq. (13). By using the main assumption stated in Section 3.1 the relationship between the elastic moduli of the virgin material and that of the damaged one becomes

$$(I-\Omega).L_o = L(\Omega) \tag{19}$$

An alternative approach was proposed in [13]. Introducing the elastic compliance $M_o=L_o^{-1}$ and $M=L^{-1}$, we define a fourth-rank tensor γ by

$$M_o.(I+\gamma) = (I+\gamma).M_o = M(\gamma) \tag{20}$$

The tensor γ is symmetrical. In the case of small damage($|\gamma|\ll 1$, $|\Omega|\ll 1$) by taking $\gamma=\Omega$ it is seen that Eqs.(19) and (20) are equivalent because $(I-\Omega)^{-1} = I+\Omega$ to the first order.
We will prove that Eq.(20) holds for elastic-brittle materials by studying a system of elastic solid containing damaged zones (§6.1).

So far in the previous sections we have only discussed the case of damage in elastic materials. Several models have also been proposed for predicting the behaviour under plastic strain, creep, fatigue or combined fatigue and creep conditions. Comprehensive reviews of these subjects exist already, see for example Lemaitre [22], Lemaitre and Chaboche [27]. For the interpretation of fatigue damage , we refer to the work of Dangvan [28].

4. PHENOMENOLOGICAL MODELS

We examine a different class of constitutive equations taking account of progressive fracturing and irreversible strain. The mathematical structure of equations is very similar to that of the theory of plasticity.

4.1. Fracturing stress

Dougill [6], Bazant [7] considered separately two irreversible mechanisms plastic strain and micro-fracturing strain. The elasto-plastic-fracturing model combines the elasto-plastic law and the elastic-brittle law, Fig.2.
   According to the first law , the strain is decomposed into elastic part ε^e and plastic (or irreversible) part ε^p. The elastic strain is defined by

$$\sigma = L.\varepsilon^e \tag{21}$$

According to the second law, the elastic moduli tensor is function of the damage state. Differentiating Eq. (21) with respect to a time like parameter gives

$$\dot{\sigma} = L.\dot{\varepsilon}^e + \dot{L}.\varepsilon^e \tag{22}$$

Therefore, the stress rate is decomposed into "elastic stress" rate and "fracturing stress rate", as shown in Fig. 2 and defined respectively by

$$\dot{\sigma}^e = L.\dot{\varepsilon}^e \tag{23}$$

$$\dot{\sigma}^f = \dot{L}.\varepsilon^e \tag{24}$$

In previous models, L depends on the damage parameters ω, the evolution law of which is given by Eq. (9). In the present models, the irreversible strain rate and the fracturing strain rate are assumed in the forms

$$\dot{\varepsilon}^p_{ij} = \lambda \partial f/\partial \sigma_{ij} \tag{25}$$

$$\dot{\sigma}^f_{ij} = -\mu \partial g/\partial \varepsilon_{ij} \tag{26}$$

where λ and μ are non negative multipliers, $f(\sigma,H)$ and $g(\varepsilon,K)$ are plastic potential and fracture potential respectively. In these equations H and K stand for internal variables. The potentials are also used for characterizing the yield criterion $f \leqslant 0$ and the fracture criterion $g \leqslant 0$. It is clear that, provided the evolution laws of H and K are known, the stress rate can be expressed in terms of the strain rate.
   The normality law (25), (26) are expressed in terms of the stress and strain respectively. Use of the strain-space was first proposed in [29] and further in [30], [6].
   The description of softening behaviour in the strain-space is the most appropriate means for obtaining uniqueness of the rate response, as shown in [29].

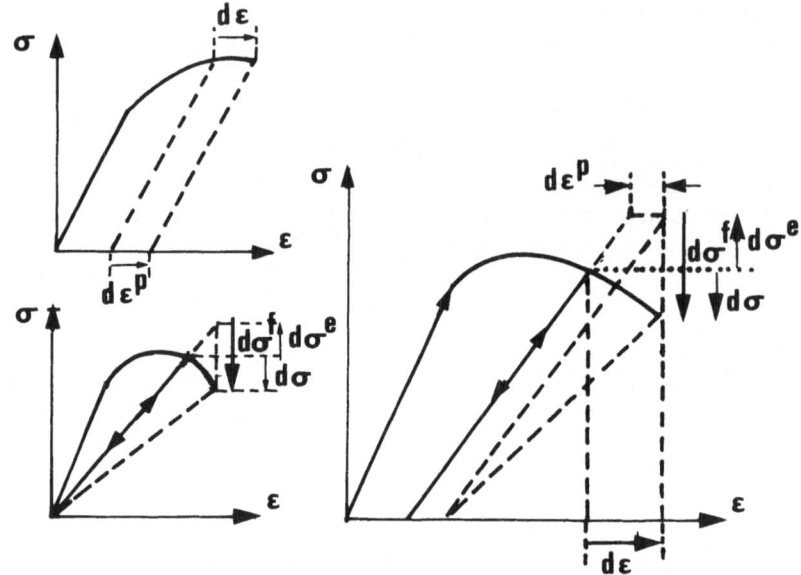

Figure 2. (a) Elasto-plastic (b) Elastic-brittle (c) Elasto-
plastic-fracturing model. Elastic stress increment
$d\sigma^e$ and fracturing stress increment $d\sigma^f$.

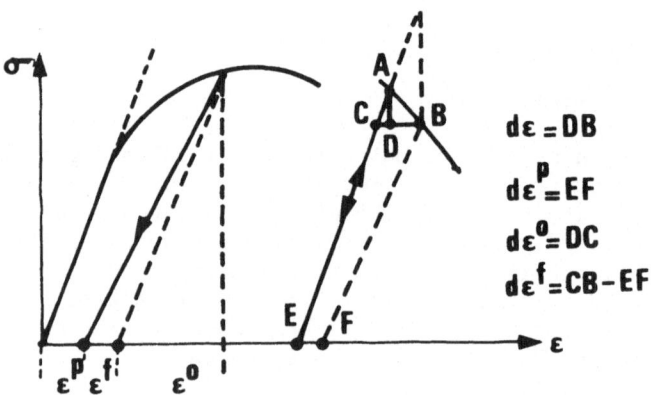

$$d\varepsilon = DB$$
$$d\varepsilon^p = EF$$
$$d\varepsilon^0 = DC$$
$$d\varepsilon^f = CB - EF$$

Figure 3. Plastic strain and fracturing strain increments.

4.2. Fracturing strain

Instead of a fracturing stress, Dragon and Mroz [8] introduced a fracturing strain ε^f. Graphical interpretations of ε^f and $d\varepsilon^f$ are shown in Fig. 3. The strain rate can be expressed as follows

$$\dot{\varepsilon} = \dot{\varepsilon}^o + \dot{\varepsilon}^P + \dot{\varepsilon}^f \tag{27}$$

where $\dot{\varepsilon}^o = M.\dot{\sigma}$. It is observed that the rates $\dot{\varepsilon}^o$ and $\dot{\varepsilon}^e$ are different, because from Eq.(22)

$$\dot{\varepsilon}^e = L^{-1}.(\dot{\sigma} - \dot{L}.\varepsilon^e) = \dot{\varepsilon}^o - L^{-1}.\dot{L}.\varepsilon^e$$

The model is completed by the evolution law, Eq.(25), for the irreversible strain rate and the equation

$$\dot{\varepsilon}^f_{ij} = \mu \partial h(\sigma_{ij},K)/\partial \sigma_{ij} \tag{28}$$

for the fracturing strain rate. An alternative form of Eq. (28) with a loading function in the strain-space $h(\varepsilon_{ij},K)$ is also given in [8]. The more general case of non-associated flow theory of plasticity and damage has been discussed in [31].

5. SIMULATIONS OF DAMAGE BEHAVIOUR

Several attempts, both numerical and theoretical, have been made to predict qualitatively the stress-strain relations for damaged materials. There is a need to use such a simulation because the identification and determination of internal variables and functions are very difficult.
     Understanding the behaviour of solids with a system of microcracks, pores or damaged zones randomly distributed in the body has been the object of some recent works.

5.1 Randomly distributed cracks and pores

Zaitsev [9], Wittman and Zaitsev [10] investigated the influence of initial distributions of flaws or cracks at pores. Knowing the behaviour of microstructures, each microstructure being characterized by a finite number of parameters such as crack length, orientation, they chose some statistical law for the initial values of parameters and generated a random macrostructure with the help of a computer. They simulated a loading curve and compared their numerical result for different initial distributions of flaws with experimental data.
     In their works, the Monte-Carlo method provides a very efficient tool for simulating the qualitative response of a macro-sample subjected to a certain loading.
     In this seminar, Roelfstra's lecture is devoted to the numerical simulation of concrete. In the preceeding lecture by Bui, the softening behaviour of a three point bend specimen of an elastic brittle material has been simulated by the propagation of one damaged zone. It is likely

that similar result holds true for a more complex system of damaged zones.

6. RELATIONSHIP BETWEEN MICROSCOPIC AND MACROSCOPIC VARIABLES

6.1. Ensemble average of microfields.

In studying polycrystalline aggregates, Hill [32], Mandel [33] established the general structures of constitutive equations and the relationship between micro-fields and the corresponding macroscopic variables.
Their methods have been applied to elastic-brittle materials in [13].
Locally, it is assumed that the damage field $\omega(x)$ takes two possible
values: either $\omega=0$ at virgin material point or $\omega=1$ at damaged zone, where
the stress tensor vanishes identically. The function $\omega(x)$ is the characteristic function of damaged zones. It is necessary to make a distinction
between the apparent volume V and the set of virgin material points,
denoted here by Ω. The damaged zones $V-\Omega$ are randomly distributed in the
macro-element in such a way that a macro-homogeneity in the sense of
Hill-Mandel can be assumed throughout the volume. Fig. 4a.

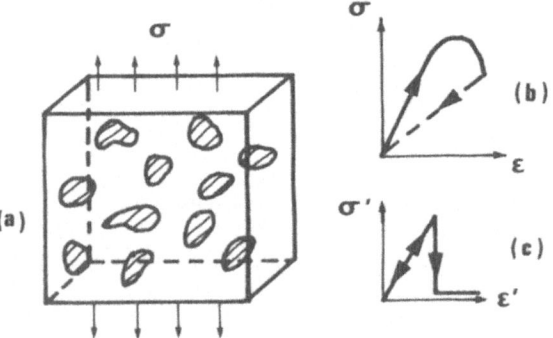

Figure 4. (a) Randomly distributed damaged zones in a macro-homoge-
neous body (b) Macroscopic behaviour
(c) Microscopic behaviour.

Let σ and ε be the homogeneous macro-stress and macro-strain respectively and let $u(x)$ be the corresponding macro-displacement field. The
corresponding micro-fields defined only in Ω are denoted by $\sigma'(x)$, $\varepsilon'(x)$
and $u'(x)$. Introduce the ensemble average over the damaged zone

$$<f> = \frac{1}{|V|} \int_\Omega f(x) \, dv \tag{29}$$

This makes it possible to relate some ensemble averages of micro-fields
to macroscopic variables.

As has been already been proved in Hill and Mandel works, we get

$$\sigma = <\sigma'> \tag{30}$$

$$\sigma_{ij} \varepsilon_{ij}^* = <\sigma'_{ij} \varepsilon_{ij}^{*\prime}> \tag{31}$$

where ε^* and $\varepsilon^{*'}(x)$ are kinematically associated and admissible fields. In the above σ and $\sigma'(x)$ are actual fields. In particular, along the boundary Γ of Ω, the micro-stress field satisfies the condition $\sigma'.n=0$.

Now, instead of the equality $\varepsilon = \langle\varepsilon'\rangle$ which holds for metal plasticity, we obtain the following one

$$\varepsilon_{ij} = \langle\varepsilon'_{ij}\rangle - \int_\Gamma 1/2(u'_i n_j + u'_j n_i) \, dS \tag{32}$$

The additional term characterizes the presence of damage. It can be used for a definition of macroscopic damage. To this end, because of the linear dependence between ε' and u', we introduce a fourth-rank tensor γ_{ijhk} such that

$$\gamma_{ijhk}\langle\varepsilon'_{hk}\rangle = - \int_\Gamma 1/2(u'_i n_j + u'_j n_i) \, ds \tag{33}$$

Combining Eqs.(32) and (33) we obtain

$$\varepsilon_{ij} = (I_{ijhk} + \gamma_{ijhk})\langle\varepsilon'_{hk}\rangle \tag{34}$$

The tensor γ provides us a mathematical measure of damage. Let us examine a consequence of Eq.(34) which confirms Eq.(20).

We consider the elastic response at two levels (microscopic and macroscopic levels)

$$\varepsilon' = M_o \cdot \sigma' \tag{35}$$

$$\varepsilon = M \cdot \sigma \tag{36}$$

By taking the ensemble average of Eq. (35) and using (34), (30) we obtain

$$(I + \gamma).M_o = M = M_o.(I + \gamma) \tag{37}$$

which confirms the equation (20).

6.2. Irreversible strain in elastic-brittle material

Micromechanics provide an interpretation of the irreversible strain which is often observed in tension loading of rocks and concrete [23]. The irreversible strain observed after an elastic unloading cannot be explained by true plastic deformation of crystals. Nor can the friction between crack surfaces be invoked in the case of tensile loading.

By studying the dissipation rate in the case where there exists an initial stress field $\sigma'_o(x)$, the field σ'_o being a self-equilibrated micro-stress, it has been shown in [14] that there are two terms

$$\dot{D} = \tfrac{1}{2}(\sigma_{ij}\dot{\varepsilon}_{ij} - \varepsilon_{ij}\dot{\sigma}_{ij}) + \sigma_{ij}\dot{\varepsilon}^{ir}_{ij}$$

The first one is the same as derived by Lemaitre in Eq. (16). The second term is similar to the plastic rate, with the irreversible strain rate

$$\dot{\varepsilon}^{ir} = \langle A^t \cdot M_o \cdot \dot{\sigma}'_o\rangle \tag{38}$$

where A^t is the transpose of the fourth-rank tensor $A(x)$ which is the
so-called "elastic stress localization tensor" introduced by Hill and
Mandel. The tensor $A(x)$ gives the elastic response $\sigma'(x)=A(x).\sigma$ under
the macrohomogeneous stress σ. The irreversible strain rate arises from
the irreversible release of the initial stress, due to the propagation
of damaged zones. The following example will illustrate the irreversible
strain in elastic-brittle material.

6.3. The three elastic-brittle bars

Consider three bars of same section area, elastic moduli and fracture
stress $\sigma_R=3$ (normalized by some reference stress). We suppose that initially the first and the third bars are in compression $\sigma'_o=-1$ while the
second one is in tension $\sigma'_o=2$. The system of bars is self-equilibrated.
We apply now the load to the system of bars. It is clear that fracture
occurs at $\sigma=3$, and is localized in the second bar. At constant applied
load $\sigma=3$, the release of stress due to rupture of the second bar will
cause a jump on deformation of the remaining bars. Because of the decrease on stiffness moduli, we obtain an irreversible strain after total
unloading. This is clearly shown in Fig.5.

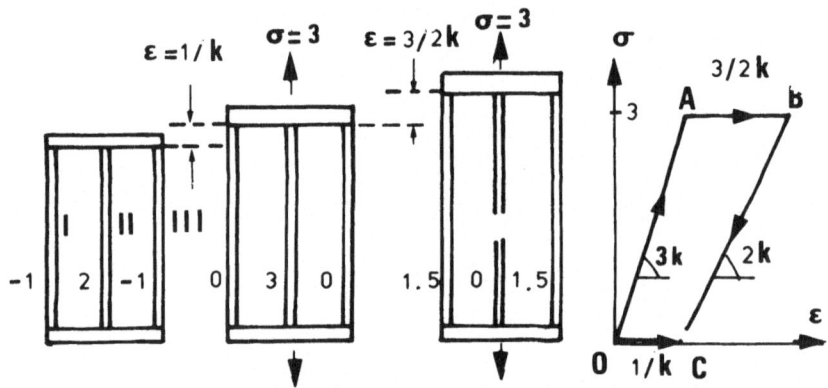

Figure 5 . The three elastic-brittle bars. The release of initial
stress at the 2nd bar induces an irreversible strain OC.

6.4. Evolution of damage

Let us reconsider the simpler case of an elastic-brittle material in its
natural state $(\sigma'_o(x)=0)$ and examine how the evolution of damage at the
macroscale, $\dot{\gamma}$, can be related to the velocity field $\phi(s)$ of damaged zones
at the microscale.

As already mentionned in the preceeding lecture by Bui, there are
additional relations between the microfield rates $\dot{\sigma}'$, $\dot{\epsilon}'$ and $\dot{\phi}(s)$
which can be used for determining these unknowns. The reader can refer
to [35] for more details on the derivation of $\phi(s)$ from external loading
rates $\dot{\sigma}.n=\dot{T}$ on S_t and $\dot{u}=v$ on S_v.

The following result was obtained in [35]. The fields $(\dot{u}',\phi)$ are solutions of a variational problem with the functional

$$J(\dot{u}',\phi) = \int_{\Omega} \tfrac{1}{2}\dot{\varepsilon}'.L_o.\dot{\varepsilon}' \; dv +$$
$$+ \int_{\Gamma} (\tfrac{1}{2}n_i W,_i \phi^2 - \phi\sigma_R\tau.\frac{d\dot{u}'}{ds}) \; ds - \int_{S_t} \dot{T}.\dot{u}'ds \qquad (39)$$

where $W=\tfrac{1}{2}\varepsilon'.L_o.\varepsilon'$, τ is the unit tangent vector to Γ, σ_R is the fracture stress. We have to seek a solution in the set of kinematically admissible fields

$$\phi \geqslant 0 \quad \text{and} \quad \dot{u}'= v \quad \text{on } S_v$$

The dependence of ϕ on the loading rates makes it possible to establish that the rate $\dot{\gamma}$ is entirely determined by solving a boundary value problem. More precisely, one can prove the following result [14]

$$\dot{M} = \int_{\Gamma} A^t.M_o.A \; \phi(s)ds \qquad (40)$$

which can be combined with Eqs. (37) for determining the rate

$$\dot{\gamma} = L_o.\int_{\Gamma} A^t.M_o.A \; \phi(s)ds \qquad (41)$$

Therefore, the micromechanical approach does not require additional evolution law such as Eq.(9).

7. CONCLUSIONS

In this lecture, we reviewed some background on continuum damage theories and focused our analysis on the dissipation rate by damage.

The merit of a phenomenological model is its simplicity. However, there are still difficult problems in the determination of internal variables and governing functions.

The approaches by micromechanical models and by numerical simulations seem to be more compléx than the first one. However, these models provide a better understanding of damage behaviour of materials.

We mainly discussed the simple model of sudden damage, as introduced in the preceeding chapter. We shown that damage can be characterized by a fourth-rank symmetrical tensor, whose rate of evolution can be theoretically determined by solving a boundary value problem.

ACKNOWLEDGEMENTS. The authors wish to thank N.Ohno for many valuable discussions. Many thanks are also due to J.L. Chaboche and J. Lemaitre for their book on the materials Sciences which gives the material for this paper.

REFERENCES

[1]   Kachanov, L.M., _Ivz. Akad Naud S.S.R. Otd Tekh Naud_, 8 (1958), 26.

[2]   Lemaitre, J., Proc. I.C.M. 1, Kyoto (1971)
[3]   Chaboche, J.L., Revue Française de Méca., n°50 (1974)
[4]   Leckie, F. and Hayhurst, D.,Proc. Roy. Soc., London,240 (1974)323
[5]   Lemaitre, J. and Chaboche, J.L., IUTAM Symposium Mech. of Visco-
      elastic media and bodies, Gothenburg, Springer Verlag (1974)
[6]   Dougill, J.W., Z.A.M.P., 27, n°4 (1976) 423.
[7]   Bazant, Z., Int.J.Solids&Struct.,(1980) 873.
[8]   dragon, A. and Mroz, Z. Int. J. Engng. Sci., 17 (1979) 121
[9]   Zaitsev, Y.V., Proc. 7th Int. Conf. Chemistry Cement, 4, (1980)176
[10] Wittmann, F.H. and Zaitsev, Y.V. Advences in Fract. Research, ICF5
      Cannes, D. François et al.(Eds), 5, (1981) 2261
[11] Andrieux, S., Thèse Univ. Paris, (1983)
[12] Marigo, J.J., Engng.Fract.Mech., 21, n°4 (1985) 861.
[13] Bui, H.D:,Dangvan,K. and Stolz, C.,Comptes Rendus Acad. Sci.Paris
      292, (1981) 863.
[14] Bui, H.D.,Dangvan,K. and Stolz, C.,Comptes Rendus Acad. Sci.Paris
      294, (1984) 1155.
[15] Germain, P., Nguyen, Q.S. and Suquet, P., J.Appl.Mech.,50, (1983)
      1013.
[16] Rousselier, G., IUTAM Symp.on Three-dimensional Constitutive Eqns.
      and Ductile Fract., Dourdan, North-Holland, (1981) 331.

[17] Lemaitre, J. and Mazars, J.,Annales Inst.Tech.Bâtiments & travaux
      publics, n° 401 (1982) 115.
[18] Nguyen, Q.S., IUTAM Symp. on Three-dimensional Constitutive Eqns.
      and Ductile Fract., Dourdan, North-Holland, (1981) 315.
[19] Bui, H.D. and Ehrlacher, A., Advances in Fract. Research, ICF5
      Cannes, D. François et al. (Eds.), 2, (1981) 553.
[20] Rabotnov, L.M., Proc. 20th Int. Cong. Appl. Mech. (1968).
[21] Andrieux, S., Comptes Rendus Acad. Sci. Paris, 293, (1981) 329.
[22] Lemaitre, J., Séminaire Matériaux & Struct. Sous chargements cy-
      cliques, Ecole Polytechnique, (1978) 133.
[23] Terrien, M., Bulletin de liaison des Ponts et Chaussées, n° 105
      (1980) 65.
[24] Murakami, S. and Ohno, N., IUTAM Symp. on creep in Structures,
      Leicester (1980)
[25] Chaboche, J.L., Euromech Colloquium 115, Grenoble (1979)
[26] Luong, M.P., Int.Conf.Fract.Mech. of Concrete, Lausanne (1985)
[27] Lemaitre, J. and Chaboche, J.L., "Mécanique des matériaux Solides"
      Dunod Ed. (1985).
[28] Dangvan, K., Sci. Tech. de l'Armement, 47, (1973).
[29] Nguyen, Q.S. and Bui, H.D., J. Méca., 13, (1974) 321.
[30] Naghdi, P.M. and Trapp, J.A., Int. J. Engng. Sci., 13, (1975) 785.
[31] Bui, H.D., Seminar Fract. Mech., St.Rémy-les-Chevreuses (1982) 1.
[32] Hill, R., J.Mech.Phys. Solids, 15, (1967) 79.
[33] Mandel, J., Proc. 11th Cong.Appl.Mech.,Munich (1964) 502.
[34] Luong, M.P., Int. Conf. Dynamics of Materials, Ecole Polytechnique
      Palaiseau (1985).
[35] Bui, H.D., Dangvan, K. and Stolz, C.,Comptes Rendus Acad. Sci.Paris
      292, (1981) 251.

NUMERICAL METHODS IN FRACTURE MECHANICS

W. Schmitt
Fraunhofer-Institut für Werkstoffmechanik
Wöhlerstrasse 11
7800 Freiburg
Federal Republic of Germany

ABSTRACT. After a short introduction sketching the continuum mechanics description, the basic equations of Fracture Mechanics will be evaluated with particular emphasis on the numerical treatment.

Among the numerical methods the Finite Element Method (FEM) will be discussed in more detail, including special crack tip elements, energy release rates and path-independent integrals.

The problem of short cracks will be touched as well as the damage function approach. A series of actual cases will be presented to demonstrate the applicability and validity of the methods introduced.

1. Introduction

The mechanical behavior of solids may be understood on the basis of the atomistic structure of the material and of the forces which cause the coherence of individual atoms or molecules. The macroscopic strength expressed e.g. as the maximum stress that can be sustained by a structural member of this material is only to a small extent influenced by theoretical atomic or molecule bond forces, as discussed by Sommer in the first lecture of this seminar. Instead, the characteristics of imperfections, their number and distribution determine the actual resistance of a material against failure.

Since information about these microscopic quantities is generally not readily available, and, if available, it could not directly be used to determine the actual behavior of a structure, as a first approximation to the real problem the concept of a continuum has been introduced. The continuum can be regarded as the mechanical model. The state of a continuous body can be described in terms of stresses, strains and displacements. Between those quantities mathematical relations exist in the form of e.g. differential equations. These relations already define the mathematical model, which also includes the boundary conditions defining an actual problem. Since exact solutions are availabe only for a very limited number of mathematical problems, in general numerical models have to be set up and solved.

It becomes now evident that this chain of simplifications from the real problem down to a mathematical or even numerical model must be

K. P. Herrmann and L. H. Larsson (eds.), Fracture of Non-Metallic Materials, 47–74.
© *1987 by ECSC, EEC, EAEC, Brussels and Luxembourg.*

regarded as a source of systematic errors. This may be demonstrated by the example of an elastic-plastic fracture mechanics test with a compact specimen which was to be analysed by the finite element method. Figure 1 demonstrates a very good agreement between analysis and experiment (here: force-displacement curve, from [1]) if a 3D-analysis is considered. Attempts to apply 2D-models lead only to some qualitative agreement in the overall performance of test and analysis. Two-dimensional simplifications in numerical analysts have been employed mainly in order to reduce the computational effort. This argument becomes less important with the rapid increase in available computer power.

In analytical and theoretical work the reduction by one dimension allows the study of singular stress and strain fields in the vicinity of crack tips resulting in a variety of near-field solutions. It must be emphasized, however, that mechanical or mathematical models incorporating only a reduced number of dimensions represent a significant deviation from the (three-dimensional) reality. This principal deficiency cannot be overcome by any numerical effort. The numerical solution will not be closer to reality than the mechanical model, except by error.

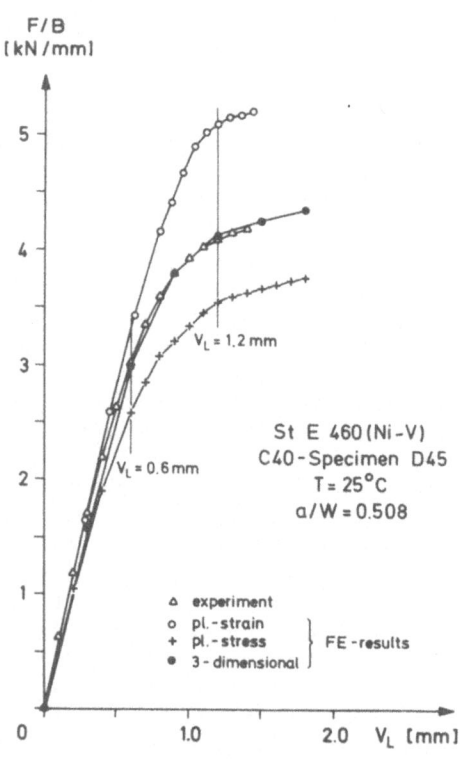

Fig. 1: Compact specimen, force-displacement diagrams, ref. 1

To demonstrate that also within numerical solutions of a given model there may be a significant amount of uncertainty, the results of one of the european round robin exercises will be shown. Here, the mechanical model to be treated was well defined: Three-point bend specimen, geometry, non-linear stress-strain curve, plane strain. Figure 2 gives the computed force-displacement curves. Possible sources of this immense scatter may only be sought in the numerical procedures employed by different participants and are explored in detail in [2].

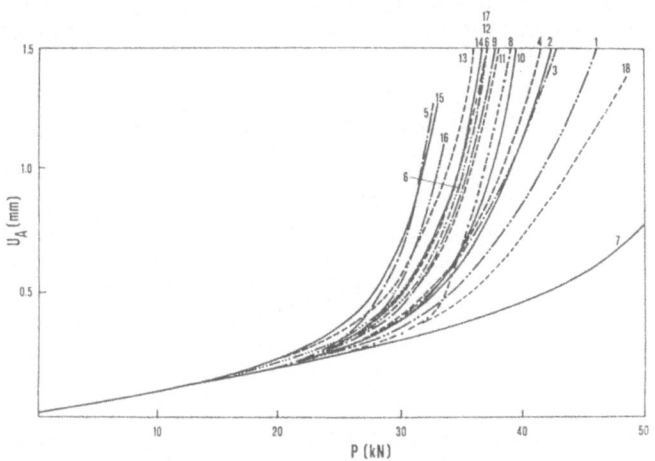

Fig. 2: 3P-Bend Specimen, force-displacement diagrams, ref. 2

From the discussions of the results it could not be concluded which solution was to be considered the "correct" solution since experimental verification is not possible and "exact" theoretical solutions of this problem (although 2D) are not available. The recognition of this fact has led to a redefinition of later stages of this round robin exercise.

The difficulties demonstrated in the last example are not typical for the majority of fracture problems of non-metallic materials, where linear elastic fracture mechanics (LEFM) is applicable. In LEFM the numerical methods and modelling techniques are well developed and reliable, if correctly applied.

2. BASIC EQUATIONS OF FRACTURE MECHANICS

In the following, the principal equations and relations of continuous fracture mechanics are compiled and discussed from the standpoint of energy conservation.

A continuous body with a crack loaded by external or internal forces will build up a stress-strain field which has singular terms when the crack tip is approached. Specifically the strain energy density

$$W(\varepsilon_{ij}) = \int_0^{\varepsilon_{ij}} \sigma_{ij} d\hat{\varepsilon}_{ij} \qquad (1)$$

has a singularity of the order r^{-1}. If the material is linear elastic the stress and strain components exhibit singularities of the order $r^{-1/2}$ and the asymptotic field around the crack tip is described by the well known Sneddon equations [3] for the mode I loading:

$$\sigma_{ij} = K_I/(2\pi r)^{1/2} f_{ij}(\beta) \qquad (2)$$

$$\varepsilon_{ij} = K_I/(2\pi r)^{1/2} g_{ij}(\beta) \qquad (3)$$

Here, r and β are polar coordinates with origin in the crack tip, K_I is the stress intensity factor, $f_{ij}(\beta)$ and $g_{ij}(\beta)$ are dimensionless functions.

If the nonlinear stress-strain curve of the material can be described by a power law

$$\sigma/\sigma_0 = (\varepsilon/\varepsilon_0)^n \qquad (4)$$

(σ_0 and ε_0 are the limits of proportionality of the uni-axial stress-strain curve, n is a hardening exponent), and if higher order terms are neglected, the (two-dimensional) asymptotic stress and strain field is given by the HRR-field after Hutchinson, Rice and Rosengren [4, 5] as

$$\sigma_{ij} = \sigma_0 [J/(r\sigma_0 \varepsilon_0 I_n)]^{n/n+1} \cdot \dot{f}_{ij}(\beta) \qquad (5)$$

$$\varepsilon_{ij} = \sigma_0 [J/(r\sigma_0 \varepsilon_0 I_n)]^{1/n+1} \cdot g'_{ij}(\beta) \qquad (6)$$

$f'_{ij}(\beta)$ and $g'_{ij}(\beta)$ are dimensionless functions, I_n is a numerical parameter which is different for plane strain and plane stress. The intensity of the singularity is characterized by the J-Integral

$$J = \oint_\Gamma (W\,dy - \vec{T}\partial\vec{u}/\partial x\,ds) \qquad (7)$$

In this original formulation by Rice the crack is oriented parallel to the x-axis, u is the displacement vector.

$\vec{T} = \{T_i\} = \sigma_{ij}n_j$:   traction vector

$\vec{n} = \{n_i\}$: outward normal vector to Γ

Γ : contour with arc element ds, surrounding the crack tip

In a more general formulation J my be interpreted as the first component of a vector of "material forces"

$$J_i = \int_\Gamma (W\delta_{ij} - \sigma_{kj}u_{k,i})n_j ds \qquad (8)$$

$_i$ denotes partial derivative with respect to x_i, δ_{ij} is Kronecker's delta.

Path-independent integrals for two-dimensional crack problems were introduced independently by Cherepanov [6] and Rice [7].

The theoretical basis for the J-concept may be found by consideration of conservation laws within the theory of elasticity. Eshelby [9, 10, 11] found that the force acting on a defect can be expressed as a surface integral of the energy momentum tensor. Buggisch et al. [12] generalized these considerations and considered also dissipative terms. Especially Rice's work gave significant impulses towards the application of the J-integral to fracture problems. He was able to demonstrate that J is equivalent to the energy release rate G which was first introduced by Griffith [8]:

$$J = G = -\partial U/\partial A \qquad (9)$$

U is the potential energy of the system, A is the area of the crack.

The application of J in experiment and in most of the analytical and numerical evaluations depends on the validity of eqn. (9). The proof was given for all materials which can be described by a deformation plasticity (i.e. the state of stress is uniquely defined by the state of strain and vice versa) or, if incremental plasticity is assumed, as long as the loading of the structure is proportional. Despite the fact that this latter assumption is violated if crack extension takes place, the error was considered tolerable if the amount of crack extension was small compared with other dimensions of the structure.

Based on the observation that in numerical analyses even with a substantial amount of stable crack growth the equivalence of eqn.(9) could be maintained, provided that the numerical solution was adequate ([13]), a reexamination of the conditions for path-independence of J was carried out, showing that eqn. (9) is strictly valid not only for elastic (linear or nonlinear) materials but also for those described by an incremental law of plasticity, provided a stress work density W is suitably defined. This can easily be demonstrated by postulating that J according to eqn. (7) or any of the components J_i according to eqn. (8) vanish identically for any closed contour not containing a singularity.

$$J_i = \int_\Gamma (\tilde{W}\delta_{ij} - \sigma_{kj}u_{k,i})n_j ds = 0$$

Application of Gauß's theorem yields

$$J_i = \iint_A (\tilde{W}\delta_{ij}\big|_{,j} - [\sigma_{kj}u_{k,i}]_{,j}) dA = 0$$

$$J_i = \iint_A (\tilde{W}_{,i} - \sigma_{kj,j}u_{k,i} - \sigma_{kj}u_{k,ij}) dA = 0$$

the second term vanishes because $\sigma_{kj,j} = 0$, and the last term yields

$$\sigma_{kj}(u_{k,j} + u_{j,k})_{,i}/2 = \sigma_{kj}\varepsilon_{kj,i}$$

Therefore,

$$J_i = \iint_A (\tilde{W}_{,i} - \sigma_{kj}\varepsilon_{kj,i}) dA = 0$$

This is identically the case if

$$\tilde{W}_{,i} = \partial\tilde{W}/\partial x_i\big|_{expl.} + \partial\tilde{W}/\partial\varepsilon_{kj}\,\varepsilon_{kj,i} = \sigma_{kj}\varepsilon_{kj,i}$$

e.g. if W does not explicitly depend on x_i. The last equation defines an algorithm to calculate W:

$$\tilde{W} = \int_0^{\varepsilon_{ij}} \sigma_{ij}d\hat{\varepsilon}_{ij} \tag{10}$$

Eqn. (9) may be replaced by

$$J = - \partial \tilde{U} / \partial A \qquad (11)$$

with

$$\tilde{U} = \int (\tilde{W} - \vec{p}\,\vec{u}\,) dV \qquad (12)$$

$\vec{p}$ and $\vec{u}$ are vectors of external forces and displacements, respectively.

The above arguments for the path-independence of J and the validity of (11) also for materials described by an incremental theory of plasticity have developed in a series of discussions. Especially Kienzler [14], Maschke and Kuna [15] gave significant contributions to the theoretical foundation.

As already stated above, the validity of eqns. (9) or (12) is the basis not only for experimental evaluation of J, but also for the most commonly used "stiffness derivative method" in FEM, which will be explained in the next chapter.

For time dependent material laws (e.g. creep) the role of the J-integral is taken over by the C*-integral [16]

$$C^* = \oint_\Gamma (\dot{W}dy - T\,\partial\dot{\vec{u}}/\partial x \; ds) = \partial U^* / \partial A \qquad (7a)$$

with

$$U^* = \int (\dot{W} - \vec{p}\,\dot{\vec{u}}\,) \, dV \qquad \dot{W} = \int_0^{\dot{\varepsilon}_{ij}} \sigma_{ij} d\dot{\hat{\varepsilon}}_{ij} \qquad (12a)$$

3. NUMERICAL METHODS, ESPECIALLY THE FINITE ELEMENT METHOD (FEM)

The limitations of the human mind are such that it cannot grasp the behavior of its complex surroundings and creations in one operation. In many situations an adequate model is obtained using a finite number of well-defined discrete components. In others the subdivision is continued indefinitely and the problem can only be defined using the mathematical fiction of an infinitesimal. This leads to differential equations or equivalent statements which imply an infinite number of elements. These continuous problems can only be solved exactly by mathematical manipulation, where the available techniques usually limit the possibilities to oversimplified situations.

To overcome this intractability of realistic problems, the discretization of continuum problems has been approached differently by mathematicians and engineers. The first have developed general techniques applicable directly to differential equations, as for instance the <u>finite difference method</u>. The engineer, on the other hand, approaches the problem by discretizing the continuum domain into a number of small <u>finite elements</u>, each of which behaves in a simplified manner (from O.C. Zienkiewicz: The finite element method, [17]).

The reader is encouraged to use Zienkiewicz's book [17] as an introduction into the FEM-method; especially non-linear and dynamic problems are treated in Bathe and Wilsons's book [18]. Here, only applications of the FEM will be discussed, after dealing with some special techniques for treating crack problems in the FEM.

3.1 Crack tip elements

The numerical treatment of crack problems requires the consideration of singular stress-strain fields. In elastic problems the stresses and strains have a $r^{-1/2}$-singularity, in elastic-plastic problems the stresses are bounded and the strains have approximately an r^{-1}-singularity. For elastic three-dimensional problems special crack tip elements have been proposed by Tracey [19]. Later, Henshell and Shaw [20] as well as Barsoum [21] discovered independently that standard quadratic isoparametric elements may be turned into singularity elements by special arrangements of the positions of the nodal points without having to modify the program. A compilation of the various possibilities and applications in fracture mechanics is given in [1] and [22].

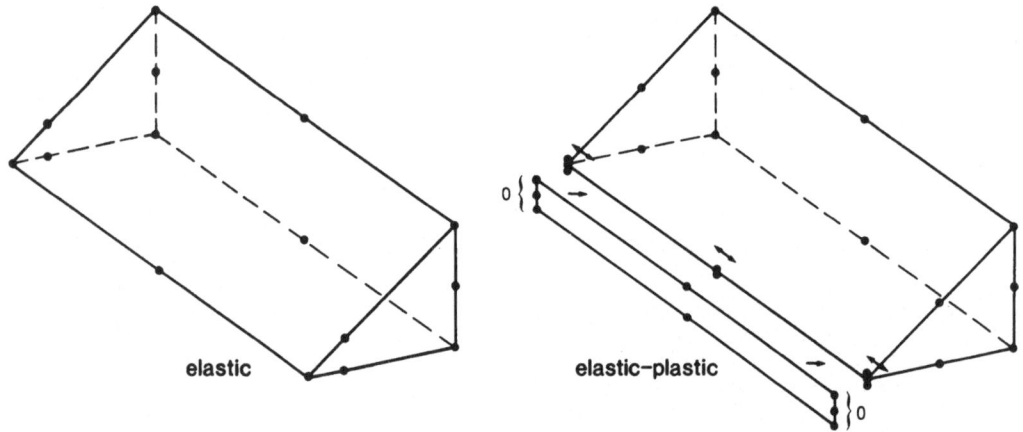

elastic   elastic-plastic

Fig. 3: The collapsed element technique

For plane crack problems the following results are important [1]:

- By degenerating one side (i.e. three nodes) on an isoparametric (quadratic) element to one point, but treating the collapsed nodes with <u>independent degrees of freedom,</u> an r^{-1} singularity of the strains is obtained.

- If the mid-side nodes of the element sides adjacent to the
  collapsed tip are shifted into positions with distances from
  the tip of one quarter of the element side, an $r^{-1/2}$-singu-
  larity in the strains is obtained.

Both types of singularities are independent from each other and
may be superposed. For practical problems it is advantageous to proceed
as follows:

- For <u>linear elastic</u> crack problems the $r^{-1/2}$-singularity in
  stresses and strains is appropriate. Therefore, the <u>quarter-</u>
  <u>point</u> element should be used with just one individual node
  point number at the crack tip.

- For <u>elastic-plastic</u> problems the r^{-1}-singularity in the
  strains is a good model, since the stresses are bounded. Therefore, collapsed elements with <u>multiple</u> (independent) <u>crack tip</u>
  <u>nodes</u> should be used. To avoid numerical problems, the mid-side
  nodes of the adjacent sides should remain in the <u>half-point</u>
  position.

Three-dimensional problems may be treated in an analogous way.
Figure 3 shows the proposed techniques in 3D.

3.2 Evaluation of energy release rates and path-independent integrals

In two-dimensional cases in the absence of body forces and initial
strains, J can be computed directly from eqn. (7). More efficiently, J
can be obtained utilizing the equivalence of J with the energy release
rate, eqn. (9), eqn. (11). Especially for FEM applications the <u>virtual</u>
<u>crack extension method</u> was introduced by Parks ([23, 24]) and later mo-
dified by DeLorenzi ([23]). Bakker ([26]) included body forces and
initial strains. It was herewith possible to extend the J-integral
concept to three-dimensional situations to obtain local values along a
three dimensional crack front.

The complete expression for J can be split into several terms:

$$J_{tot} = J_{stat} + J_T + J_p + J_d \qquad (13)$$

Here, J_{stat} refers to the static case (eqn. 7) when the external
loads do not vary with the virtual crack extension, as would for instance be the case if the crack flanks were loaded by pressure.

IWM has implemented J- and C*-routines based on the contour integral and on the virtual crack extension method for 2D and 3D situations, arbitrary material and loading ([1,27]).

3.3 Evaluation of the linear elastic stress intensity factor K_I

From Sneddon's equations in the vicinity of a crack tip the stress intensity factor may be derived, if stresses, strains or displacements are known with sufficient accuracy. In numerical solutions the accuracy usually decreases if a singularity is approached. Therefore extrapolation to r=0 at constant angle ß usually improves the accuracy.

$$K_I = \lim_{r \to 0} \sigma_y(r)\,(2\pi r)^{1/2} \qquad\qquad (14)$$

from the ligament stresses (ß=0)

$$K_I = \lim_{r \to 0} \frac{v(r)\,E}{4(1-f\nu^2)}\,(2\pi/r)^{1/2} \qquad (15)$$

from COD (ß=π)

Most advantageous and most accurate is the evaluation of K_I from the J-Integral

$$K_I = [J\,E/(1-f\nu^2)]^{1/2} \qquad\qquad (16)$$

with σ_y: stress normal to the ligament

v: crack opening displacement

E: Young's modulus

ν: Poissons's ratio

f: factor (f=1 for plane strain, f=0 for plane stress)

According to [17] the discretisation error in the FEM for second order elements is $O(h^2)$ for stresses and strains, $O(h^3)$ for displacements and $O(h^4)$ for energy terms. Therefore, the evaluation of K_I from eqn. (16) is most preferable, followed by eqn. (15) and, last, eqn. (14).

3.4 Simulation of crack growth

In order to include crack propagation in the finite element model a method developed by DeLorenzi [28] was modified and extended to three-dimensional applications. The node shifting and releasing technique is explained in Figure 4 for 2D. The crack propagation law of the model is either an experimental observation, e.g. a measured v(Δa) curve, or a material resistance curve, e.g. J(Δa). According to this law the crack tip nodes and the respective side nodes are shifted by small increments. In the following load step the stiffness matrix is reformed.

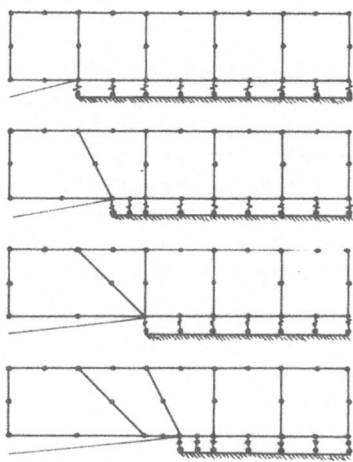

Fig. 4: The node shifting and releasing technique

The stresses and strains of the preceding load step are used as starting values for the equilibrium iterations. The convergence of these iterations may be substantially impeded by the fact that local unloading occurs as the crack propagates. In this case it is helpful to use an elastic stiffness matrix for the iterations. When the corner node of the next element is reached the two nodes of the first element are released. This procedure allows small amounts of crack extension to be produced even with relatively coarse finite element grids, which is important for 3D-applications.

3.5 The problem of short cracks

The concept of fracture mechanics - analytical or numerical - assumes that the behavior of the crack is determined by a singular stress or strain field surrounding the crack tip. This field is obtained under the assumption of homogeneous material. The singular field is embedded by a non-singular far field. The process zone is small compared with the crack size.

In particular in cases where critical crack sizes are very small (say, in metals of the same order of magnitude as grain sizes) the validity of these assumptions becomes questionable. It is therefore a matter of principle whether fracture mechanics based e.g. on the K-concept is applicable ([29]).

If the answer is positive, the numerical treatment of short cracks poses additional difficulties. Even if it is assumed that the same type of singularity elements is applicable and that the virtual crack extension method works, sufficiently accurate FEM models may require so many degrees of freedom that they can be treated only with very powerful computers, especially for 3D problems.

3.6 The damage function approach

The numerical procedures discussed so far assume one distinct crack embedded in a continuum. The material law of the continuum does not change. If the crack propagates the geometry of the model is adjusted. An alternate approach originating from microscopic considerations models the local damage at a crack tip (e.g. fractional void volume) as a function e.g. of the equivalent strain. This damage modifies the constitutive equations of the material in a continuum mechanics sense. The damage function approach is the object of the lecture by Bui and Stolz in this seminar.

4. EXAMPLES

In the first example it will be demonstrated that proper application of the FEM leads to accurate results in cases where the exact solution is known. In [1] the convergence behavior of the K_I-analysis of a compact specimen was examined, compared with the solution of [30]. The relative error in K_I with respect to the exact solution is given in table 1 for different mesh refinements, indicated by the total number of elements, NEL. The crack tip was modelled by quarter-point singularity elements.

NEL	Eqn.(14)	Eqn.(15)	Eqn.(16)	
4	9	16	15	%
16	13	5	1	
64	9	3	1	

Table 1: Accuracy of K_I-solutions

Except the case with only 4 elements the superiority of eqn.16 is clearly demonstrated.
    To demonstrate that the FEM is flexibly adaptable to complex geometry and loading situations, the analysis of axial surface flaws in a hollow cylinder (steel) subject to pressure and thermal loading is presented ([31]). Three-dimensional elastic FEM-calculations were performed to study the variation of K_I along the crack front of two semi-elliptical surface flaws in a pressure vessel.
    Fig. 5 shows the stress intensity factor K_I for one of the crack geometries of [31] over the position along the crack front for pure pressure loading. The evaluation of K_I is done according to eqn.(16) and (15) and compared with a reference solution. Fig. 6 shows K_I for temperature loading (thermal shock) for the same geometry.

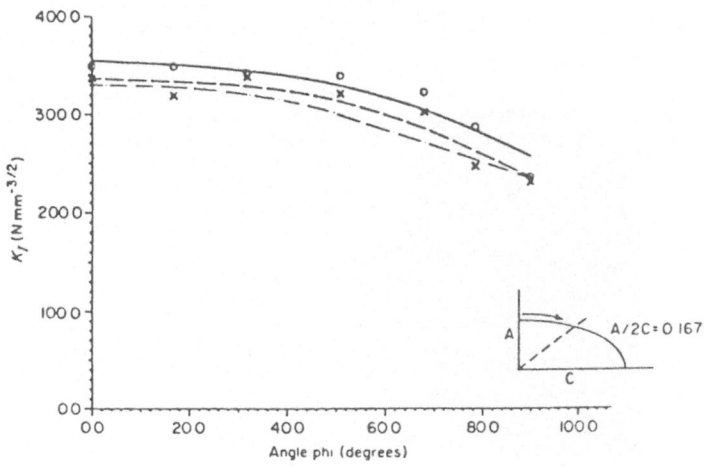

Fig. 5: Vessel with surface flaw: K_I for pressure loading, ref. 31

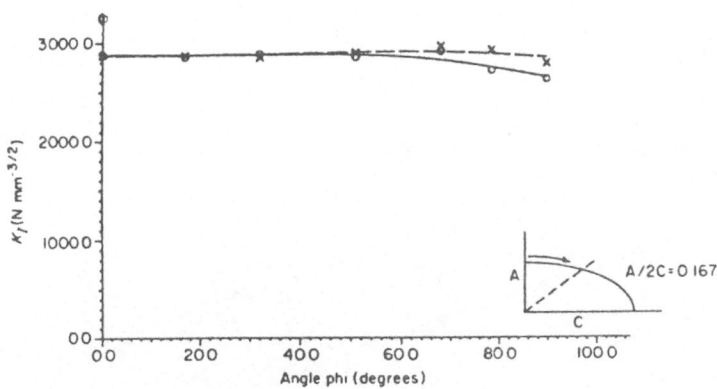

Fig. 6: Vessel with surface flaw: K_I for temperature loading, ref. 31

The last example on steel is meant to give an idea of the power of the numerical techniques if crack propagation and material non-linearities are involved. This is certainly not representative for most of the applications of FM in non-metallic materials, but at elevated temperatures non-linear effects may also become important here.

In this study ([32]) it was attempted to apply a J(Δa)-curve obtained from a side-grooved compact specimen to a smooth specimen of the same material tested at the same temperature. Fig. 7 shows the frac-

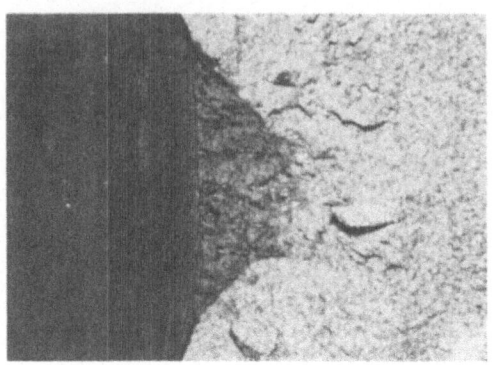

Fig. 7: Compact specimen, Steel A 542, fracture surface

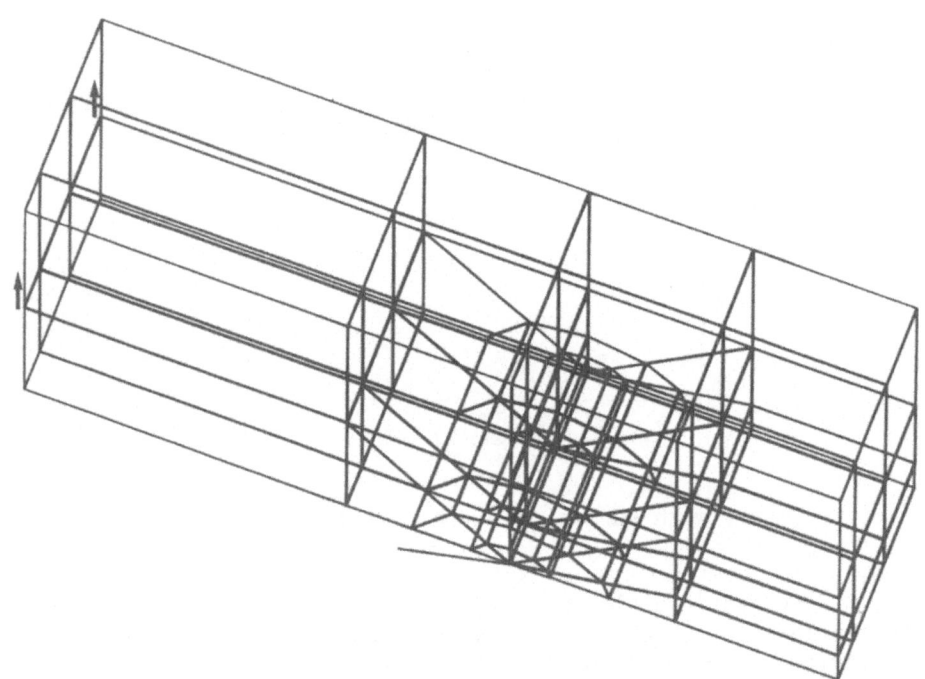

Fig. 8: Compact specimen, steel A 542, FEM-mesh 3D

ture surface of this specimen. From the pronounced tunneling of the crack front it is evident that only a 3D analysis including stable crack extension would be appropriate. Figure 8 shows the FEM-mesh employed (one quarter of the structure due to symmetry). The simulation is done by applying constant displacement increments along the load line of the specimen. The reactions of the specimen are calculated considering a multi-linear stresss-strain curve assuming an incremental law of plasticity, i.e. the unloading path of the stress-strain curve is elastic.

The J-integral (Figure 9) was evaluated at four points across the (half-)specimen thickness according to eqn. 11. The crack front was locally advanced (Figure 10) according to the J-resistance curve determined from the side-grooved specimen. The overall agreement of analysis and experiment is surprisingly good, as can be seen from the force-displacement diagrams, Figure 11.

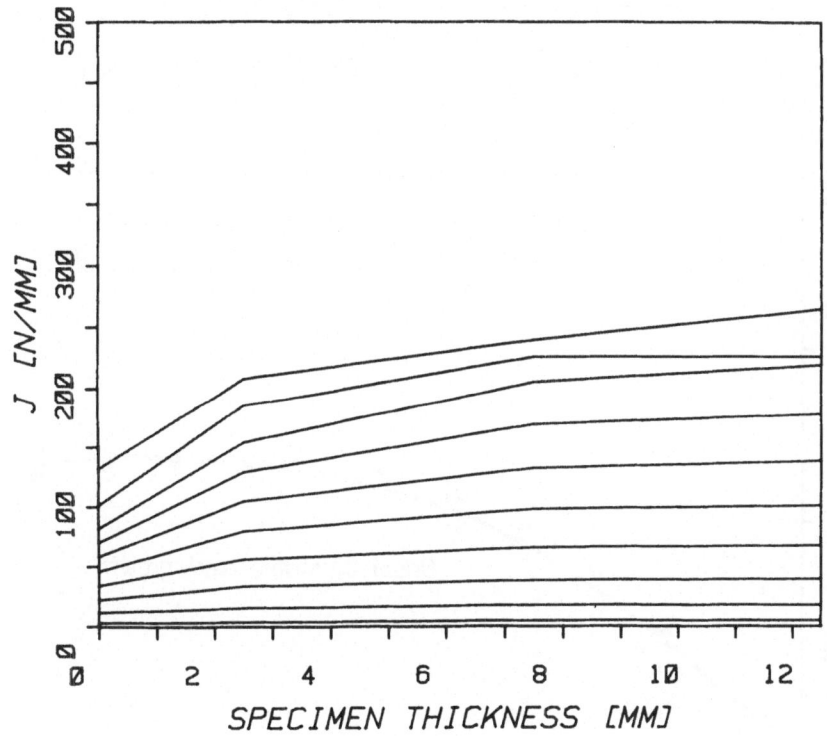

Fig. 9: Compact specimen, steel A 542, J-integral vs. thickness, parameter: load

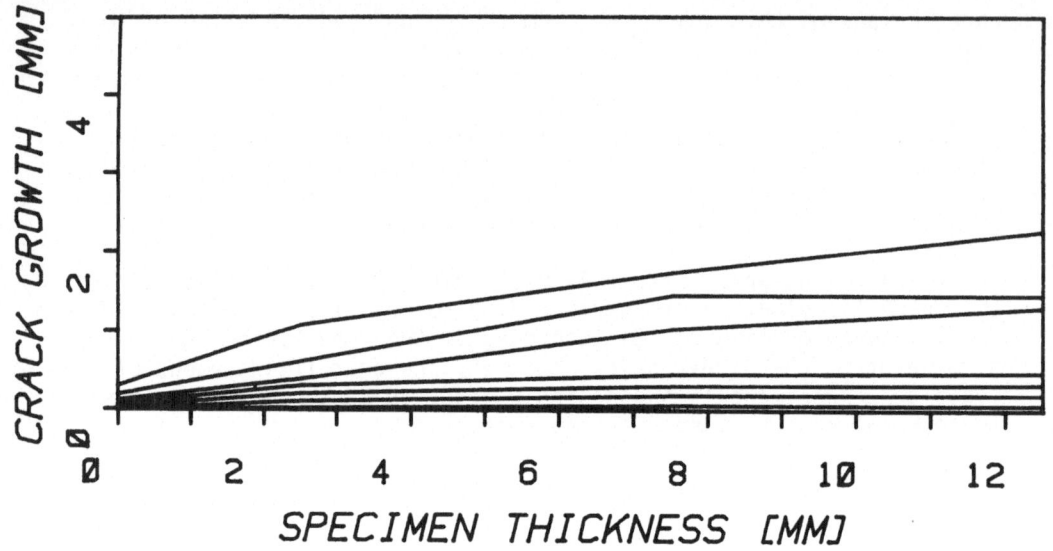

Fig. 10: Crack front position vs. thickness, parameter: load

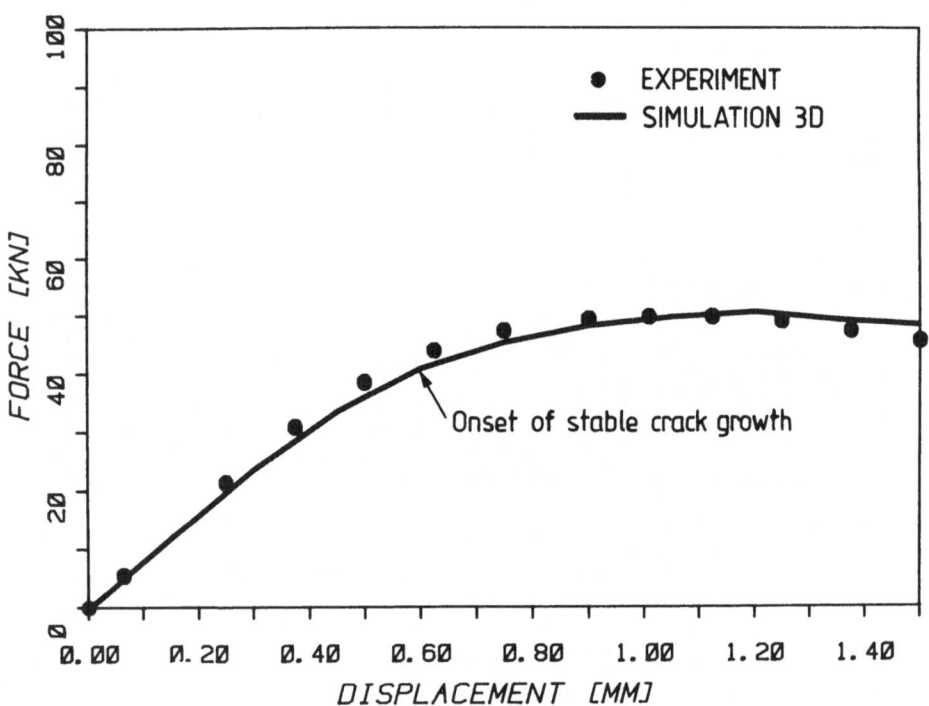

Fig. 11: Compact specimen, steel A 542, force-displacement diagrams
experiment and simulation

A more detailed evaluation of the J-resistance curves, Fig. 12, reveals a series of problems concerning the transferability of specimen results to real structures.

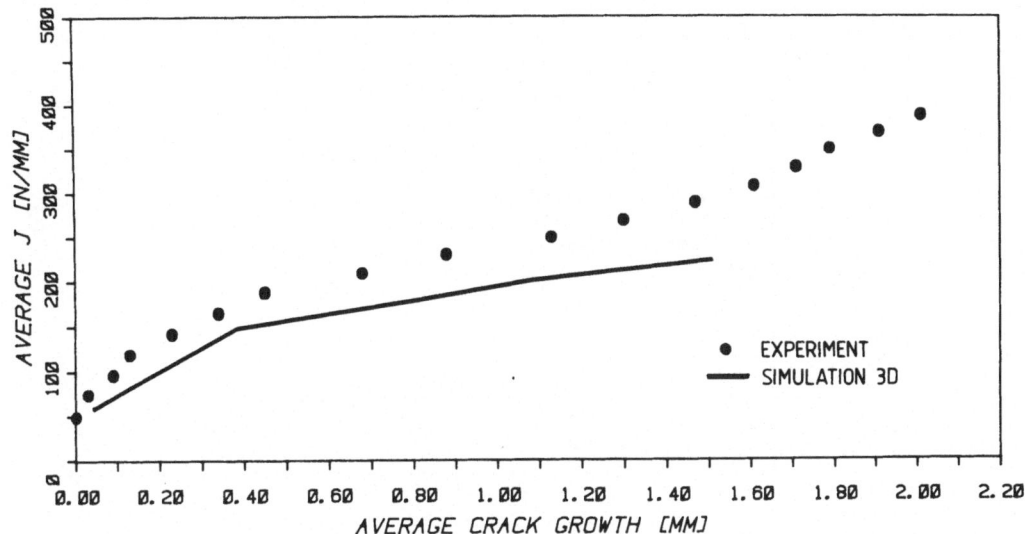

Fig. 12: Compact specimen, steel A 542, J-resistance curves
FEM and experiment

In the next example the numerical simulation of crack arrest experiments is demonstrated utilising the finite difference method. The experiments([33]) were performed with double-cantilever beam specimens (RDCB) made of ARALDITE B. The numerical simulations were performed several years after the experiments in order to verify numerical procedures and the sensitivity of different models ([34, 35]).
Figure 13 shows a schematic view of the experimental set-up, Figure 14 gives the finite difference grid. The crack extension is modelled by gradually releasing the stresses along the crack path by applying a weighted mean of the equation holding for the stress-free state and that of zero y-displacement. In the "generation mode" the crack length as a function of time a(t) is input as depicted from the experiment. The dynamic J-integral is evaluated following eqn. (7) but applying a dynamic correction and consequently converted into K_I. In the "prediction mode" an experimental relation between stress intensity factor and crack velocity is used to control the crack propagation in the model.

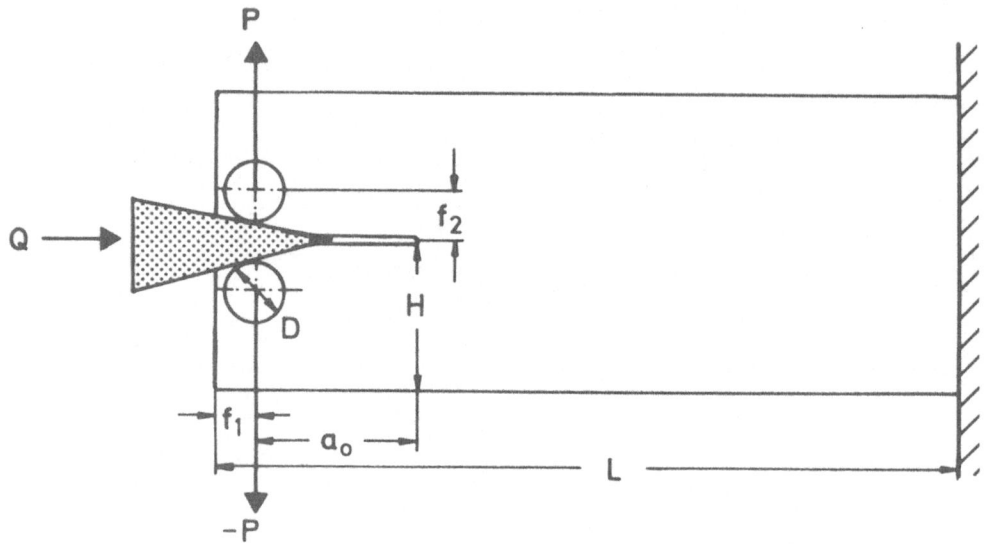

Fig. 13: RDCB-specimen, experimental set-up, Ref. 33

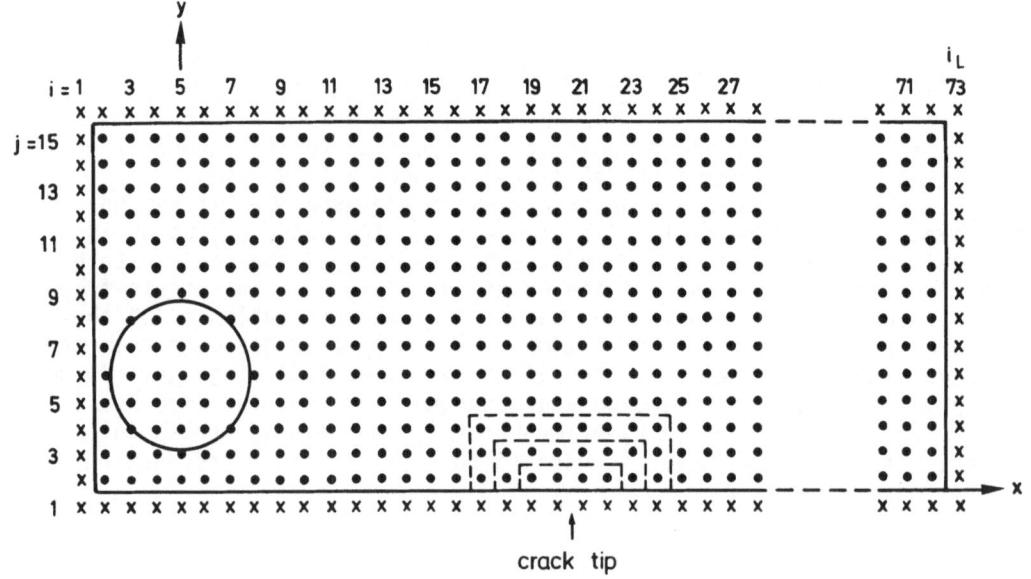

Fig. 14: RDCB-specimen, finite difference grid

Figure 15 gives a comparison of measured and simulated $K_I(a)$ and $v(a)$ curves. The agreement is considered quite satisfactory.

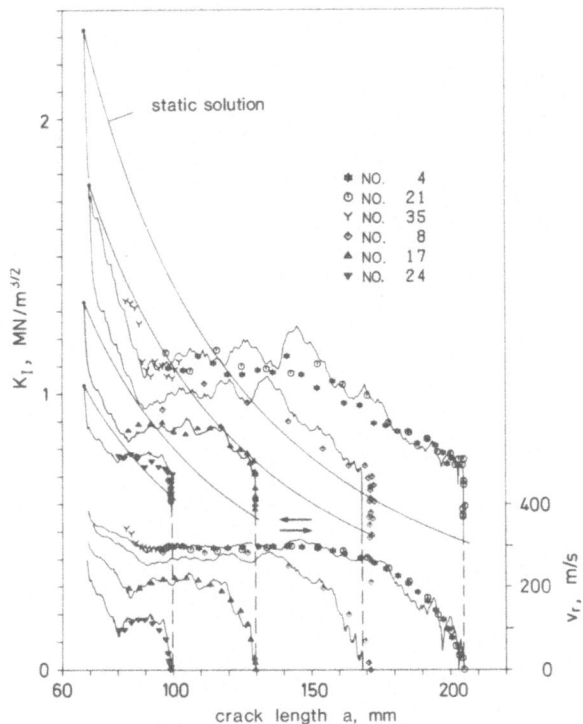

Fig. 15: RDCB-specimens, simulation and experiments

One branch of the comminution research is concerned with the behavior of single particles under defined loading conditions. Shape and size of these particles have significant influence on the result of the comminution process. So, numerical studies have been performed on the behavior of spheres from different materials under different loading conditions [36]. From these results a series of fracture phenomena could be explained which were also found in experiments. Here, the behavior of elastic spheres with meridional flaws are examined numerically.

Figure 16 gives a view of the 3D-FEM mesh with a meridional flaw. The sphere was assumed to be pressed between coplanar planes at both poles. To examine the number of parts to occur most likely under these circumstances different numbers of flaws were examined: n = 1,2,3,4, 6,12. An "energy release rate" was calculated by comparing the total energy of the uncracked sphere with the total energy of the same sphere loaded to the same displacement but having a certain number of cracks. The result of this calculation is plotted in Figure 17, indicating that a number of flaws n=3 should be the most likely situation to occur under these conditions.

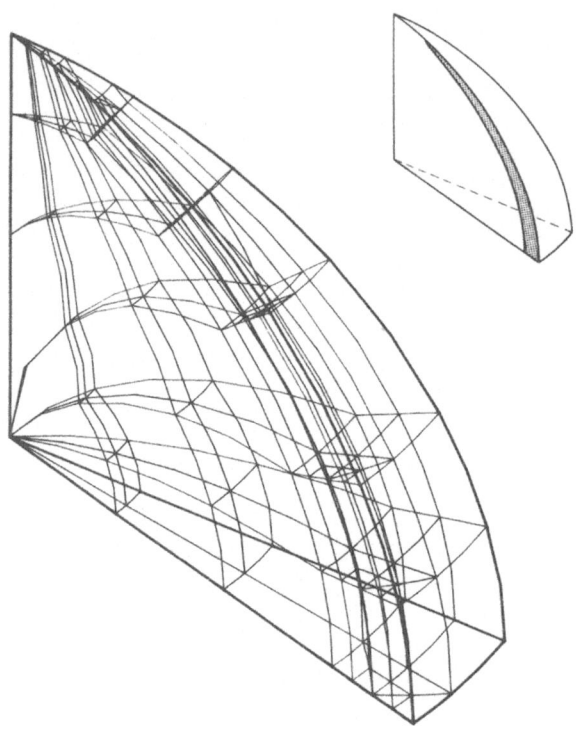

Fig. 16: Elastic sphere with meridional flaws: Part of FEM-mesh,
ref. 36

This is in agreement with experimental results for brittle
material, e.g. glass or PMMA at low temperature [37].
The situation changes when ductile materials are under consideration. Six and more meridional cracks are observed. The numerical
investigation on elastic-plastic spheres with meridional cracks has not
yet been performed.
The problems discussed so far included the modelling of a definite
crack. In many applications the FEM is applied just to do a stress analysis which is sometimes followed by a fracture mechanics evaluation.
The last two examples deal with the design of proof tests for ceramic
materials, where FEM is needed to ensure that the desired stress distribution is obtained in the test samples.

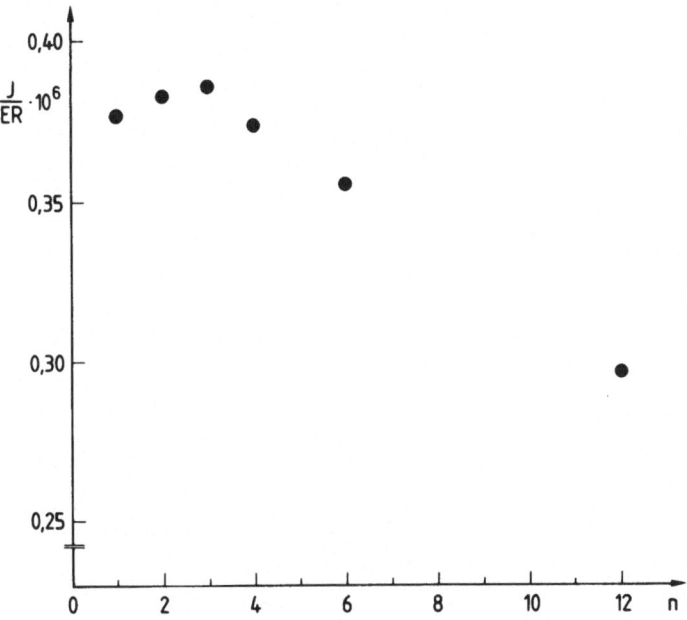

Fig. 17. Elastic sphere with meridional flaws: energy release rates

The strength of ceramic materials is strongly dependent on the condition of the surface, i.e. on flaws generated during the machining of the surface. The common strength testing procedures using flexural bars, however, are additionally dependent on the flaws at the edges of the bar which are specially sensitive to the particular shaping and machining procedure. Therefore, the concentric-ring test which has already been used for testing glasses, was adapted to ceramic materials (fig.18, [38]).

Since the specimen size and the testing arrangement had to be modified, analytical and numerical investigations have been carried out in order to check whether the advantages of the concentric-ring test remain valid, if applied to glass. If the effect of the applied load is modelled properly (via Hertz's compression) and geometrical nonlinearities are taken into account it turns out that the homogenous biaxial stress distribution generated inside the inner ring is conserved, as can be seen in fig.19. Therefore, the test arrangement has been proved to deliver reliable strength values.

In order to improve the reliability of ceramic hip-joint heads by applying a fracture mechanics concept to define the conditions of a proof test, the stresses acting in an implanted component e.g. during walking must be known. Therefore, a combined numerical and experimental investigation ([39]) was performed to determine these stresses.

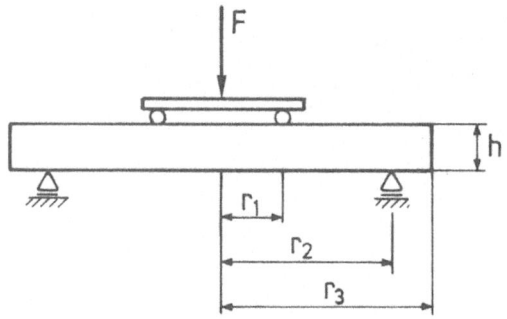

Fig. 18: The concentric-ring test, experimental set-up, schematic

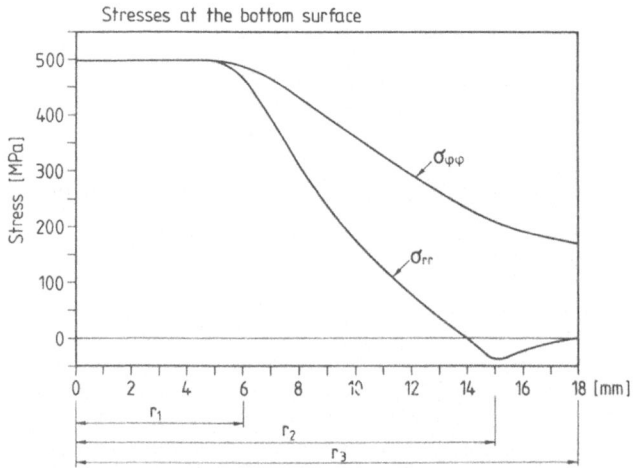

Fig. 19: The concentric-ring test, biaxial stress distribution

Figure 20 demonstrates the experimental configuration, Figure 21 gives the finite element mesh. Since the ceramic head is fixed on the stem by friction forces along the conical interface between head and stem, in addition to the external loads also friction and sliding must be simulated. This was done by simplified FEM models with varying friction coefficients. The stresses at the outside differed significantly from the measured ones. It was concluded that the displacement pattern along the sliding stretch could not be linear. Therefore an improved model was set up where displacement profiles at the inner surface of the head were preset in radial direction and varied until accordance between calculated and measured stresses at the outside of the head was found.

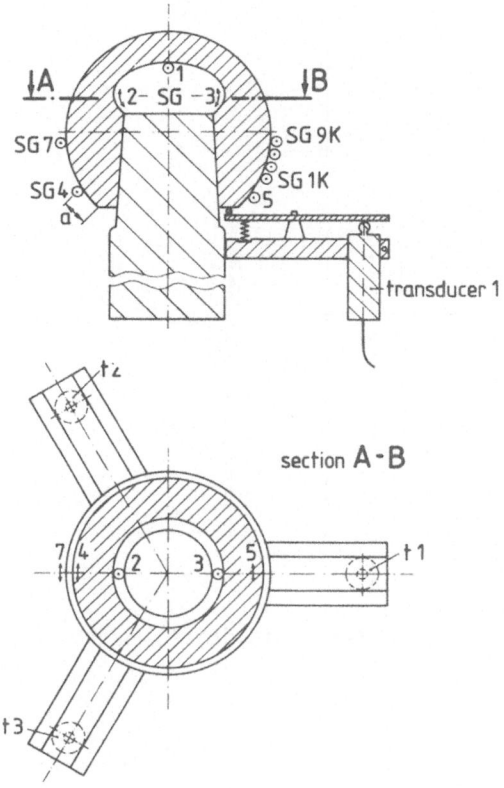

Fig. 20: Reliability of ceramic hip-joint heads:
experimental set-up, Ref. 39

Figure 22 shows the final result of this procedure. The total stress
pattern including also the upper part was obtained by superposition of
this result with that of fixed bonds, i.e. infinite friction coefficient
(Figure 23). With this FEM model adjusted to experimental findings conditions for a proof test could be defined.

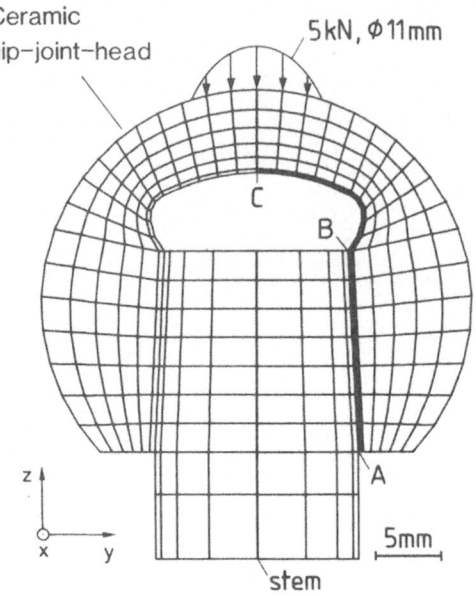

Fig. 21: Reliability of ceramic hip-joint heads:
FEM-model, ref. 34

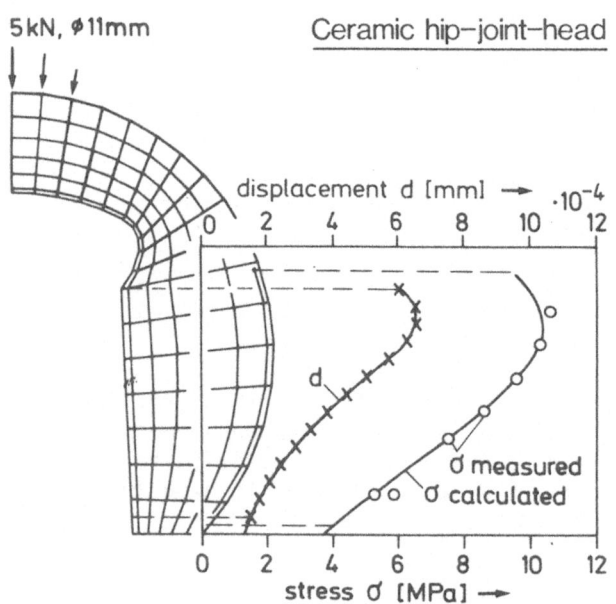

Fig. 22: Reliability of ceramic hip-joint heads:
displacement profiles, ref. 34

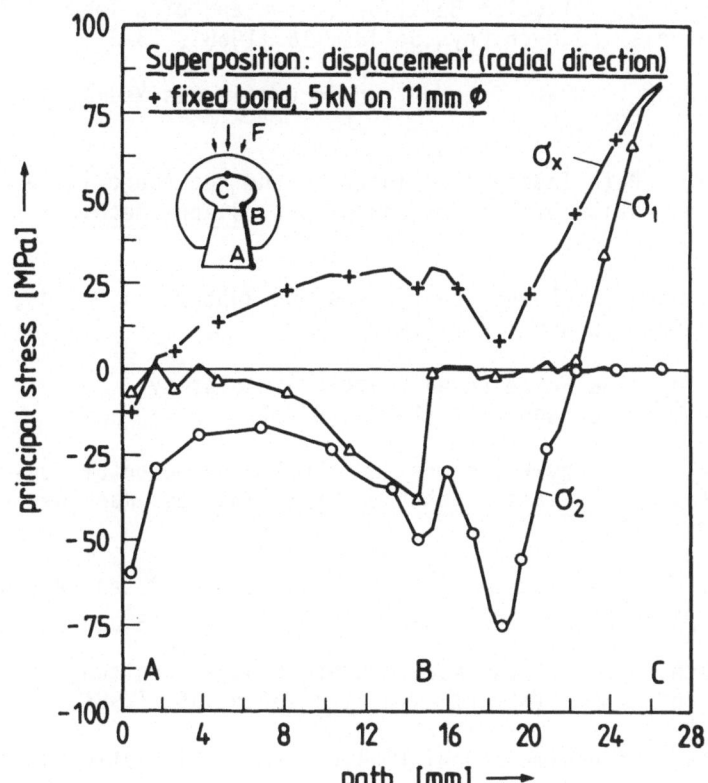

Fig. 23: Reliability of ceramic hip-joint-heads:
final stress distribution, Ref. 34

6. REFERENCES

[1] W. Schmitt, Anwendung der Methode der finiten Elemente in der Bruchmechanik unter besonderer Berücksichtigung dreidimensionaler und elastisch-plastischer Probleme, IWM-Bericht W2/82

[2] L.H. Larsson, A Calculational Round-Robin in Elastic-Plastic Fracture Mechanics, 6th International Conference on Structural Mechanics in Reactor Technology G 3/2, Paris (1981)

[3] J.N. Sneddon, The Distribution of Stress in the Neighborhood of a Crack in an Elastic Solid, Proc. Phys. Soc. (London), 187 (1946), 229-260

[4] J.R. Rice and G.F. Rosengren, Plane Strain Deformation Near a Crack Tip in a Power-Law Hardening Material, J. Mech.Phys.Solids, 16 (1986), 1-12

[5] J.W. Hutchinson, Singular Behavior at the End of a Tensile Crack in a Hardening Material, J.Mech.Phys.Solids, 16 (1968), 13-31

[6] G.P. Cherepanov, Crack Propagation in Continuous Media, J.Appl.Math. Mech., 31 (1967), 503-512

[7] J.R. Rice, A Path Independent Integral and the Approximate Analysis of Strain Concentration by Notches and Cracks, J.Appl.Mech., 35 (1968) 379-386

[8] A.A. Griffith, The Phenomena of Flow and Rupture in Solids, Phil.Trans.Roy.Soc. (London) 221 (1921)

[9] J.D. Eshelby, The Force on an Elastic Singularity, Phil.Trans.Roy.Soc. (London) 244 (1951), 87-112

[10] J.D. Eshelby, Energy Relations and the Energy-momentum Tensor in Continuum Mechanics, Inelastic Behavior of Solids (ed. Kanninen et al.), New York (1970) 77-115

[11] J.D. Eshelby, The Calculation of Energy Release Rates, Prospects of Fracture Mechanics (ed. Sih et al.), Leyden (1975) 69-84

[12] H. Buggisch, D. Gross und K.H, Krüger, Einige Erhaltungssätze der Kontinuumsmechanik vom J-Integral-Typ, Ing. Arch. 50, (1981) 103-111

[13] W. Schmitt, Three-Dimensional Finite Element Simulation of Post-Yield Fracture Experiments, to appear in Int. J .Solids Struct.

[14] R. Kienzler, Fraunhofer-Institut für Werkstoffmechanik, private communication, 1985

[15] H. Maschke and M. Kuna, Akademie der Wissenschaften der DDR, private communication, 1986

[16] H. Riedel and J.R. Rice, Tensile Cracks in Creeping Solids, Fracture Mechanics: Twelfth Conference, ASTM STP 700 (1980) 112-130

[17] O.C. Zienkiewicz, The Finite Element Method, McGraw-Hill, London (1977)

[18] K.J. Bathe and E.L. Wilson, Numerical Methods in Finite Element Analysis, Prentice-Hall, New Jersey (1976)

[19] D.M. Tracey, Finite Elements for the Three-Dimensional Elastic Crack Analysis, Nuclear Engineering and Design, 26 (1974) 282-290

[20] R.D. Henshell and K.G. Shaw, Crack Tip Finite Elements are Unnecessary, Int. J. Num. Meth. Engn. 9 (1975) 496-507

[21] R.S. Barsoum, Triangular Quarter-Point Elements as Elastic and Perfectly-Plastic Crack Tip Elements, Int. J. Num. Meth. Engn. 11 (1977) 85-98

[22] L. Bank-Sills and Y. Bortman, Reappraisal of the Quarter-Point Quadrilateral Element in Fracture Mechanics, Tel-Aviv University

[23] D.M. Parks, A Stiffness Derivative Finite Element Technique for Determination of Elastic Crack Tip Stress Intensity Factors, Int. J. Fracture, 10 (1974) 487-502

[24] D.M. Parks, The Virtual Crack Extension Method for Nonlinear Material Behavior, Comp. Methods Appl. Mech. Eng. 12 (1977) 353-364

[25] H.G. DeLorenzi, Energy Release Rate Calculations by the Finite Element Method, General Electric Technical Information Series, Report No. 82CRD205 (1982)

[26] A. Bakker, The J-Concept: Theoretical Basic and its Use in EPFM, in: Elastic-Plastic Fracture Mechanics, L.H. Larsson (Ed.), D. Reidel Publishing Co., Dordrecht / Boston / Lancaster (1985), 13-53

[27] R. Kienzler, H. Kordisch, W. Schmitt und D. Siegele, Analogien zwischen den Beanspruchungsparametern der Zähbruchmechanik J und der Kriechbruchmechanik C*, 17. Sitzung des Arbeitskreises Bruchvorgänge im DVM, Basel (1985)

[28] H.G. DeLorenzi, J-Integral and Crack Growth Calculations with the Finite Element Program ADINA, Methodology for Plastic Fracture, EPRI Report SRD-78-124 (1976)

[29] E. Sommer, Bruchmechanische Bewertung von Oberflächenrissen: Grundlagen, Experimente, Anwendungen, Springer (1984)

[30] ASTM E 399-78, Standard Method of Test for Plane Strain Fracture Toughness of Metallic Materials, Annual Book of ASTM Standards (1978), 512-533

[31] W. Schmitt and E. Keim, Linear Elastic Analysis of Semi-Elliptical Axial Surface Cracks in a Hollow Cylinder, Int.J.Pres.Ves. and Piping (7) (1979)

[32] W. Schmitt, Three-Dimensional Finite Element Simulation of Post-Yield Fracture Experiments, Computational Fracture Mechanics-Nonlinear and 3-D-Problems, PVP-Vol. 85 AMD Vol. 61 (1984) 119-131

[33] J.F. Kalthoff, J. Beinert and S. Winkler, Measurements of Dynamic Stress Intensity Factors for Fast Running and Arresting Cracks in Double-Cantilever-Beam Specimens, Fast Fracture and Arrest ASTM STP 627

[34] H. Stöckl, Numerische Simulation von Rißarrestversuchen an ARALDIT B mit dem Verfahren der finiten Differenzen, IWM-Bericht W 9/82 (1982)

[35] W. Schmitt and H. Stöckl, Numerical Simulation of Crack Arrest Experiments, CSNI Specialists' Meeting on Crack Arrest, Freiburg (1984)

[36] R. Kienzler, Numerische Beanspruchungsanalyse einer definiert belasteten Kugel 3, IWM-Bericht W 6/85 (1985)

[37] M. Stieß: Die Druckbeanspruchung von elastischen und inelastischen Kugeln bis zum Bruch, Dissertation TH Karlsruhe (1976)

[38] U. Soltesz, H. Richter and R. Kienzler, The Concentric-Ring Test and its Application for Determining the Surface Strength of Ceramics, to be presented at the World Congress on High Tech Ceramics, Milan, June 1986

[39] G. Kleer, D. Siegele and U. Soltesz, Evaluation of Stresses in Ceramic Hip-Joint Heads, Fourth Meeting of the European Society of Biomechanics, Davos (September 1984)

MECHANICAL BEHAVIOR OF CERAMICS

Philippe BOCH
ENSCI, 47 Av. A. Thomas
87065 LIMOGES FRANCE

ABSTRACT. The main definitions of what ceramic materials and ceramic processes are have been recalled first, and the characteristics of the brittle behaviour of ceramics have been reviewed. The case of alumina has been exemplified, to show the influence of microstructural parameters.

1. CERAMIC MATERIALS

1.1. Ceramic compounds

From the Materials Science point of view, ceramics are inorganic and non – metallic compounds:

INORGANIC MATERIALS   ORGANIC MATERIALS

METALS / NON- METALS = CERAMICS

This definition is very wide, and it does not distinguish natural materials (e.g. rocks) from synthetic materials ; it does not give information on the structural state (glass, single crystal, polycrystal) ; it does not consider the case of composites (e.g. inorganic fibers and organic matrix).
  Materials Science considers two limit cases of inorganic materials:
  * Metals (metallic bonds, with "free" electrons) ;
  * Ceramics (iono-covalent bonds, without "free" electrons). However, it must be noticed that some compounds (e.g. carbides) are difficult to classify because they have mixed bonds (partially metallic and partially ceramic bonds).

K. P. Herrmann and L. H. Larsson (eds.), Fracture of Non-Metallic Materials, 75–93.
© *1987 by ECSC, EEC, EAEC, Brussels and Luxembourg.*

1.2. Ceramic processes /ceramic materials

Industry considers *articles,* and Materials Engineering allows us to write:
 Ceramic articles are made by applying *ceramic technology* to *ceramic materials.*
    The most common ceramic technology is the sintering of powders, and we can classify the materials made of ceramic compounds into three broad classes :

* *Ceramic technology ---> ceramics*
Powders   --->shaping   --->thermal   treatments   and consolidation by sintering (for hot-pressing, shaping is made during thermal treatments) ;

* *Glass technology ---> industrial glasses*
Powders ---> thermal treatments and melting --->shaping , and consolidation during cooling ;

* *Cement technology ---> cements, lime, plasters*
Thermal treatments --->reactive powders --->shaping, and consolidation at room temperature by chemical reactions.

    Such a classification eliminates the natural materials (rocks), and distinguishes ceramics from glasses not by structural considerations but by the nature of the process. It must be noticed that it is over simplified, and that it does not consider many important techniques (e.g. chemical vapor deposition).

1.3. Ceramic products

"Traditional" ceramics are mainly silicateous compounds, elaborated by using natural raw materials (e.g. clays, with kaolinite as main mineral, of equivalent formula : $Al_2O_3-2SiO_2-2H_2O$, quartz, i.e. crystallised silica: SiO_2, fusing additions, e.g. feldspar orthose: $K_2O-Al_2O_3-6SiO_2$). Their chemical composition can be plotted in a ternary diagram with "elementary" oxides at each corner: Al_2O_3, SiO_2, M_xO_y (e.g. K_2O) (ref. Fig.1)
    "Technical" ceramics are elaborated from starting materials which were already synthesized (e.g. Bayer alumina). From the chemical point of view the main classes are: OXIDES (Al_2O_3, SiO_2, MgO, UO_2, BeO...) and NON-OXIDES (nitrides, carbides, borides, silicides...). The particularity of non- oxides is they are not stable in normal atmosphere, where they oxydize. Such an oxydation is very pronounced at high temperatures, and can lead to a complete destruction of the material. Moreover, silicon compounds (e.g. silicon nitride: Si_3N_4 or silicon carbide:

SiC) can be protected by a superficial layer of air-tight silica : this explains, for a part, the fact that "new" non-oxide ceramics are mainly silicon compounds.

A common classification of ceramic products is as follows:
* (Glass)
* (Cement, Lime, and Plaster)
* Building ceramics (heavy clays, bricks, tiles)
* Sanitarywares
* Tablewares   CHIEFLY SILICATEOUS CERAMICS
* Enamels
* Refractories

* Abrasives
* High Tech. Ceramics
- Electric functions
- Magnetic functions   CHIEFLY NON-SILICATEOUS
- Optical functions   CERAMICS
- Chemical functions
- Thermal functions
- Mechanical functions
- Biological functions
- Nuclear functions

1.4. Ceramic microstructures / microcracks

Most of industrial ceramics have complex microstructures characterized by :
 * Several phases, crystalline and / or amorphous ;
 * A residual porosity due to the fact that sintering generally does not lead to a perfectly densified state ;
 * Various defects (collectively named: *"Microcracks"*, e.g. voids, inclusions, exageratingly large grains, cracks...). These flaws will control brittle fracture.
    It must be noticed that in the simple case where linear fracture mechanics is valid :

$$\sigma_v = (1/Y) \; K_c \; (a)^{-1/2} \qquad (1)$$

(with σ_v the mechanical strength, K_c the fracture toughness, a the equivalent size of the critical microcrack, and Y a numerical term), toughness can be considered as a material parameter, chiefly depending on the nature of the compound, whereas "a", the size of the critical flaw, chiefly depends on the processes which have been used. Thus, it follows that the improvement of the mechanical properties, particularly the strength of ceramic *articles* , can generally be obtained more by improving the *processes* than by improving the *materials* if by this word we understand "compounds".

Silicateous ceramics often contain an abondant glassy phase (porcelain, for instance, is constituted of a silica- rich glassy matrix in which crystalline needles of mullite are embedded). On the contrary, most of non-silicateous ceramics are crystalline. However, they generally contain small amounts of glassy segregated phase along the grain boundaries. Although in a very small quantity those segregations play an important role because their softening point mainly controls the mechanical properties at high temperatures, e.g. the brittle --->ductile transition.(ref. Fig.2)

Fracture toughness of ceramics is typically one order of magnitude below toughness of metals (K_c is in the range of 1-5 MPa$\sqrt{}$ m instead of some tenths of MPa$\sqrt{}$ m). This means that the critical microcracks in ceramic articles are very small : two orders of magnitude below what they are in metallic pieces (they will be expressed in μm, instead of mm). Therefore the prevention of defects in processing articles, or the detection of defects in as-delivered pieces, will be more difficult in ceramics than in metals. In the former case the scale of the microstructure is similar to the scale of the flaws (e.g. 50 μm), whereas in the latter case critical flaws are expressed in cm, microstructure remaining expressed in μm.

1.5. Sintering and shrinkage / Associated flaws

The refractoriness of ceramic materials (high melting temperatures or decomposition before melting) and their brittleness generally prevent the use of the processes which are usual for metals, i.e. casting, machining, shaping by plastic forming etc. So, a ceramic article is generally processed by *sintering* compacted powders, in order to obtain the article in its final size and final shape, with very limited post-firing operations (e.g. polishing with diamond abrasive wheels). We must therefore stress the two following points:

* Contrary to metallurgists, for whom there is a frequent separation of function between *producing materials* (wires, bars, plates...) and *making articles* (in their finished form) ceramists only know *making materials AND articles.*

* Pressing powders leads to a mean "green" relative density of about 60% : if the sintering process is performed up to the final densification i.e. 100% the porosity which must be eliminated is about 40%, hence a *Shrinkage* of about 40% (in volume) or 15% (in linear dimensions). So, it will be very difficult to avoid differential strains, able to provoke macroscopic warping and also microscopic CRACKS. Such cracks will often be the origin where brittle fracture will initiate.

2. MECHANICAL BEHAVIOR OF CERAMICS / BRITTLENESS

2.1. Structure of ceramics and brittleness [1 to 5]

In metals, plasticity can develop due to the easy moving of numerous dislocations, which is allowed by the metallic bonding : no directivity, no local electrostatic restrictions. However, some limitations to the plastic behaviour of polycrystals arise when the number of the independent slip systems is not enough (von Mises' criterion requires 5 independent systems). This explains the low temperature brittleness of B.C.C. or hexagonal metals.

In ceramics, iono-covalent bonds are directive and have local electrostatic restrictions. There are very few dislocations (typically 4 orders of magnitude below metals) and those dislocations have a limited mobility. Moreover, ceramic compounds have often a low symmetry, and von Mises' criterion is not satisfied. Finally, the following statements can be made :

* Polycrystalline ceramics are brittle at low and medium temperatures ; limited plasticity only develops at high temperature i.e. $T > Tm/2$, where Tm is the melting temperature (e.g. 2050°C for alumina, 2750°C for zirconia)

* Limited plasticity can be observed at room temperature in cubic crystals (e.g. alkaline halides : LiF, NaCl...) when they are stressed in favourable directions.

* Glassy phases become viscous hence allow a plastic flow when diffusion is fast enough. The glassy transition may be at rather low temperatures, which means the refractoriness of multiphase ceramics is primarily depending on their glassy segregations.

2.2. Linear elasticity / low fracture strain

The stress - strain curve (e.g. as determined by a tensile test) is a straight line up to the fracture : the ultimate strength, σ_u is equal to the yield limit, σ_y There is no plastic domain (see Fig.3). Other comments are :

* Young's moduli of ceramics are generally higher than Young's moduli of metals (e.g. E is 400 GPa in alumina and 200 GPa in steel).

* Strength may be of the same order as in metals (e.g. 100 MPa for glasses, 300 MPa for alumina, 800 MPa for silicon nitride, more than 1000 MPa for partially stabilized zirconia). This is the tensile strength ; the compressive strength is about an order of magnitude higher.

* Fracture strain, $(\Delta l/l)u$ is always very low : for

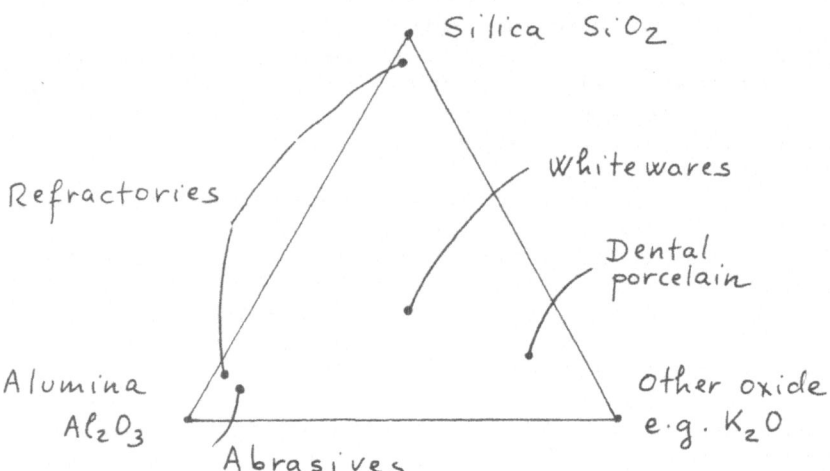

Fig.1 Schematic of the compositions of silicateous (traditional) ceramics

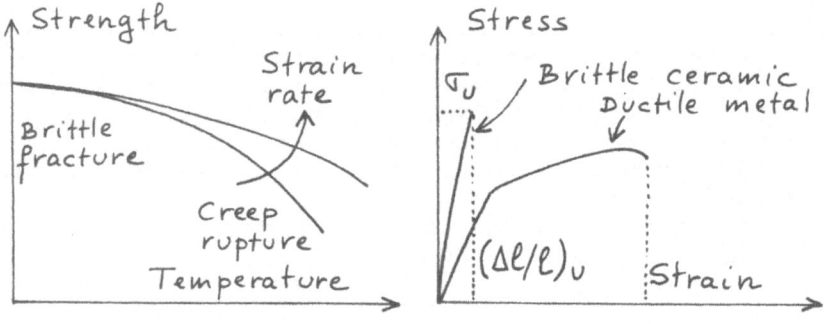

Fig.2 Influence of temperature on fracture

Fig.3 Stress / strain diagram

instance if τ_u is 300 MPa and E is 300 GPa, $(\Delta l/l)u$ is 0.001. ---> Ceramic pieces do not lead to major problems under imposed stresses; difficulties arise under imposed strains.

2.3. Sensitivity to differential strains / thermal shocks

Due to their very low ultimate strain , ceramics do not accept local plasticity, which means they are sensitive to differential strains. Thereby, in the case of "thermal shocks" there can exist :

Non-zero thermal expansion coefficient
↓
Uniform thermal expansion coef. but temperature gradients
or
Uniform temperature but heterogeneous / anisotropic
thermal exp. coef.
↓
Differential strains, which lead to local fracture if $(\Delta l/l) > (\Delta l/l)u$

The expression "thermal shocks" corresponds to the common case where fast thermal exchanges induce thermal gradients in a piece (such gradients are facilitated by the generally low thermal conductivity of ceramics). However, a more subtle case corresponds to a uniform temperature but to an anisotropic (crystals of non-cubic symmetry) or to an heterogeneous (multiphase materials) thermal expansion. A rather high anisotropy can lead to grain-boundary microcracks which act to initiate the fracture : it is the case in aluminium titanate, or in cordierite, for which the low mean dilatation of polycrystals results from the strong anisotropy of the single crystals (grains). Even in weakly anisotropic crystals (e.g. alumina) microcracking can develop when grains are large : this will be discussed in a next paragraph. However, the most spectacular cases are related with the phase transitions, e.g.: quartz / cristobalite / tridymite or tetragonal zirconia / monoclinic zirconia.
Thermal (i.e.dilatometric) strains are not always detrimental : they are used to produce tempered glass (wind-screens). Phase transitions are responsible for the toughening of partially-stabilized zirconia (as explained in the lecture by N. Claussen).

2.4. Low toughness / sensitivity to small microcracks

Ceramics have a low toughness : the critical stress intensity factor, K_c, is less than 1 MPa$\sqrt{m}$ for glasses, about 3.5 MPa$\sqrt{m}$ for alumina, 4.5 MPa$\sqrt{m}$ for silicon nitride, and could reach 10 MPa$\sqrt{m}$ for some PSZ. (For such PSZ non linear effects related to the toughening mechanism of stress-induced transformation T -->M are such that the linear relationship between σ_u and K_c is no longer valid) [6]. Cermets, particularly Wc-Co for cutting tools, have a better toughness due to the bonding metallic phase ; fiber composites also have a noticeable toughness : Kc = 20 MPa$\sqrt{m}$ has been indicated for SiC / SiC composites. However, questions arise about the meaning of K_{Ic} in anisotropic composites, when the K_{Ic} measurements are made by using classic tests, e.g. bending of notched beams, which do not lead to a crack propagation in mode 1. If toughness is expressed in fracture energy, it ranges from < 1 N/m for glasses to > 100-1000 N/m for cermets, polycrystalline ceramics having intermediate values of 10-100 N/m. In comparison, metals frequently reach 100,000 N/m.

The sensitivity to small microcracks has already been indicated : it is the main drawback associated with a low toughness. Thus, for ceramic pieces submitted to loads in the ranges which are usual in mechanical engineering, the critical size of microcracks is of some tenths of micrometers!

2.5. Low impact resistance

Impact resistance of ceramics is an important problem. which has not received enough attention. Basically, local fracture occurs during impact when the stress intensity factor K_c is reached at the tip of a crack. This value can be reached for a short moment only which leads to a small propagation of the crack, not sufficient for breaking the piece but sufficient for weakening it.

The low toughness and the low fracture resistance of ceramics implies a low impact resistance. However:

* The high hardness and the good resistance to abrasion of ceramics can help them to endure abrasive shocks, where metals do not succeed due to their surface degradation;

* Dynamical effects associated with shock-waves can lead to a unusual behaviour, e.g. when compressive shocks on one side of a piece provoke a degradation near the opposite side, due to tensile waves going back after reflection onto that side.

2.6. Scattering of strength values

The most damaging characteristic is strength scattering. It directly results from the scattering of severity of flaws:

$$\sigma_u = (1/Y)K_c(a)^{-1/2} \qquad (2)$$

The usual statistics is Weibull's statistics, which relates the probality of rupture, Pf, to the strength, σ_u, by the expression:

$$Pf = 1- \exp - [(\sigma_u - \sigma')/(\sigma o)^m \qquad (3)$$

where σ' is the value for which Pf = 0 (which is often chosen equal to zero), m is "Weibull's modulus"(which characterizes the width of the distribution), and σo is a normalizing factor.

Weibull's diagrams are drawn by plotting :

$$Ln\, Ln\, 1 / (1-Pf) = m\, Ln\, (\sigma_u - \sigma') - m\, Ln\, \sigma_o \qquad (4)$$

which gives a straigth line of slope "m".(ref. Fig. 4).

Many studies have been devoted to Weibull'statistics; we need only to mention the main results:

* The volume (and / or the surface) of the loaded piece must be taken into account : Pf decreases when the volume increases, because the probability of finding a large flaw increases in a similar manner. The nature of the load also influes on Pf, because it modifies the volume under stresses. In the simple case of a tensile test of two pieces of respective volume V1 and V2:

$$\sigma_u,1 / \sigma_u,2 = (V2 / V1)^{1/m} \qquad (5)$$

Thus, the strength values determined by breaking small samples, e.g. by using a 3-point bend test which leads to noticeable stresses in a small volume only, do NOT be directly transposed to calculate large pieces. Those values must be corrected (toward lower ones) by taking into account the differences in volume and nature of load.

For the previous example of homogeneous traction, if m = 10 and V2 = 1000 V1 (i.e. a ratio for the linear dimensions of 10) , thus: $\sigma_u,2 \simeq 0.5\ \sigma_u,1$ which shows that Weibull's correction is far from being negligible.

* Increasing Weibull's modulus for their products is one of the main preoccupations of ceramists. Typical values are m = 6 for industrial glasses, m = 10 for technical ceramics of medium quality, m > 15 for fine-grain, high-quality materials.

However, what does Weibull's modulus exactly mean for a "material"? The value of σ', which is related to the most severe crack, and Weibull's modulus, m, which characterizes the distribution width, depend on microstructural features, hence on both the nature of the compound and the type of process which has been used to produce the piece. In what follows that the "same"ceramic material can exhibit a low Weibull modulus (e.g. for large pieces which have been produced by injection molding, with a bad control of the binder pyrolysis) or a high Weibull modulus (e.g. for small samples, processed by extrusion and polishing).

The"basic" relation: *"Ceramic Piece = Ceramic Material + Ceramic Technology"* must be always remembered : it is of no interest to improve *"Materials"* if *"Technology"* cannot be correlatively improved or, worst, if is degraded.

2.6. Time and environment effects

Ceramics are sensitive to static fatigue which can lead to the delayed fracture of a piece submitted to a medium load for a long time. Basically, static fatigue is a chemically activated phenomenon which corresponds to the *slow crack growth*, and which depends on the environment. For instance humidity is known to increase the S.C.G. in nearly all ceramics.

S.C.G. is usually studied by drawing the "v = f(K)" curves, where v is the S.C.G. rate and K the stress intensity factor. The typical aspect of such curves is shown in Fig. 5. In zone 1 which corresponds to the zone where materials are normally loaded v depends on K according to a power law, as it was detailled in Sommer's lecture (ref. Eq. 2-20) :

$$v = A K^n \qquad (6)$$

The values of "n" are very high (e.g. 15 to 50) and the accuracy on their determination is bad. In particular, there is generally a discrepancy between the values which are measured by using samples with "macroscopic" notches (e.g. double torsion of notched plates) and those obtained from the S.C.G. of "natural" flaws (e.g. tensile tests at various stressing rates)[7].

Further, delayed fracture is another reason, besides Weibull's effects, to confirm that nominal values of strength, determined from short tests in "clean" atmosphere, are noticeably higher than the static fatigue limit, which is often the parameter of main interest (long time, especially in polluted atmospheres). Cyclic fatigue (from "mechanical" origin or from "thermal" origin) has not been studied in a sufficiently detailed manner. The possible development of ceramic pieces in thermal engines will certainly focus on this topic.

2.7. Life-time and reliability

It is possible to take into account both Weibull's scattering and S.C.G. by combining data from Weibull's diagram and from v = f(K) curves. This leads to a *Probabilistic* treatment of life-time, expressed in the so-called S.P.T.(Strength or Stress − Probability − Time) diagrams (ref. Fig. 6). Various representations are possible to link fracture probability, Pf, to the stress level and to the time during which the loads are applied.

It would be a mistake to be perfectly confident in the quantitative applicabilty of such diagrams : their accuracy is limited due to the uncertainties on Weibull's modulus, m, and on the stress exponent in the $v = f(K)$ curves, n. However, it would also be a mistake to underestimate their utility : if the extrapolations on volume (size of the laboratory samples--->size of the real pieces) and on time (time of the experiments--->duration of the real service) are not too large (e.g. a ratio < 10), the use of S.P.T. diagrams is safer than the rather arbitrary choice of a "safety factor" chosen from the data of "inert" strength. The principles that the S.P.T. diagrams stand for have been used for many years in the testing of carbonated beverage bottles.

When a probabilistic treatment cannot be acceptable (e.g. when the need of a very high safety requires the choice of a very low value for Pf, which leads to an overestimation of pieces) a *determinative* treatment can be chosen. Two methods are applicable:

* The *non- destructive evaluation* of all pieces in order to detect the defects which are more severe than a given level. Visual inspection, dye penetrant, X-rays, and ultrasonic waves are the main techniques.

N.D.E. has been successful for the control of metallic structures (e.g. space industries, atomic industries...) but it is doubtful whether its applicability will be the same for ceramics. On the one hand because the very small size of microcracks dramatically increases the difficulty and the cost of control and on the other hand because the similarity of scale of microstructure and defects makes identification of defects very difficult.

* The *proof-test* of all pieces in order to break the ones which will not resist the service conditions. *Proof testing* consists in applying an overload, σ. Max, during a short time, t.min, to a piece which must resist a load, σ. min, during a long time, t.Max. This method allows us to be sure of a minimal lifetime of a piece which is not broken after overloading :

$$tf \geqslant \frac{2 K_c^{\,2-n}}{Y^2 A \sigma.min^2 (n-2)} \left[\left(\frac{\sigma.Max}{\sigma.min}\right)^{n-2} - 1 \right] \qquad (7)$$

where K_c, n, and A are the parameters determined by the v=f(K) curves, and Y the usual geometric term.

Many studies (e.g. ref.[8]) have disscussed the main advantages and difficulties of proof test. In particular, the chemical environment must be the same for the determination of the v=f(K) curve and for the real in-service conditions. The extrapolation in the scale of time (t.min --->t.Max) must not be too large (maybe inferior to 10).

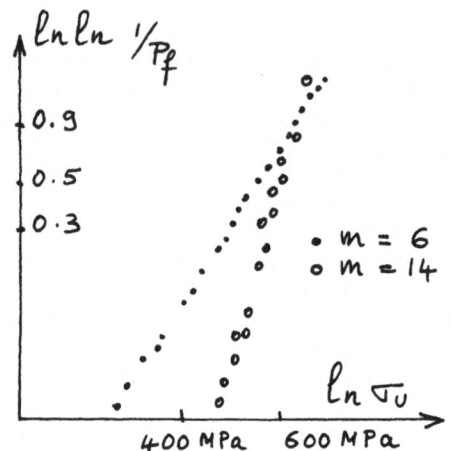

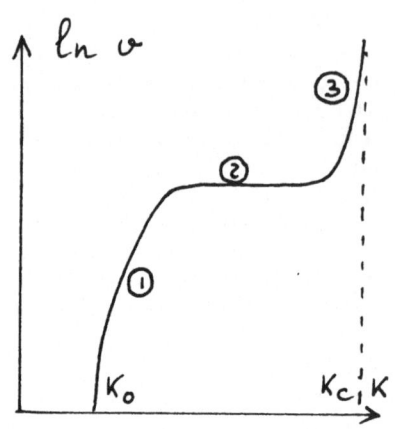

Fig.4 Weibull's
diagram for two alumina
ceramics of different
quality

Fig.5 Typical aspect
of a v = f(K) curve

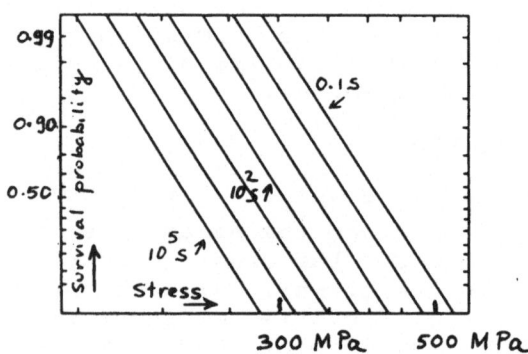

Fig.6 SPT diagram (after Davidge, in Ref. [4])

A very well known example of an application of proof test is the control of the windows of the American space shuttle; an older and a more common example is the control of abrasive grinding wheels by rotating them at speed in excess of the allowed speed.

3. TEMPERATURE AND MICROSTRUCTURE INFLUENCE : THE CASE OF ALUMINA

The discussion will be made on the case of alumina, by summarizing the study of A.G. Evans and Y. Fu *"The mechanical behaviour of alumina a model anisotropic brittle solid"* in Ref. [9]. Alumina is the archetype of technical ceramics, and ranges from low-purity, low-density, coarse-grain materials (e.g. refractories) to high-purity, high-density, fine-grain materials (e.g. substrates for electronics). Its symetry is trigonal, wich means it has an anisotropic thermal expansion, even if the anisotropy is rather low : a//c = 9 ; a $\perp$ c = 8 (in K $^{-1}$ between 20 to 1000°C).

So, alumina is the best model material to study the influence of the main physical parameters (temperature and microstructure features) on the fracture properties of ceramics.

The general trends in mechanical behaviour is as follows :

* at "low" temperatures (typically < Tm/2, if Tm is the melting temperature) alumina is a linearly elastic, purely brittle material. Strength is relatively insensitive to temperature and time (except for slow crack growth),

* at *high temperatures* (T > Tm/2), an appreciable permanent deformation can be observed before rupture. Strength is strongly sensitive to temperature and time, and creep phenomena occur.

It should be mentioned that the brittle --> non-brittle transition temperature is very sensitive to the alumina purity. The value of Tm/2 corresponds to a purity of about 99 % . For impure aluminas, which generally have vitreous segregations, the temperature dependence is related to the amount and to the properties (e.g. viscosity) of those segregations.

* At *low temperatures* the brittle behaviour depends on grain size :

- for *fine-grain* materials, strength is relatively insensitive to grain size, because fracture is initiated by microcracks (voids, inclusions, shrinkage flaws...) ;

- for *coarse-grain* materials, strength is roughly proportional to the inverse of the square root of the grain size, and fracture is related to microcrack coalescence phenomena (ref. Fig. 7).

3.1. The initial microcracking

Microcracks in brittle polycrystals of non-cubic symmetry are due to the anisotropic thermal expansion of material which leads to residual stresses with a wavelength of the order of the grain size oscillating between tension and compression along adjoining facets.

The derivation of the residual stress amplitude, σa , shows it can be written as :

$$\sigma a \sim E \ da \ dT \ [1+A \ \ln \ (\ell/x)] \ F \tag{8}$$

with E Young's modulus, da the thermal expansion anisotropy, dT the cooling range below the equilibrium temperature (often the sintering temperature), ℓ the grain size, x a distance, A. a constant and F a function of orientation. This shows σa increases when ℓ increases, therefore microcracking becomes more probable as grain size increases.

* In the *absence* of external loads a spontaneous microcracking occurs when the grain size is above a critical grain size [10] :

$$\ell c \sim (\ Kb \ / \ E \ da \ dT \ ) \ ^2 \tag{9}$$

with Kb the fracture resistance of grain boundaries.

* In the *presence* of external loads the "external" stresses are superposed on the residual ones. Microcracking begins when a critical stress is reached :

$$\sigma_c \sim E \ da \ dT \ [\ (\ell/\ell c) \ ^{1/2} - 1] \tag{10}$$

and the fraction of facets which are microcracked, f, increases when the external stress, σ, increases :

$$f \sim (\sigma - \sigma_c)/ \ E \ da \ dT \tag{11}$$

3.2. Microcracking : influence of the elastic properties

A microcracked material has its Young's modulus, E, and its Poisson's ratio, v , which have decreased in comparison with the non-microcracked material (Eo and vo) [11] :

$$E/Eo \sim [1 + \frac{16 \ (10-3v)(1-v^2)}{45(2-v)} N b^3]^{-1} \tag{12}$$

$$v/vo \sim [1 + \frac{16(3-v)(1-v^2)}{15(2-v)} N b^3]^{-1} \tag{13}$$

(for N randomly oriented cracks per unit volume and a crack shape as ellipsoids of revolution of low aspect ratio, b being the radius of revolution of the cracks).

This leads to a non-linear stress-strain curve, with an hysteretic loop, and a permanent dilatation because the microcracks relieve the residual stress on the facets subject to tension (ref. Fig. 8).

3.3. Microcracking : the process zone

The propagation of the major crack can lead to its growth, with a kink along the contiguous grain boundary subject to tension, or to the initiation of secondary cracks, if the grain boundary contiguous with the primary crack is subject to compression while the grain boundary where the secondary cracks initiate is subject to tension (ref. Fig.9). However, secondary cracks only develop if the facet length (which is on the order of the grain size) is superior to a critical length, $\ell'c$:

$$\ell'c \sim 2/5 \; \ell c \qquad (14)$$

Evans' analysis explains the transition between the two regions of the brittle behaviour :

* For *small* grains ($\ell \ll \ell'c$) no secondary cracks develop and fracture occurs by the progressive growth of a microcrack which is large in comparison whith the grains.

The mechanical properties (strength and toughness) do not sensibly depend on the grain size and on the grain shape. However, toughness can be increased by crack deflecting mechanisms, sensitive to the grain shape. Equiaxed grains lead to K_c about 3 MPa$\sqrt{m}$ while elongated grains can lead to K_c about 5 MPa$\sqrt{m}$ (similar effects are observed in silicon nitride, where a control of the α /β transition allows the obtention of bimodal microstructures, with an improved toughness).

* For *coarse* grains ($\ell \gg \ell'c$) secondary cracks develop : there is a microcrack process zone and fracture occurs by microcrack coalescence. The mechanical properties strongly depend on grain morphology, with possible toughening.

3.4. Microcrack coalescence

A semi — quantitative analysis considers that fracture proceeds when sufficient microcracks occur on contiguous grain boundaries that a critical microcrack develops. This leads to the fracture strength, σu :

$$\sigma_U = \frac{\pi K_c}{2 \sqrt{\ell}} \left[\frac{\ln(4(\sigma_U - \sigma_c)/ EdadT]}{\ln(\ell^3 /4V)} \right]^{1/2} \qquad (15)$$

where .. is always the critical stress and V is the volume of the stressed piece.

For $\sigma_U \gg \sigma_c$ the grain size dependence of σ_U scales with $\ell^{-1/2}$, which corresponds to the experimental observations.

3.5. Microcrack toughening

The process zone can result in crack shielding, which increases toughness and leads to an "R-curve", i.e. to an increase of the resistance to crack growth, Kr, when the crack develops (ref. Fig. 10).

The frontal process zone does not noticeably increase toughness, because shielding is counteracted by the degradation of the material ahead of the crack tip. However, an extended zone over the crack surface increases toughness. Such an increase is proportional to the square root of the microcrack zone width, h, and the R-curve is due to the fact that h increases when the crack grows, which increases Kr.

The grain dependence of toughness is shown in Fig. 11. For small grains (e.g. $< 10 \mu$m) there is no toughening, because no secondary cracks develop, whereas for coarse grains (e.g. $> 100 \mu$m) the spontaneous thermal microcracking reduces shielding. The maximum toughness should be observed for intermediate grain sizes (e.g. about 50μm), with a possibility of an increase up too twice the value of the unmicrocracked material. An increase in temperature increases lc and decreases h, thus decreasing the toughening capabilities.

Microcrack toughening (M.T.) is of special interest in zirconia - toughened materials, where also exists stress-induced-transformation toughening (S.I.T.T.). It is worth noticing that M.T. generally does not increase strength, whereas S.I.T.T. increases both toughness and strength (More details can be found in the lecture by Claussen).

Further, Evans' comment about the meaning of toughness measurements is important :
* For notched beam tests (e.g. single edge notched beam), with no initial wake along the notch, the measured toughness is the intrinsic value, K_{co} at the left of the R curve ;
* For cantilever tests (e.g. double cantilever beam), in which an initial wake has been created during precracking, the measured toughness is the steady state value K_c at the right of the R curve.

3.6. The high temperature behaviour

At high temperatures alumina fails by creep mechanisms, which involve three main steps : nucleation, growth, and coalescence of cracks.
* The nucleation of cracks generally occurs at microstructural and chemical heterogeneities ;
* The crack growth includes regions of slow crack growth,

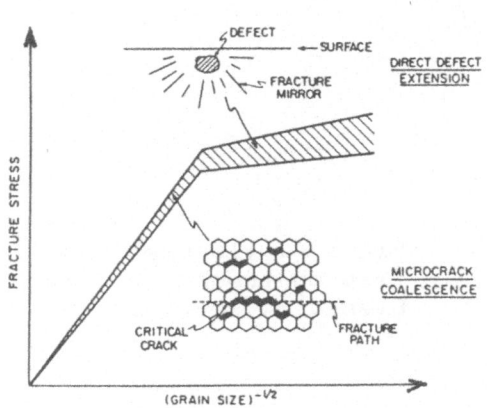

Fig.7 Influence of grain size on fracture (after Evans, in Ref. [9])

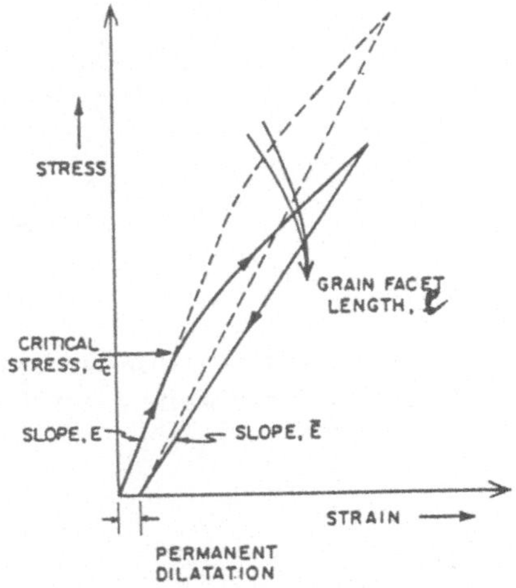

Fig.8 Hysteresis due to microcracking (after Evans, in Ref. [9])

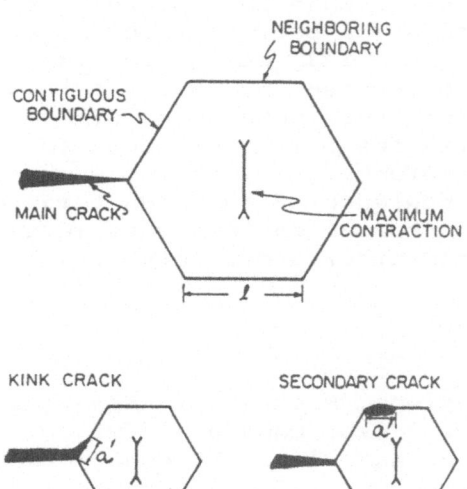

Fig.9 Conditions of nucleation of secondary cracks (after Evans, in Ref. [9])

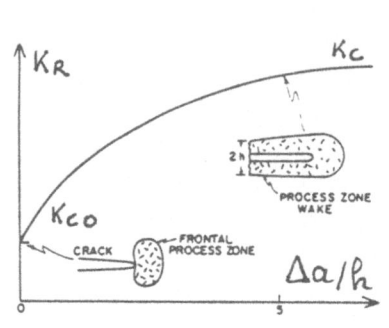

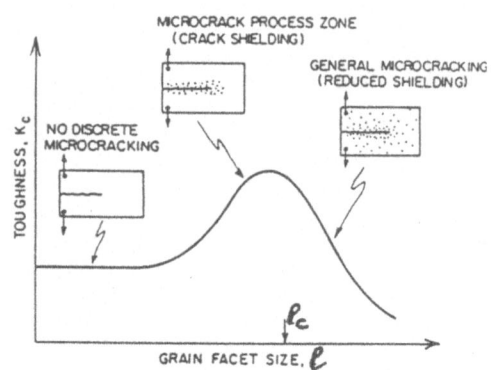

Fig.10 R curve effect
after Evans, in Ref.[9])

Fig.11 Influence of grain
size on toughness (after
Evans, in Ref.[9])

blunting and crack healing ;
 * The coalescence of *microcracks* to give a *macrocrack* is
the last step, but the more time-consuming one, in the
failure process.
   The transition between brittle fracture and creep
rupture takes place when the steady state creep stress
becomes less than the brittle fracture strength, σ_U.
   Fracture maps [12] delineate the fields in which
fracture is controlled by different mechanisms, which
helps to predict the behaviour of a certain material
(purity, grain-size...) submitted to given conditions
(stress, temperature...). A lot of work remains to be done
for a better knowledge of the high temperature properties,
chiefly when corrosive environments (e.g. air in the case
of non-oxide ceramics) are involved.

4. REFERENCES

[1] R.W. Davidge, *Mechanical Behaviour of Ceramics*,
Cambridge Univ. Press, London (1979)
   [2] B.R. Lawn and T.R. Wilshaw, *Fracture of Brittle
Solids*, Cambridge Univ. Press, London (1975)
   [3] A. de S. Jayatilaka, *Fracture of Engineering Brittle
Materials*, Appl. Sc. Publ., London (1978)
   [4] W.E.C. Creyke, I.E.J. Sainsbury and R. Morrell,
Design with Non- ductile Materials, Appl. Sc. Publ. London
(1982)

[5] *Fracture Mechanics of Ceramics,* Vol 1-6 ; Vol 7-8 to be published in 1986, Plenum press, New York (1974, 1978, 1983 /1986)

[6] M.V. Swain, 'Limitation of Maximum Strength of Zirconia-Toughened Ceramics by Transformation Toughening Increment', *J.Am. Ceram. Soc.,* 68, 4, (1985) C97

[7] J.E. Ritter, J.P. Fahey and P.L. Duffy, 'Slow Stressing Rate Testing of Ceramic Materials', *Am, Ceram. Soc. Bull.,* 12 (1984) 1517

[8] J.E. Ritter, P.B. Oates, E.R. Fuller and S.M. Wiederhorn, 'Proof Testing of Ceramics' *J. Mat. Sc.,* 15 (1980) 2275

[9] A.G. Evans, *Fracture in Ceramic Materials,* Noyes Publications, Park Ridge (1984)

[10] R.W. Davidge, 'Cracking at Grain Boundaries in Polycrystalline Brittle Materials, *Acta Met.,* 29 (1981) 1695

[11] D.P.H. Hasselman and J.P. Singh, 'Analysis of Thermal Stress Resistance of Microcracked Brittle Ceramics', *Am. Ceram. Soc. Bull.,* 58 (1979) 856

[12] M.F. Ashby, C. Ghandhi and D.M.R. Taplin, 'Fracture Maps...', *Acta Met.,* 27 (1979) 699 and 1565

ENGINEERING PERFORMANCE PREDICTION FOR CERAMICS

R.W. Davidge
Materials Development Division
AERE Harwell
OX11 ORA UK

ABSTRACT. Fracture mechanics provides the fundamental understanding for the mechanical properties of ceramics. Although the basic Griffith equation is applicable in simple cases there are complications because the fracture surface energy or fracture toughness are not independent of crack size and vary with time; crack interaction effects are also significant. Weibull statistics are used to discuss statistical variations in strength. Practical examples for grinding wheels and prosthetic hip joints illustrate the principles of performance prediction. Some problems associated with high temperature deformation are introduced.

1. INTRODUCTION

Engineering ceramics are being exploited increasingly in a wide variety of applications. Engineering components can fail through a variety of causes including mechanical wear (as in a bearing), chemical degradation (as in a hostile chemical environment, often at elevated temperature), or through mechanical failure (in a component under stress). We shall concentrate on this last aspect and the property of paramount importance is the tensile strength. The compressive strength is usually an order of magnitude greater than the tensile strength. The toughness or fracture resistance of the material is generally of secondary importance (except in how it affects strength) because once failure is initiated fracture is usually catastrophic.

The basic framework for understanding the strength and toughness of ceramics is provided by fracture mechanics [1]. Fracture occurs in a brittle manner and originates from some flaw in the material, Fig. 1. The tensile fracture strength of a ceramic σ_f can be understood in terms of the Griffith equation

$$\sigma_f = \frac{1}{Y}\left(\frac{2E\gamma_i}{a}\right)^{\frac{1}{2}} = \frac{K_{IC}}{Ya^{\frac{1}{2}}} \tag{1}$$

where Y is a geometrical constant, E is the Young's modulus, γ_i is an

95

K. P. Herrmann and L. H. Larsson (eds.), Fracture of Non-Metallic Materials, 95–115.

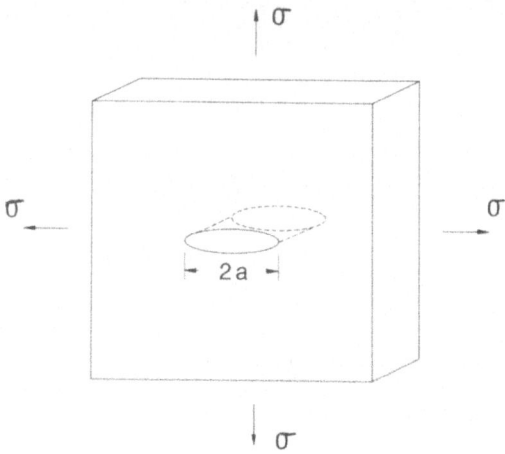

Figure 1.   The Griffith crack.

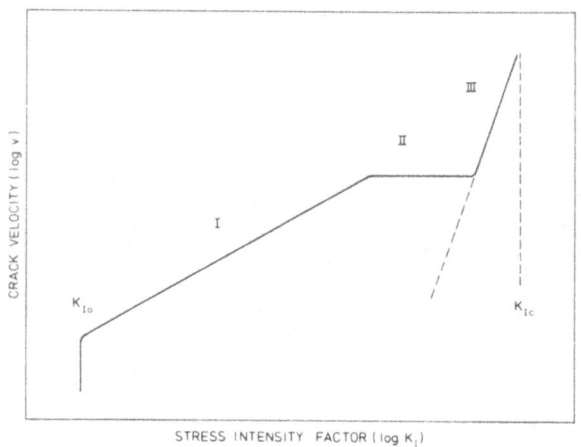

Figure 2.   Typical K-v curves for ceramics showing threshold value K_{IO}, critical value K_{IC}, and the three stages of crack growth.

effective surface energy, a is a flaw size, and K_{IC} is a stress intensity factor or fracture toughness. This equation implies that the surface energy or fracture toughness of the material is a constant, so that provided that the flaw size is known, the fracture strength can be estimated.

The effective surface energy and the stress intensity factor are both measures of the toughness of the ceramic. According to Eq. (1) any increase in toughness leads ideally to an increase in strength. Toughness however is not useful independently, to prevent catastrophic failure once fracture is initiated, unless it is very high.

The argument can be quantified by considering a cube of material of dimension L stressed in tension to the fracture stress σ_f. The elastic energy in the cube is thus $\sigma_f^2 L^3/2E$. The energy to fracture the cube is that required to create two fracture surfaces of area L^2, that is $2L^2\gamma_f$, where γ_f is a work of fracture. The critical condition to fracture the cube completely is therefore

$$L \geqslant \frac{4E\gamma_f}{\sigma_f^2} \tag{2}$$

assuming that all the elastic energy is converted to surface energy. For typical values E = 400 GPa, γ_f = 20 J m^{-2}, σ_f = 400 MPa, L $\geqslant$ 200 μm. Cubes larger than 200 μm could fracture catastrophically under these conditions. Noting that L is proportional to γ_f, a cube of 10 mm would require a work of fracture of 1 kJm^{-2} to prevent catastrophic failure; this value is considerably greater than that observed for most ceramics.

Tensile strength (conveniently measured in a bend test) is thus the key parameter. In an ideal experiment (such as a large sharp flaw in a piece of glass tested at low temperature in an inert atmosphere) the Griffith Eq. (1) in its simple form is adequate to explain quantitatively strength in terms of a flaw size and a surface energy. Generally, things are not so simple because a number of basic assumptions are not upheld in realistic situations:

- γ_i and K_{IC} are not independent of time.
- γ_i and K_{IC} are not independent of crack size.
- Several (small) cracks can interact cooperatively.

We discuss these problems below.

2. NON-IDEAL BEHAVIOUR: TIME AND MICROSTRUCTURAL EFFECTS

2.1 Time Dependent Effects

When a brittle ceramic specimen, with a large preformed crack, is fractured in tension, in vacuum or inert atmosphere, its behaviour follows classical expectations of the Griffith Eq. (1). The fracture surface energy is essentially constant, and specimens with similar-sized cracks fail at a similar force.

Under more aggressive conditions, for example wet environments for oxides, or at high temperatures for all ceramics, the situation is very different. Data relating crack velocity to stress intensity factor have been obtained for a wide range of ceramic materials by using standard fracture mechanics tests, particularly the double torsion test and the double cantilever beam test. The classical behaviour observed is shown in Fig. 2 where there are three well defined regions. Above a threshold stress intensity factor the crack velocity (v) in region I is proportional to the nth power of the stress intensity factor

$$v \propto K_I^n \qquad\qquad\qquad (3)$$

as already mentioned in the lecture by Sommer (ref. Eq. 2.20). In region II the crack velocity is independent of stress intensity factor, and in region III the behaviour is similar to that in region I but with a significantly higher slope. Fracture finally occurs at the critical stress intensity factor K_{IC}. The rate of crack growth in region I is reaction rate controlled, whereas in region II it depends on the diffusion of a corrosive species to the crack tip. In most situations the corrosive agent is water and region II occurs at higher crack velocities with increase in water content. Region III represents an inherent behaviour of the material and is the only stage observed when testing is done in vacuum.

The important effects relating to time dependence of strength are generally concerned almost solely with region I in that the behaviour associated with other regions occurs during very short times. This has led to the emergence of the parameter n as all-important in determining the time dependent behaviour of ceramics.

2.2   Crack Size Effects

Data for the surface energy are obtained generally from specimens with large preformed cracks, but there are complications at flaw sizes comparable in size to microstructural features. The surface energy for a grain size crack should approximate to the cleavage or grain boundary surface energy (γ_o), which is usually at least an order of magnitude less than the effective surface energy for macroscopic fracture (γ_i). Further propagation of the crack beyond a grain size dimension has a greater surface energy requirement and this may be a simple step function, or more complicated as indicated in Fig. 3, where it is envisaged that γ rises to γ_i over a length of a few grain dimensions. Strength is controlled by the maximum value of γ/a and in Fig. 3 the value at points P, O and R, for the three curves drawn. Fracture may thus be preceded by a period of subcritical crack growth based on microstructural considerations. This is in addition to the environmental stress corrosion effects discussed above. Thus for inherent flaws it is conceptually difficult to discuss strength in terms of an independent crack size and stress intensity factor in that these two parameters interact strongly.

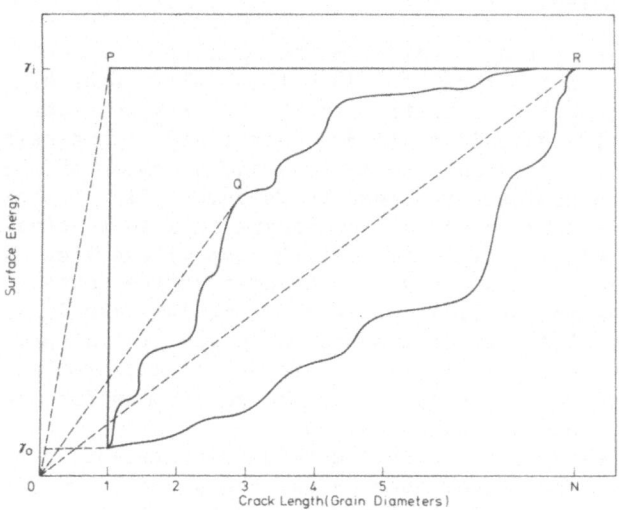

Figure 3. Variation in effective surface energy with grain
size. Strength is controlled by the maximum value of the ratio
γ/a i.e. the slope of lines OP, OQ, OR.

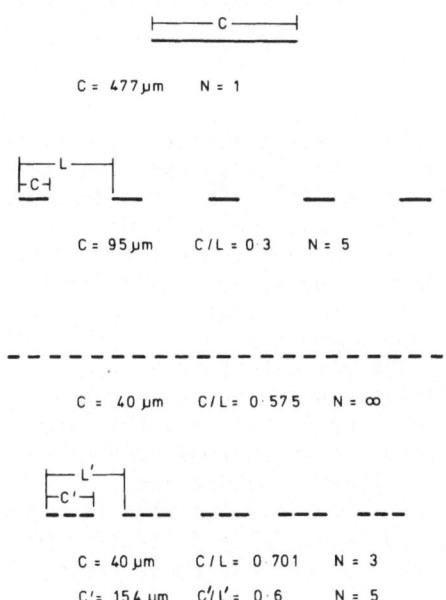

Figure 4. Various arrays of linear cracks that would all
propagate at the same stress [2].

2.3   Crack Density Effects

A further complication arises from the density of inherent cracks. The theory assumes that these flaws do not mutually interact. However, when the crack spacing L-C is similar to the crack size C this is not true. Okada & Sines [2] have identified a relatively high density of active microcracks in polycrystalline alumina and a number of close but small microcracks can coalesce and lead to failure. Fig. 4 shows three such arrays of cracks that could all propagate in a short time at the same stress as the single large crack at the top of the figure. Note that this has serious consequences for non-destructive testing techniques. The nil-observation of flaws above a particular size in a sample could lead to an optimistic prediction for strength if fracture was initiated from closely-spaced smaller flaws. An additional problem is that the times to failure under delayed fracture conditions for the various crack arrays vary by an order of magnitude.

Related effects are expected to be significant in porous materials. For porosity values larger than 5-10%, the porosity in ceramics is of the open, connected type. In highly porous materials the pores, which act as flaws, are all interconnected and the entire specimen can be regarded as being permeated by a single large flaw.

2.4   Combined Effects

We see then that fracture toughness is a function of both crack velocity and crack length

$$K_{IC} = f(v,a) \qquad\qquad (4)$$

The K_{IC}/v relation is clearly established for large cracks (Eq. (3)). But there is no simple way of relating K_{IC} and a; a strength test simply indicates the maximum value of $K_{IC}/a^{\frac{1}{2}}$. The time dependent crack growth parameter n is crucial in predicting the performance of engineering ceramics but because of the K_{IC}/a dependence the n value obtained from large cracks is not necessarily the same as that for small cracks. For example, in alumina it has been found that the value of n obtained from strain rate variations is approximately one half that obtained from double torsion tests [3]. The K/v data from macroscopic cracks would thus lead to a highly optimistic performance of components.

Fortunately a different approach is possible which automatically compensates for these effects. Conventional mechanical testing of ceramics can be done under different conditions:
  • constant strain rate ($\dot{\varepsilon}$) (e.g. modulus of rupture bend test).
  • constant stress (σ) (e.g. delayed fracture bend test).
Note that there is an equivalence between these tests, Fig. 5, according to

$$t_{\dot{\varepsilon}} = (n + 1)t_{\sigma} \qquad\qquad (5)$$

When $t_{\dot{\varepsilon}}$ and t_{σ} are the times to failure in a constant strain rate or constant stress test.

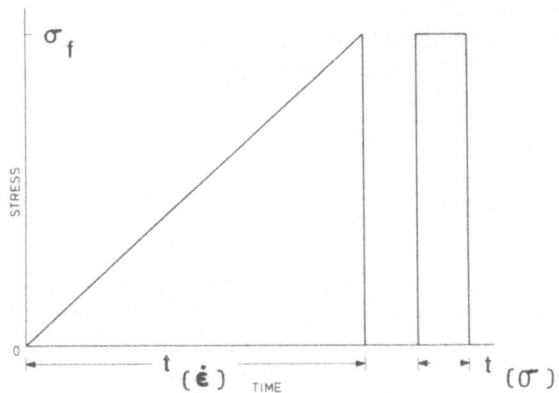

Figure 5. Comparison of constant strain rate and constant stress tests.

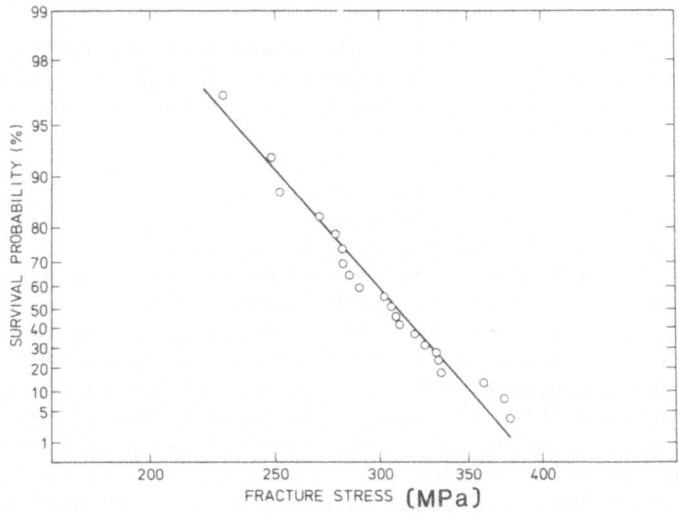

Figure 6. Statistical variation in strength of an engineering ceramic plotted according to the Weibull function.

Thus in the constant strain rate test, recalling the form of the K/v curve, significant crack growth occurs only at stresses near the fracture stress. From tests conducted under a range of either strain rate or stress it is thus possible to estimate n from respectively

$$\left(\frac{\sigma_{\dot{\varepsilon}_1}}{\sigma_{\dot{\varepsilon}_2}}\right)^{n+1} = \frac{\dot{\varepsilon}_1}{\dot{\varepsilon}_2} \tag{6}$$

$$t_f \propto \sigma^{-n} \tag{7}$$

Estimation of n from these tests is relevant to the crack growth from inherent flaws and automatically includes both the time and microstructural effects. This enables sound prediction of performance to be made.

3.   STATISTICAL EFFECTS

When a property of interest such as strength varies statistically it is obviously important to determine precisely how the property varies, in a quantitative mathematically explicit form if possible, so that some probabilistic prediction of performance can be made. Strength depends primarily on the particular flaw in a specimen suffering the highest stress intensity ($\sigma \sqrt{a}$), or, in the case of a uniform tensile stress, the largest flaw. If strength varies in the same functional way for different ceramics then a meaningful comparison can be made between materials.

In recent years it has become almost universally popular to analyse statistical variations in the strength of ceramics in terms of Weibull statistics. This has obvious advantages as we shall show later. However we cannot stress too strongly that
- this is an empirical approach,
- it does not hold for all situations,
- prediction of failure probabilities to low stress levels may have poorly defined confidence limits.

The statistical treatments used to describe the strength of ceramics are based on weakest link theory and on assumptions referred to in statistical theory as the Poisson postulates. The general theory of statistics of extremes is given in the book of Gumbel [4] and the weakest link theory of strength, pioneered by Weibull [5] is one aspect of this. In essence the problem is: given a particular distribution of property values (e.g. strength), what is the minimum value expected for a particular number of samples? For example one might be required to estimate the highest operating stress that could be applied to a component to give a failure rate of less than 1 in 10^6 from a number of observations of strength (typically < 100).

The theory assumes that the material can be divided into a number of small regions (which could be volume dV or area dA depending on whether volume or surface flaws are controlling) that act independently, each containing a flaw associated with a particular strength σ. The strength of the whole is then determined by the strength of the weakest region, as in the links of a chain. The theory as applied to ceramics [6-10] considers a function $f(\sigma)$ such that the number of flaws per unit volume (or area) associated with a strength between σ and $\sigma + d\sigma$ is $f(\sigma)d\sigma$. If $d\phi$ is the probability of failure of an element

$$d\phi(\sigma) = dV \int_0^\sigma f(\sigma)d\sigma \qquad (8)$$

with a survival probability $1 - d\phi(\sigma)$. In material of volume V there are V/dV volumes dV. The probability $(1-\phi)$ that no volume has a strength lower than σ is the product of the separate probabilities and

$$1-\phi = (1-d\phi)^{V/dV} = (1-dV \int_0^\sigma f(\sigma)d\sigma)^{V/dV} \qquad (9)$$

This reduces to, as $V/dV \rightarrow \infty$,

$$1-\phi = \exp V \int_0^\sigma f(\sigma)d\sigma \qquad (10)$$

The theory of the statistics of extremes considers three kinds of asymptotic function for $f(\sigma)$: the first two refer to variates that are unlimited in extreme values; for distributions that are bounded at one extreme the so-called asymptotic function of the third kind holds. This third function is thus of most relevance to the strength of engineering ceramics which must have a tensile strength > 0. Analysis shows that the simplest function for $f(\sigma)$, relevant to a lower limit σ_u, is given by

$$\int_0^\sigma f(\sigma) = (\frac{\sigma-\sigma_u}{\sigma_0})^m \qquad (11)$$

where σ_0 is a 'scale' parameter and m the 'shape' parameter. This will be recognized as the well-known Weibull distribution; m is usually known as the Weibull modulus and increases with decreasing variability in strength.

Weibull statistics are widely applied to the strength of ceramics, and for many data there is good agreement between experimental results and the above function. In many cases the lower limit of strength σ_u is taken as zero and thus, rearranging Eqs. (10) and (11) gives a proportionality between survival probability $P_s = 1-\phi$ as

$$\ln \ln (1/P_s) \propto m \ln \sigma \qquad (12)$$

Typical data fitting this relation are shown in Fig. 6. The prediction of stress levels for high survival probabilities is possible but this should be treated with some caution as extrapolations way beyond the experimental data are involved and there is little evidence to indicate whether the predictions are accurate.

An additional attraction of Weibull statistics is that, for materials that are consistent with the theory, it is possible to predict effects on strength of the state of stress (e.g. uniaxial, biaxial), the distribution of stress, and the specimen size. In this casr however we are concerned more with the average rather than extreme values of strength. Both these effects are a manifestation of the fact that the greater the volume of material under stress, or the greater the number of flaws perpendicular to the tensile stress, the higher the chance of finding a large flaw.

These effects are quantified in the examples below for a range of values of Weibull modulus of 5, 10 and 20. For specimens under a tensile stress, assuming a volume distribution of flaws, the mean strengths σ_{V1} and σ_{V2} for specimens of volume V_1 and V_2 are

$$\frac{\sigma_{V1}}{\sigma_{V2}} = (\frac{V_2}{V_1})^{1/m} \tag{13}$$

Thus for various values of V_1/V_2 and m we have values for σ_{V1}/σ_{V2} of:

V_1/V_2	m		
	5	10	20
10	0.63	0.79	0.89
100	0.40	0.63	0.79
1000	0.25	0.50	0.71

For a bar under a tensile stress, three point, four point (quarter length knife edges) and pure bending, assuming a volume distribution of flaws, the relative mean strengths are:

$$1 \quad : \quad [2\,(m+1)^2]^{1/m} \quad : \quad \frac{4(m+1)^2}{m+2}^{1/m} \quad : \quad [2(m+1)]^{1/m},$$

or for particular values of m:

Stress state	m		
	5	10	20
Tension	1	1	1
3-point bend	2.35	1.73	1.40
4-point bend	1.83	1.45	1.25
Pure bend	1.64	1.36	1.21

4. STRENGTH PROBABILITY TIME DIAGRAMS

It is convenient to combine the time dependent and statistical strength features into a single design concept, the SPT diagram [3]. Fig. 7 shows data for the strain rate dependence of alumina specimens for three strain rates, each a decade apart, which broke the specimens in average times of about 5, 50 and 500 s. Application of Eqs. (6) and (12) enables estimates of n and m to be made and hence generate an SPT diagram, Fig. 8. The confidence in the results for long times can be increased through using data from constant stress tests which relate to much longer failure times.

The application of the diagram is straightforward. Suppose a 99% reliability is required with a lifetime of 10^5 s then the maximum stress that can be applied is 160 MPa - considerably less than the mean bend strength 400 MPa. Corrections for stress state and distribution, and for component size can be made according to the procedures in section 3. This enables performance to be predicted only on a probabilistic basis but a technique with some guarantee of performance is often more useful such as proof testing.

5. PROOF TESTING

This is a technique whereby individual samples from a batch, that are not up to specification can be removed before entering service. In its simplest form a proof test would subject components to particular conditions relating to performance requirements, whereupon some would fail, with the remainder expected to perform satisfactorily in service.

During a realistic proof test only a small fraction of the specimens would be rejected through an appropriate selection of proof-test conditions. In spite of the weakening effect of the proof test on the weaker survivors, the strength distribution of the survivors is greatly improved compared with that of the original population. Furthermore, it is possible under **ideal** conditions to give a **guarantee** for the performance of all the survivors. The stress-intensity factor K_{Ip} on the surviving specimens for an **instantaneous** removal of the proof stress, is given by

$$K_{Ip} = Y\sigma_p a_p^{\frac{1}{2}} < K_{IC},$$ (14)

where a_p is the crack size after proof-testing (the crack size may have increased during the test). Note that $K_{Ip} < K_{IC}$ otherwise failure would have occurred. In subsequent service under an applied stress σ_a ($< \sigma_p$) the corresponding stress-intensity factor value K_{Ia} is

$$K_{Ia} = Y\sigma_a a_p^{\frac{1}{2}}.$$ (15)

Thus

$$K_{Ia} < \frac{\sigma_a}{\sigma_p} K_{IC}.$$ (16)

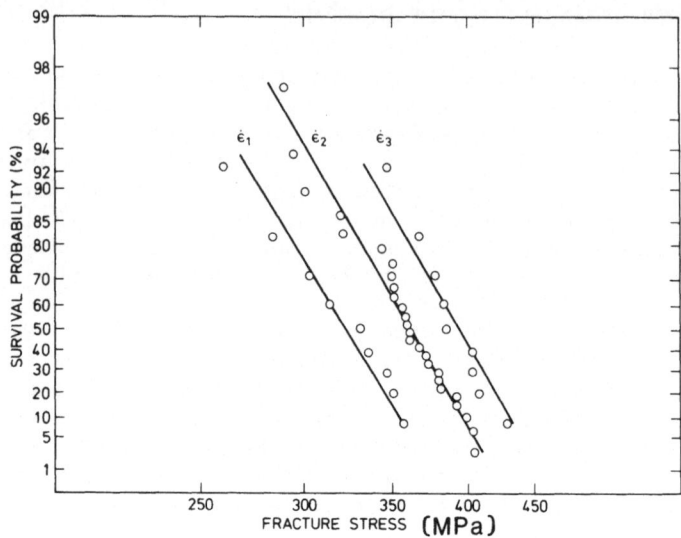

Figure 7. Strain rate dependence of strength of alumina tested at three strain rates each 10x apart.

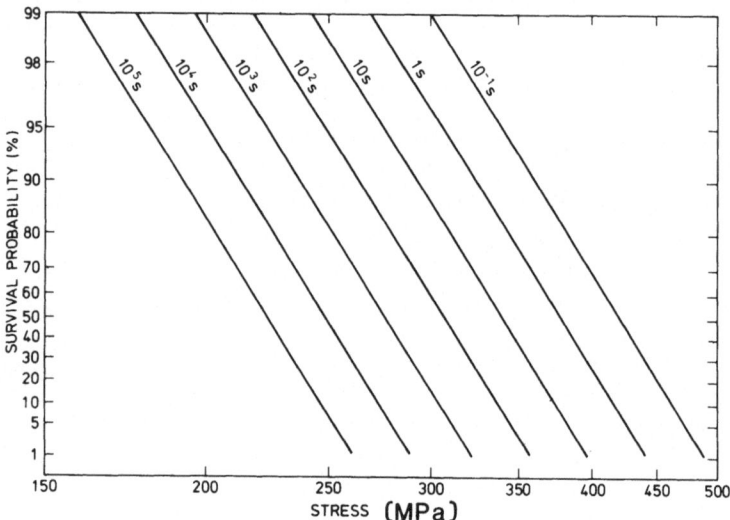

Figure 8. Strength-probability-time diagram for alumina.

Using the maximum value of K_{Ia} it can be shown that the minimum time to failure t_{min} at σ_a is

$$t_{min} \propto \frac{(\sigma_a/\sigma_p)^{2-n}}{\sigma_a^2} \qquad (17)$$

The practical significance of this equation is best appreicated by noting that for a given ratio of the proof stress to the required working stress (proof-test ratio) the minimum time to failure is proportional to σ_a^{-2}. Plots of t_{min} v. σ_a thus give a series of straight lines, on logarithmic paper, of slope -2. An example for glass is shown in fig. 9. Superimposed are failure-probability data obtained from independent measurements. Suppose for example it is required to operate a component for 10^6 s at a stress of 50 MN/m^2. Application of a proof-test ratio of 2.4 ($\sigma_p = 120$ MN/m^2) would give the required performance of the survivors, $\sim 9\%$ of the specimens having failed the proof-test.

Ideally proof-testing should be conducted under conditions where sub-critical crack growth is absent. The effects of environment could be eliminated by testing in vacuum or at very low temperatures but this would still leave the intrinsic crack growth mechanisms (region III) which are however, less significant. When crack growth occurs during the proof test, growth at the proof stress is accounted for automatically by the above analysis. Unfortunately any crack growth on unloading from the proof stress is not accounted for and, although this effect may be small, it effectively nullifies the guarantee of performance, which can now only be expressed in statistical terms.

6. SPECIFIC EXAMPLES

6.1 Grinding Wheels

The grinding wheel is simply a solid disc with a central hole and is thus a readily analysed component in terms of stress, Fig. 10. The major stresses are a consequence of the rotation of the wheel, whether or not any work is being done at its periphery where the stresses due to interaction with the workpiece are relatively small. The maximum stress due to rotation is located at the central hole in a tangential direction and is given by

$$\sigma_{max} = \frac{(3+\nu)}{4} \rho\omega^2 \left(b^2 + \frac{(1-\nu)}{(3+\nu)} a^2\right) \qquad (18)$$

This stress falls with increase in radius to zero at $r = b$ (the maximum radial stress is less than half this value and is located away from the hole). Note that the stress is proportional to the square of the velocity. ν is Poisson's ratio and ρ the density. The other quantities are indicated in Fig. 10.

Some of the principles involved [11] in performance prediction for grinding wheels are illustrated in Fig. 11. This SPT (strength

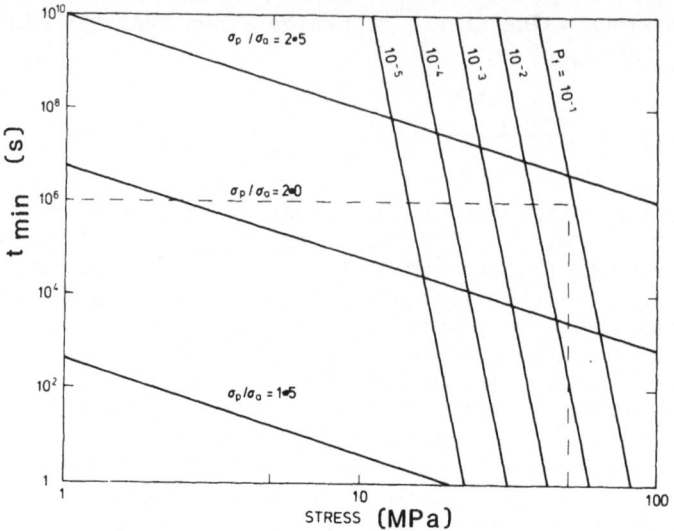

Figure 9.   Proof test diagram for glass.

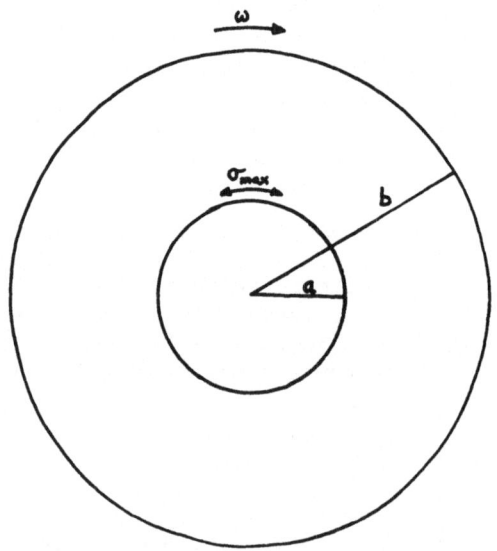

Figure 10.   Sketch of grinding wheel.

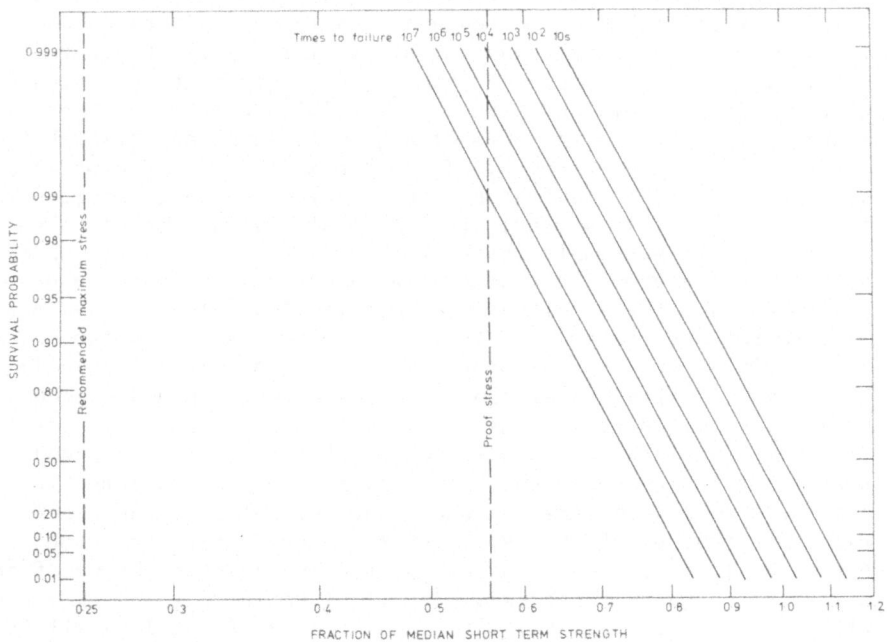

Figure 11. SPT diagram for grinding wheel.

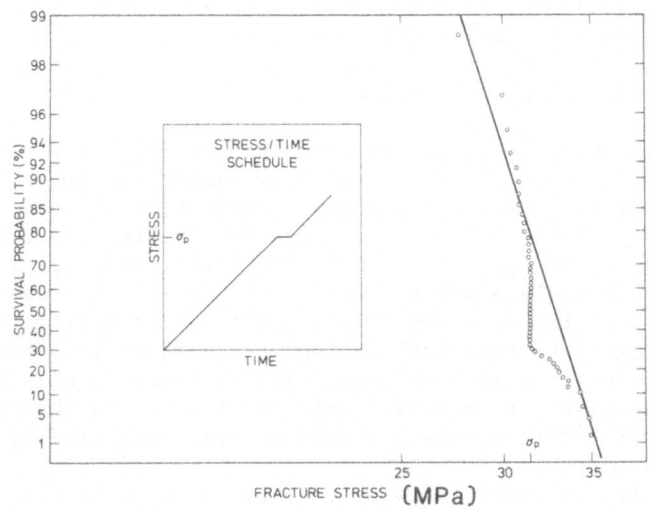

Figure 12. Variation in strength of vitreous bonded alumina grinding wheel: solid line – tested at constant strain rate; experimental points – tested according to stress–time cycle in insert.

probability time) diagram was produced from measurements on test bars of grinding wheel material tested under various strain rate conditions and under delayed fracture (constant stress loading). This permitted estimation of the constants n and m and a reference stress, and the diagram was then plotted from the principles discussed above. The diagram is expressed in terms of a normalised strength which is the average short time strength. The maximum stress on the wheel during service (limited by maximum permitted velocity) is generally around one-quarter of this average short term strength. One can thus see that any failure is extremely unlikely. Furthermore, grinding wheels are subjected to a proof test procedure before use whereby the wheel is rotated to 1.5 its permitted use velocity (equivalent to 2.25 times the maximum service stress). This procedure generally fractures 1 in 10^3 to 10^4 of wheels so that the weaker wheels are automatically eliminated.

Fig. 12 illustrates some of these effects in a different way. The solid line refers to a series of wheels broken under spin test conditions (for which no data points are included for clarity). A second set of wheels was broken under the stress time schedule indicated in the inset. The conditions were arranged such that one-third of the wheels failed below a chosen proof stress, one-third broke during the proof stress held for about 10 s, and the remaining third broke after the proof stress with a further increase in stress. The first third of the data are coincident with the original results. The second third of the data all failed at the constant stress value, the remaining third broke at stresses which increasingly approached the original curve with increasing stress. This shows that those specimens which just survived the proof stress are weakened to a small extent as indicated. The elimination of the weaker specimens through the proof test procedure gives an enhanced performance of the survivors. Under ideal conditions this can give a guaranteed minimum life-time for the component.

6.2   Prosthetic Alumina Hip Joints

The use of high quality alumina ceramics as the ball in prosthetic hip joint replacements [12-13] is increasingly widespread and so far between 10^5-10^6 implants have been made. In a common design, the alumina ball with a closely machined tapered hole as shown in Fig. 13 is attached through a metal coupling to the femur. The ball generally locates into a high density polyethylene socket, but in some cases this may also be alumina. Although the wear resistance and chemical inertness of alumina are being exploited in this application, the ceramic is stressed through its attachment to the metal coupling. This produces a circumferential stress in the ceramic, in many ways similar to that in the grinding wheel. In service this stress can only increase dictated by the normal range of human activity.

An SPT diagram for a biograde alumina is shown in Fig. 14. The high reliability can be appreciated bearing in mind that the maximum working stress is 70 MPa. Similarly a proof procedure can also be applied and this information is summarised in Fig. 15. This indicates that by overstressing the component by a factor of 1.6 by a proof test will give a guaranteed minimum lifetime under normal use of 30 years.

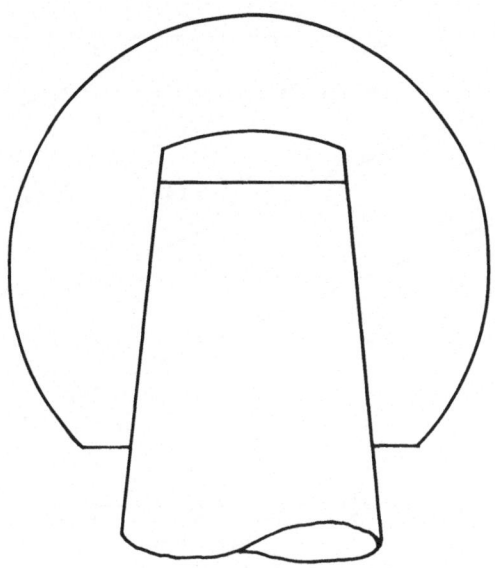

Figure 13. Geometry of cross section of alumina femoral head.

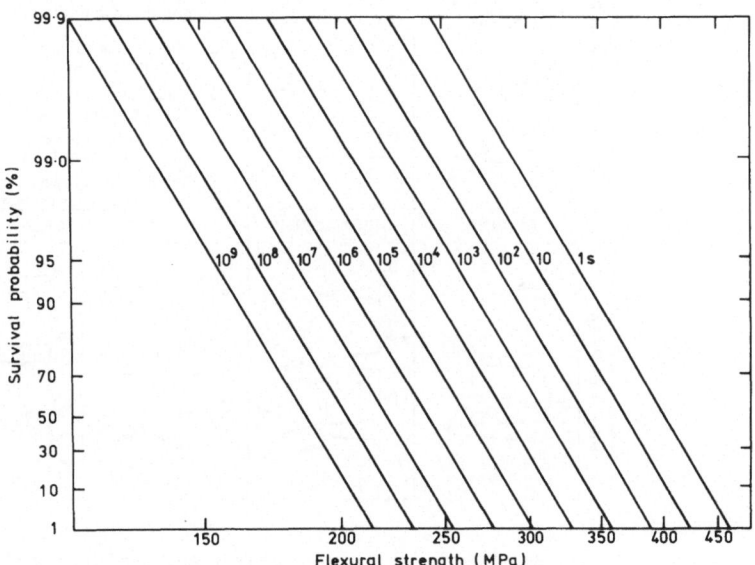

Figure 14. SPT diagram for bio grade alumina.

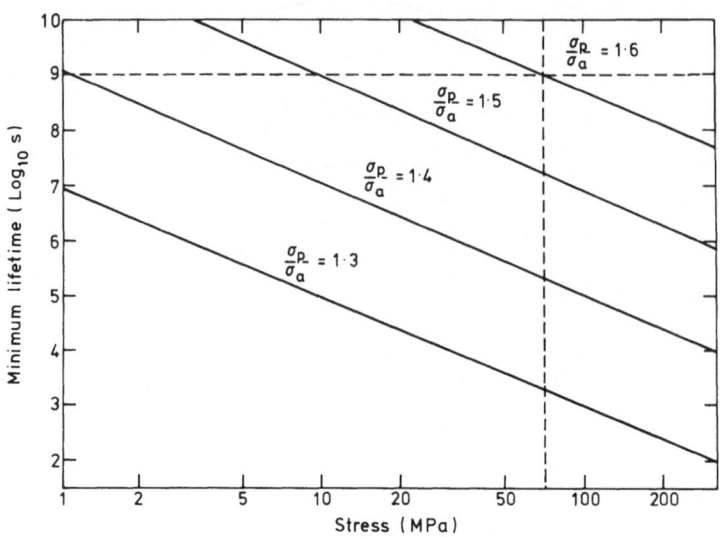

Figure 15.   Proof test diagram for bio grade alumina.

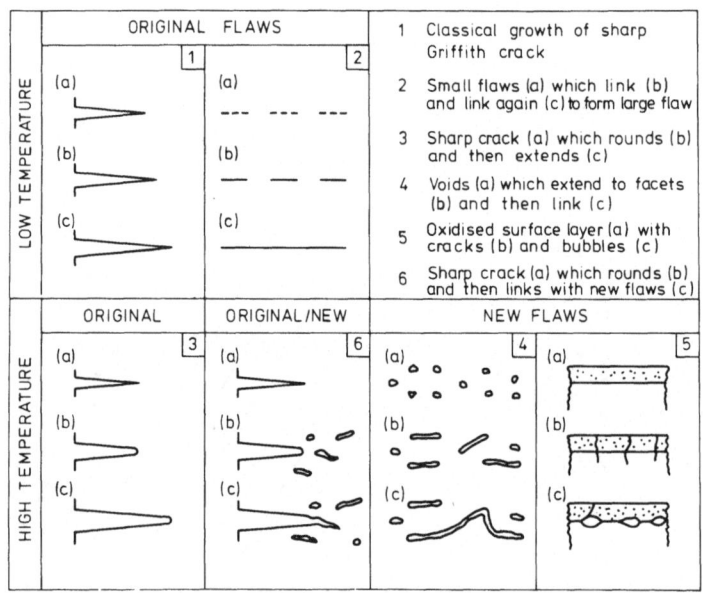

·Figure 16.   Summary of crack nucleation and growth mechanisms.

6.2   High Temperature Applications

The number of current high temperature engineering applciations for
ceramics is relatively small but the high current interest for use as
engine components makes this application area potentially enormous.
Performance prediction at high temperature is poor [14].   Although the
initial inherent flaws present after fabrication affect the initial
properties, additional flaws are generated due to either physical or
chemical effects.   Fig. 16 illustrates some of the possibilities.   The
simplest effect (example 3) is localised plastic deformation particularly
under the high stresses near the tips of cracks which can lead to crack
rounding.   This can lead to an initial enhancement of strength but after
this the crack can then grow in its rounded configuration.
      Example 4 indicates how creep processes can generate new defects
particularly at grain boundaries.   This is initially manifested as small
voids which eventually grow to defects on grain boundary facets which may
then link together.   This process can either extend the original flaws or
create new flaws which are larger than the original flaws.
      Example 5 shows some chemical effects, say, in a non-oxide ceramic
like silicon nitride which is used in an oxidising envornment.   This
leads to the formation of a silica surface layer and more complex
diffusional effects involving impurities withint he ceramic or in the
operating environment.   This can lead to additional cracking and a
reduction in strength.
      There is a great deal of active work in progress to study many of
these and related effects but at present the situation is too complicated
to predict failure in a way that was possible for low temperature
operation.   Fortunately for high stresses and short times, the same

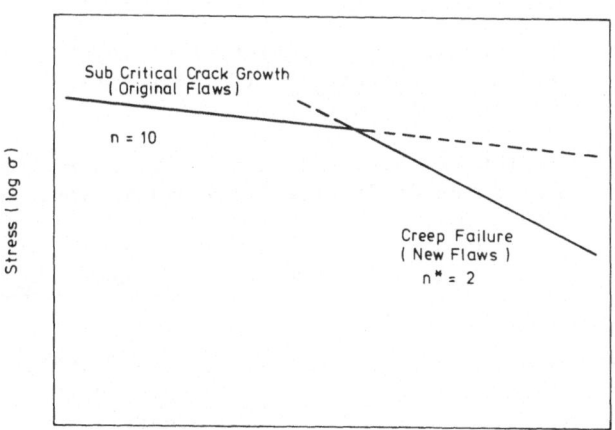

Figure 17.   Stress time endurance limit diagram for engineering
ceramics at high temperature.

functional relationship (Eq. 5) holds as for low temperatures. For lower stresses and longer times, however, the situation is not amenable to such analysis. In some cases, however, a different failure criterion may be used based on a limiting creep strain. In this case a STEL diagram (Stress Time Endurance Limit) may be constructed as indicated in Fig. 17. A lot more work is required in this area before high temperature engineering ceramic components can be operated reliably in a predicted manner.

7. REFERENCES

[1]   Davidge, R.W., The Mechanical Behaviour of Ceramics, Cambridge University Press (1979).

[2]   Okada, T. ánd Sines, G., Crack Coalescence and Microscopic Crack Growth in the Delayed Fracture of Alumina, J. Amer. Ceram. Soc. 66 (1983) 719.

[3]   Davidge, R.W., McLaren, J.R. and Tappin, G., Strength – Probability – Time Relationships in Ceramics, J. Mater. Sci. 8 (1973) 1699.

[4]   Gumbel, E.J., Statistics of Extremes, (1958), New York, Columbia University Press.

[5]   Weibull, W., A Statistical Distribution Function of Wide Applicability, J. Appl. Mech. 18 (1951) 293.

[6]   Matthews, J.R., McClintock, F.A. and Shack, W.J., Statistical Determination of Surface Flaw Density in Brittle Materials, J. Amer. Ceram. Soc. 59 (1976) 304.

[7]   McClintock, F.A., Statistics of Brittle Fracture, Fracture mechanics of ceramics, (Ed. R.C. Bradt et al), 1 (1974) 93; New York, Plenum Press.

[8]   McClintock, F.A. and Zaverl, F., An Analysis of the Mechanics and Statistics of Brittle Crack Initiation, Int. J. Fract. 15 (1979) 107.

[9]   Evans, A.G. and Jones, R.L., J. Amer. Ceram. Soc. 61 (1978) 156.

[10]   Batdorf, S., Fundamentals of the Statistical Theory of Fracture, Fracture mechanics of ceramics, (ed. R.C. Bradt et al), 3 (1978) 1; New York, Plenum Press.

[11]   McLaren, J.R., Davidge, R.W., Cotton, D.C.L., Haywood, S.A. and Robson, M.E., Predicting the Effect of Proof Tests on the Safe Operation of Grinding Wheels, Proc. Br. Ceram. Soc. 26 (1978) 67.

[12] Davidge, R.W., Structural Degradation of Ceramics, <u>Biomaterials</u> 5 (1984) 37.

[13] Real, M.W., Cooper, D.R., Morrell, R., Rawlings, R., Weightman, B. and Davidge, R.W., Mechanical Assessment of Biograde Alumina, <u>Science of Ceramics</u> **13** (1986) in press.

[14] Davidge, R.W., Durability of Engineering Ceramics for High Temperature Engineering Applications, Proc. 2nd Irish Durability and Fracture Congress, Limerick, <u>Mechanically, chemically and thermally induced failure in engineering materials</u>, Ed. J.D. Bolton and S. Hampshire, Irish Elsevier, Shannon (1984) 291.

THERMAL SHOCK PROPERTIES OF CERAMICS

P. BOCH
Ecole Nationale Supérieure de Céramique Industrielle

47 Av. A. Thomas
87065 Limoges
FRANCE

ABSTRACT. Ceramic materials are brittle solids, which have a very low fracture deformation. They are sensitive to differential strains such as those which are induced by temperature gradients or by thermal expansion heterogeneities or anisotropies. The two main theories of thermal shocks are the thermoelastic theory, which considers that fracture is controlled by the initiation of new cracks, and the energetic (Hasselman 's) theory, which considers that fracture is controlled by the propagation of preexisting flaws.

1. BRITTLENESS OF CERAMICS AND THERMAL SHOCK SENSITIVITY

1.1. Sensitivity to differential strains

Ceramic materials are brittle solids which have a linearly elastic behaviour at low and medium temperatures. Their maximum deformation is very low (e.g. 0.001), which means they are sensitive to differential strains which would not be dangerous in ductile metals where plastic flow releases local stress peaks.
---> Even high strength ceramics can break in a catastrophic manner if they are loaded regardless of stress concentrations.

The main cause of differential strains is THERMAL EXPANSION, and ceramic pieces are frequently destroyed by "thermal shock" effects. This is true for domestic items (e.g. dinnerwares) and for industrial refractories (where corrosion is also operative). Thereby the development of "structural ceramics", to be used in thermal machines, reinforces the need of a better understanding of the thermal shock resistance of ceramics.
---> A very low thermal expansion is the perfect answer to the thermal shock problem. This is obtained for

117

K. P. Herrmann and L. H. Larsson (eds.), Fracture of Non-Metallic Materials, 117–135.
© *1987 by ECSC, EEC, EAEC, Brussels and Luxembourg.*

instance in pure silica glass, where α is only about
$0.5\ 10^{-6}\ K^{-1}$. However, it must be recalled that even
ceramic pieces with a very low thermal expansion can
endure thermal stresses if their "environment" expands.

1.2. Thermal expansion of solids

Basically, thermal expansion is related to the imperfect
harmonicity of interatomic potentials. For crystals, it
can be shown that :

$$\alpha = A\ C_v$$

where C_v is the specific heat and A a proportionality
constant. At low temperature C_v varies as T^3 and at high
temperature (T > Debye temperature) C_v is constant:
---> It is a common error to consider the thermal
expansion coefficient of a material at room temperature
even for uses at high temperatures. In particular for
ceramics which have high Debye temperature the variation
of α vs. T can be very pronounced.
        The amplitude of α is inversely proportional to the
strength of atomic bonds, which means that low values of α
are associated with high melting points and high elastic
constants.

1.3. Thermal expansion of crystalline ceramics

   * Only crystals of cubic symmetry are isotropic ; lower
symmetries lead to an anisotropic expansion, geometrically
figured by a revolution ellipsoid (trigonal, tetragonal,
and hexagonal) or a non-revolution ellipsoid
(orthorhombic, monoclinic, and triclinic). Some crystals
have a very strong anisotropy with a positive value of α
along some directions and a negative one along others: it
is the case of calcite $CaCO3$, cordierite
$2MgO-2Al2O3-5SiO2$, or aluminium titanate $Al2TiO5$.
   * "Pure" polycrystalline aggregates have a mean thermal
coefficient which is isotropic if grains are randomly
oriented but which is anisotropic in case of crystalline
orientation texture and non-cubic symmetry. Cracks can
develop at grain boundaries due to expansion mismatch (See
the lecture on the Mechanical Properties of Ceramics).
   * Multiphase materials are constituted of phases which
generally have different thermal coefficient. The mean
value of α is related to the coefficient of each phase
with a weight which is proportional to its volumic
fraction multiplied by its elastic modulus. Cracks can
develop at interphase boundaries. An important case is the
one of phase transitions (e.g. quartz->cristobalite, or
tetragonal zirconia ->monoclinic zirconia).

1.4. Thermal expansion of glasses

Glasses have isotropic thermal expansion. However, they are out of equilibrium solids which can lead to a peculiar thermal behaviour: density depends on the cooling rate and decreases when the cooling rate increases. When heated, glasses begin to expand but they can contract at temperatures where diffusion is fast enough to allow the atomic arrangement to modify into a more stable configuration. Thermal expansion may thus be considered as the sum of two terms: the first being related to the variation of temperature for a given structure and the second to the variation of structure at a given temperature. This can be expressed by the introduction of a "fictive temperature" , T*, which characterizes the structural variations :

$$dl/l = \alpha_1 \, dT + \alpha_2 \, dT*$$

1.5. Numerical values

Oxides have a mean thermal expansion coefficient, between 20°C and 1000°C, $\bar{\alpha}$,of about 10 (in $10^{-6} \, K^{-1}$): for instance this value is 11 for zirconia, 9 for alumina etc. Silicates have lower expansion: 5.5 for mullite, 5 for zirconium silicate. Ceramics with covalent bonds (e.g. silicon carbide or silicon nitride) have a coefficient of about 3-4. Such a low expansion helps them to whistand thermal shocks. Silica-lime-soda glasses have an expansion of about 10 ; borosilicate glasses (e.g."Pyrex") and fused silica have $\bar{\alpha}$ values which can decrease to 0.3 (whereas quartz, i.e. crystalline silica has a coefficient of nearly 10).

2. THERMAL STRESSES ; THERMOELASTIC THEORY OF THERMAL SHOCKS

2.1. Thermal stresses

For the simple case of a bar uniformely heated with its ends constrained to avoid expansion (ref. Fig. 1) Hooke's law gives:

$$dl/l = \alpha \, dT \qquad\qquad (1)$$

$$\sigma = - E \, dl/l = - E \, \alpha \, dT \qquad\qquad (2)$$

where dT is the temperature variation, E is Young's modulus and σ is the normal stress in a cross-section area.

For alumina ($E = 400$GPa , $\alpha = 10^{-6}$ K^{-1}), dT = 100°C leads to σ = 400 MPa , which is in excess of the usual tensile strength of about 250 MPa. This shows that rather low temperature gradients can provoke fracture, such gradients being facilitated by the generally low thermal conductivity of ceramics (with some exceptions : diamond, beryllium oxide, and to a lesser extent silicon carbide).

2.2 Temperature gradients

Thermal stresses can be found in a piece even at uniform temperature, due to heterogeneity or anisotropy effects. However, the most important cases remain the ones where stresses are due to temperature gradients, particularly if those gradients result from a fast cooling (or heating).
        For an infinite plate submitted to a steady-state heat flow there exists a parabolic temperature distribution, which leads to a parabolic stress distribution. Thereby for cooling the surface suffers tensile stresses whereas the core is under compression. Thus, because the tensile strength is lower than the compressive one fracture will initiate at the surface. The reverse case is valid for heating.

REMARK. Glass specimens (e.g. windscreens) can be strengthened by treatments leading to residual compressive stresses. That is the case of tempered glass. The initial temperature must be high enough to allow the glass to be in a viscous state. Then fast cooling leads to the solidification of the surface and to its contraction, which is not opposed by the hot and viscous core. The ulterior cooling of the core, thus its solidification and its contraction, leads to the final state of compression on the surface and tension in the core. (ref. Fig. 2).

2.3. The thermoelastic theory

The temperature map inside a piece has been calculated for various geometries and various heat-exchange conditions. In all cases the stress in a point and for a given facet, σ, can be written (in incorporating the minus sign in dT):

$$\sigma = E \alpha \, dT \, f(\nu) \qquad (3)$$

where $f(\nu)$ is a function which depends on geometry, heat exchange, and Poisson's ratio, ν.
For a thin plate of infinite dimensions uniformly cooled:

$$f(\nu) = 1/(1-\nu) \qquad (4)$$
$$\sigma_M = (E \alpha \, dT)/(1-\nu) \qquad (5)$$

σ_M is the MAXIMAL stress which would appear at the surface of the plate by cooling down the surface to the lower temperature T1, whereas the core maintained at the initial temperature T2. This is a limit case which is never reached and the actual stress, σ, is only a fraction of σ_M:

$$\sigma = S* \sigma_M = S* (E \alpha dT)/(1-\nu) \qquad (6)$$

where S* is a dimensionless coefficient, inferior to one, which characterizes the shock severity.

The value of S* depends on Biot's number, β:

$$\beta = L h / k \qquad (7)$$

where L is a "mean" dimension of the piece, h the heat-exchange coefficient, and k the thermal conductivity.

Thereby valids : $S*(\beta) ---> 1$ if $\beta ---> $ infinity (very fast thermal exchanges) and $S*(\beta) ---> 0$ if $\beta ---> 0$ (very low thermal exchanges). Figure 3 shows the typical aspect of the $S*(\beta)$ curves vs. time for various shock severities. It can be noticed that the stress peak is not obtained at t = 0 but at a time which increases when the shock severity decreases. Local fracture will develop when the stress σ reaches the local strength σu:

$$\sigma u = (S*(\beta) E \alpha dT)/(1-\nu) \qquad (8)$$

which corresponds to a CRITICAL TEMPERATURE GRADIENT:

$$dTc = \sigma u (1-\nu) /E \alpha S*(\beta) \qquad (9)$$

The FIRST THERMAL SHOCK PARAMETER R is thus defined as :

$$R = \sigma u (1-\nu) / E \alpha \qquad (9)$$

Thereby R is expressed in Kelvin and represents the maximal temperature gradient the material can resist without failure.

For a given piece submitted to given heat exchanges, characterized by a certain Biot's number and thus by a certain $S*(\beta)$ value, the maximal temperature variation allowed before fracture will be:

$$dTc = R / S*(\beta) \qquad (11)$$

which is always greater than R.

R characterizes the resistance to thermal cracking of the material. It increases whith increasing strength σu and decreases with increasing Young's modulus E and thermal expansion coefficient α , respectively. Therefore

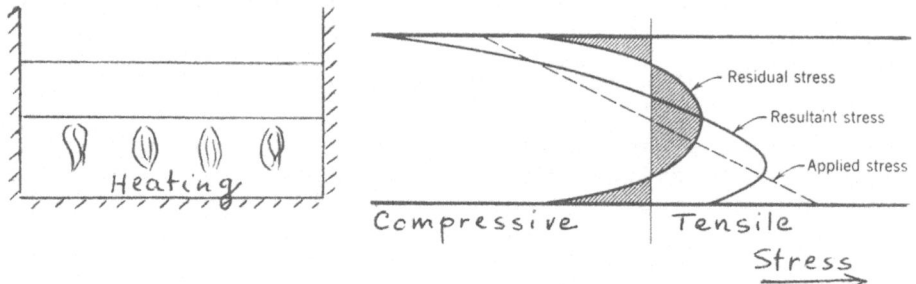

Fig. 1 Uniformly heated
bar with constrained ends
to avoid thermal expansion

Fig. 2 Residual stress,
applied stress, and resultant
stress distribution for
transverse loading of a
tempered glass plate
(from ref. [6])

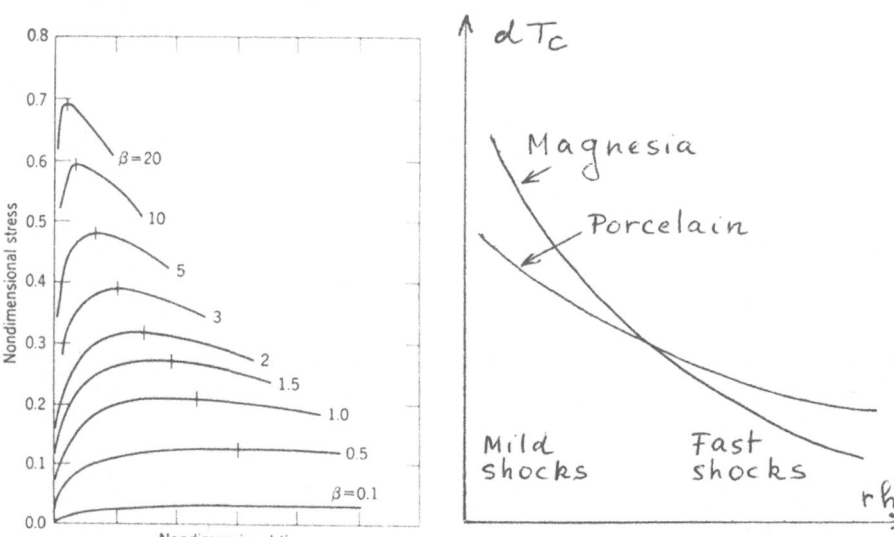

Fig. 3 Variation of non-
dimensional surface stress
(S*) with dimensionless
time for an infinite flat
plate with different
values of Biot's modulus β
(from ref. [6])

Fig. 4 Variation in quench
temperature (dTc) causing
fracture for different
materials under different
conditions of heat transfer
(expressed by their rh value)
(from ref. [6])

R is a *materials parameter*. However, it can be pointed out that the thermal cracking resistance of a *piece* also depends on S*(β) which is linked to both *materials parameters* (e.g. conductivity k) and *non-materials parameters* (e.g. heat exchanges).

R is generally considered for FAST thermal shocks. For MILD thermal shocks it is usual to consider another parameter R'. For mild shocks, indeed, S*(β) appears to be roughly proportional to β:

$$S*(\beta) = A \beta \qquad (12)$$

where A is a constant.

From equation (6) it follows :

$$dT = \sigma(1-\nu) / (E \propto A \beta) \qquad (13)$$
$$dT = \sigma(1-\nu) k / E \propto L A h \qquad (14)$$
$$dT = (k \sigma (1-\nu)/E \propto)/(1/ALh) \qquad (15)$$

by using β = Lh/k.

Fracture will develop when the temperature variation leads to the stress equal to the strength:

$$dTc = (k \sigma u (1-\nu)/E \propto)/(1/ALh) \qquad (16)$$

i.e.:

$$dTc = R' K/h \qquad (17)$$

where R' = kR is the SECOND THERMAL SHOCK RESISTANCE PARAMETER, k is a geometrical parameter equal to 1/AL and k is always the conductivity. R' is expressed in Wm^{-1} . Figure 4 shows the variation of dTc vs. Lh for different materials. The interesting point is that there is a crossing of the curves. Thus, a comparison between porcelain and magnesia, for instance, shows that for mild thermal shocks (left hand side of the diagram) MgO resists better than porcelain whereas it is the contrary for fast thermal shocks (right hand side of the diagram). This means there is no possibility to establish a hierarchy of the capability of materials to resist thermal cracking without taking into account the shock severity. Thus, the best materials for fast shocks are those with a high R value whereas the best materials for mild shocks are those with high R' values. The "same" shock (e.g. water quenching) will be a fast one for a material having a low conductivity (---> high Biot's number) and a mild one for a material having a high conductivity (---> low Biot's number).

The quantitative use of the thermoelastic theory is difficult for two reasons:

1) The thermal exchanges are generally known with a very low accuracy. Biot's number, for example, can vary on an order of magnitude in the case of water quenching, depending on the water temperature, the velocity of the

quenched piece when entering the bath, the shape of the piece, etc. This induces an error of S*(β) which is very large (>50%).

2) The criterion of damage , i.e. fracture develops when $\sigma = \sigma u$, raises the question of the meaning of the strength σu. It is known that strength is a "secondary" parameter only, related to the toughness , Kc, and to the equivalent size of the most critical flaw, a:

$$\sigma u = (1/Y) \; Kc \; (a)^{-1/2} \tag{18}$$

This shows that the level of the maximum stress, σ, at a certain point, M, is not enough to know whether the fracture will develop. Besides, σu may vary depending on its experimental determination: size of samples (Weibull's effect), type of test (tensile test, or bend test...),etc. The relative error of the value of σu which should be chosen is also large (>25%).

REMARK. R and R' are generally written in the case of a biaxial state of stress which leads to the factor (1-ν) in the numerator. Such a factor is equal to unity for an uniaxial stress state and is equal to (1-2ν) for a triaxial stress state.

SUMMARY. The *Thermoelastic theory* of thermal shock resistance is based on the assumption that brittle fracture is controlled by the *initiation of new cracks,* and does not take into account the propagation of preexisting flaws in an explicit manner. It leads to the definition of two main parameters, R and R', which correspond to fast shocks and mild shocks, respectively. The increase of the maximum temperature variation a specimen can resist is obtained by increasing the strength and the thermal conductivity and by decreasing the thermal expansion coefficient, Young's modulus, and Poisson's ratio of the material.

3. THE ENERGETIC THEORY OF THERMAL SHOCK RESISTANCE (D.P.H. HASSELMAN)

This theory is based on the assumption that the brittle fracture of thermally shocked ceramic specimens is controlled by the *extension of preexisting microcracks.* The corresponding calculations are derived from an energetic criterion, analogous to Griffith's one. It is proposed that the driving force for crack propagation can be derived from the elastic energy stored in the specimen at the moment of fracture. Thermal stress fracture will not be catastrophic if the total elastic energy is less than the total fracture energy required to propagate a crack over an area equivalent to the cross sectional area of the specimen.

Crack propagation under thermal stress conditions generally occurs in the absence of external forces. The driving energy for crack propagation is solely due to the stresses inside the shocked specimen. This is analogous to crack propagation under conditions of "fixed grips". Thereby the propagation of a crack decreases the effective Young's modulus which leads to a relaxation of stresses, and thus to a decrease of the driving energy for crack propagation. Hasselman has initially chosen the case of a specimen uniformly cooled (or heated) through a temperature difference of dT. If the external boundaries are rigidly restrained the triaxial state of stress will be:

$$\sigma = \alpha\, E_o dT\, /\, (1-2\nu) \tag{19}$$

The material is supposed perfectly brittle with no stress relaxation by creep mechanisms. Further, it contains microcracks, uniformly distributed throughout the material, with a density of N per unit volume. All cracks have a "penny shape" of uniform size, l being the radius. They are supposed to propagate all in the same time, in radial direction and with no interaction between them.

The total energy per unit volume, Wt, is the sum of the elastic energy, We, plus the fracture energy of the cracks Wf. For penny-shaped cracks, the total fracture energy is :

$$Wf = 2\pi N [l]^2\, G \tag{20}$$

where G is the fracture surface energy, the factor 2 being due to the two surfaces of the cracks.

The elastic energy has a rather complicated form:

$$We = \frac{3\, E_o (\alpha\, dT)^2}{2(1-2\nu_o)} \left(1+\frac{16(1-\nu_o^2)\, N\, l^3}{9(1-2\nu_o)} \right)^{-1} \tag{21}$$

The crack stability depends on the sign of the derivative :

$$dWt/dl \tag{22}$$

If dWt/dl is > 0 the crack propagation increases the total energy of the system and the cracks do not propagate ; if dWt/dl is < 0 the crack propagation decreases the total energy and the cracks propagate.

The critical temperature difference, dTc, corresponds to the limit case where dWt/dc = 0:

$$dTc = \left(\frac{G(1-2\nu)^2 \pi}{2E\alpha^2 I(1-\nu_o^2)} \right)^{1/2} \left(1 + \frac{16(1-\nu_o^2)Nl^3}{9(1-2\nu)} \right) \qquad (23)$$

This expression is graphically shown in Fig.5. The curve
is "U-shaped" and the interesting comments are as follows:
   * The dTc value begins to decrease if l increases, and
after passing a mimimum increases again with increasing l.
   * For small cracks, i.e. on the left hand side of the
curve, the slope of the curve is < 0. This means that if
dT becomes equal to dTc the rate of the decrease in
elastic energy associated with the crack extension exceeds
the rate of increase of fracture energy : so the cracks
will propagate in a "dynamic" or " catastrophic" manner.
   * For large initial cracks, i.e. on the right hand side
of the minimum, the slope is > 0. The increase of dT
beyond dTc results in a "quasi-static" propagation only.
   * For small cracks, the excess energy is transformed
into kinetic energy of the propagating cracks. So the
final length of such cracks is greater than the length
which would correspond to the initial dTc value on the
RIGHT HAND SIDE of the U curve : it is lf and not l'f.
Those cracks are now "undercritical" with respect to the
dTc value which initiated them and they cannot propagate
again unless dT is increased about a finite quantity in
excess of dTc (ref. Fig. 6).
   The difference between the previsions of the
thermoelastic theory and of the energetic one can be
clearly seen in the case of short cracks. For such cracks,
which will show a dynamic propagation, the final length,
lf, is generally much greater than the initial length, lo.
The calculation leads to:

$$lf \approx (3 (1-2\nu) /8 (1-\nu^2) \ lo \ N)^{1/2} \qquad (24)$$

However, this equation can be modified by taking into
account the value of the critical fracture stress, σ_U,
which is for a penny-shaped crack:

$$\sigma_U = (G E /2 lo (1-\nu^2))^{1/2} \qquad (25)$$

Further, by introducing equ. (25) into equ. (24) gives :

$$lf \approx (3 (1-2\nu) \sigma_U^2 /(4\pi NGE))^{1/2} \qquad (26)$$

This shows that the area, A, traversed by each propagating
crack (A = π lf^2) is inversely proportional to the factor:

$$G E /\sigma_U^2 (1-2\nu) \qquad (27)$$

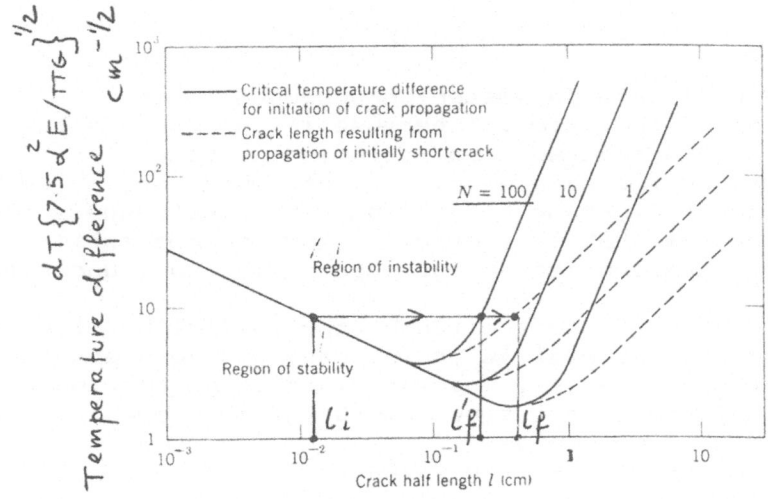

Fig. 5 Temperature difference (dTc) required to initiate crack propagation as a function of crack length (l) and crack density (N) (from ref. [1])

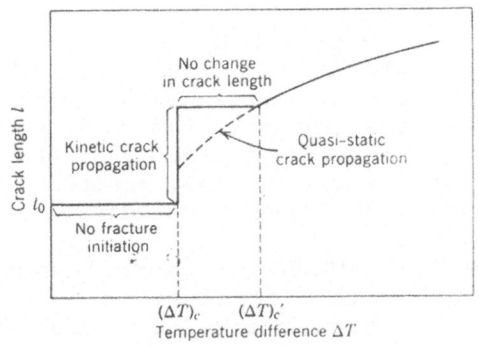

Fig. 6 Crack length as a function of temperature difference dT (from ref [1])

Apart from the term $(1-2\nu)$ the factor (27) is identical to the so-called FOURTH THERMAL SHOCK RESISTANCE PARAMETER:

$$R'''' = G \, E \, / \, \sigma_U^2 \; (1-\nu) \tag{28}$$

An increase of this parameter corresponds to an improvement of the material because it corresponds to a decrease of the cracked area. Such an increase of R'''' is obtained by an increase of G, i.e. of toughness, as well as of E, and a decrease of σ_U. Thus the requirements are the contrary of what they were for the thermoelastic theory which called for an increase of σ_U and a decrease of E.
---> If fracture is controlled by the INITIATION of newly created cracks the goal is to increase strength and to reduce thermal stresses, whereas if it is controlled by the EXTENSION of preexisting cracks the main point is to reduce the driving energy.
For long cracks dTc has the value :

$$dTc = (128 \pi G (1-\nu^2) \, N^2 \, l^5 \, / 81 \alpha^2 \, E)^{1/2} \tag{29}$$

which leads to the THERMAL STRESS STABILITY PARAMETER Rst:

$$Rst = (G/\alpha^2 \, E)^{1/2} \tag{30}$$

Rst can be compared to the R parameter of the thermoelastic theory which also suggests the introduction of R'st, analogous to R':

$$R'st = k \, R \tag{31}$$

REMARK. The analytical values which have been given correspond to a particular case, namely penny-shaped cracks of the same size, propagating simultaneously without interaction. Realistic cases are flaws of various shapes and sizes with interactions. However, the qualitative trends of the theory remain valid.

SUMMARY. The ENERGETIC THEORY of thermal shock is based on the assumption that brittle fracture is controlled by the propagation of preexisting cracks. It leads to the definition of several parameters (R'''', Rst, etc.) which focus on the reduction of the driving energy to limit thermal cracking.

4. THERMAL SHOCK RESISTANCE PARAMETERS

Many parameters have been defined with respect to various cases of thermal exchanges (convection, radiation...) and of characteristics of materials, even in the case where some stress relaxation may occur (e.g. by creep). The most important parameters are:

* Resistance to the initiation of cracks (short cracks) :

$R = \sigma_u (1-\nu)/\alpha E$: the maximum allowable temperature difference to which a body can be subjected in convective heat exchanges for $\beta \gg 1$.

$R'= kR$: same definition as for R but for $\beta \leq 1$.

* Resistance to the extent of thermal damage:

Dynamical propagation (short cracks):

$R'''= E/(1-\nu)\sigma_u{}^2$: the minimum in the elastic energy at fracture available for crack propagation.

$R''''= G R'''$: the minimum in the extent of crack propagation on initiation of thermal cracking.

Quasi-static propagation (long cracks):

$Rst = (G/\alpha^2 E)^{1/2}$: the maximum allowable temperature difference required to propagate long cracks, for $\beta \gg 1$.

$R'st = k Rst$: same definition as for Rst but for $\beta \leq 1$.

5. THE CHOICE OF MATERIALS

The determination of dTc for samples of a given geometry consisting of a certain material, is usually made by submitting a set of samples to thermal shocks of increasing severity dTi (i=1,2,3...) until cracks develop. The control of cracking is frequently obtained by breaking the shocked samples, which means the number of samples must be equal to the number of points on the $\sigma_u = f(dTi)$ curve. A non destructive evaluation allows the use of a few samples only, each of them being submitted to several successive shocks of increasing severity.

Figure 6 shows two curves corresponding to two limit cases:

(1) For samples containing *short cracks*, the initial strength $(\sigma_u)i$ is high, and dTc is also high (as can be foreseen by using the thermoelastic theory as well as the energetic theory). However, if dTc is exceeded there is a sudden drop, $(\sigma_u)i - (\sigma_u)f$ and the final strength, $(\sigma_u)f$, is very low. This is due to the dynamic propagation of cracks in a highly stressed body. It can be noticed that $(\sigma_u)f$ remains on a plateau beyond dTc, because of the kinetic energy effects previously mentionned, which lead to final cracks undercritical with respect to dTc.

(2) For samples containing *long cracks,* the initial
strength is low and dTc is usually low — which corresponds
to a crack length just at the right hand side of the
minimum of the U curve — but may be high — for a crack
length far away from the minimum.

These two limit cases help us to explain the possible
way of performing a good choice of an appropriate
material :

(1) If high initial strength is required (e.g. for
optical, magnetical or mechanical uses) the choice will be
of fine grain, dense ceramics (e.g. glasses, or
dense-sintered polycrystalline ceramics). However, the
thermal exchanges must not lead to values of dT > dTc,
which would provoke unacceptable damage to the shocked
specimen.

(2) If the main goal is to obtain a damage resistant
specimen (e.g. for furnace refractories), the choice will
be of a coarse, porous and microcracked material. The
accidental increase of dT beyond dTc will not lead to
noticeable damage. Fibrous refractories are the best
examples of ceramic materials with a very low initial
strength but a nearly complete insensitivity to thermal
shocks.

6. MICROSTRUCTURAL ASPECTS OF THERMAL SHOCK RESISTANCE AND
THERMAL CRACKING

Most of the physical quantities which are taken into
account in the thermal shock parameters are sensitive to
microstructural features, and consequently the thermal
shock behaviour of materials is very sensitive to
microstructure. The equivalent size of the "microcracks",
l, is the main parameter, because it influences not only
the mechanical stability of the system and the kinetics of
cracking but also other properties such as Young's
modulus, Poisson's ratio, or thermal conductivity.

* For thermal conductivity, a statistical random
orientation of cracks in the form of ellipsoids of
revolution leads to:

$$k = k_0 \ (1+8Nr^3 \ /9) \qquad (32)$$

where k and k_0 are the thermal conductivity of the
microcracked and non-microcracked material, respectively.
Further, r is the radius of the ellipsoid which is
supposed to be very elongated. The interesting quantity
here is Nr^3 .

* For elastic properties, i.e. Young's modulus and
Poisson's ratio, the crack dependence is also expressed in
Nr^3 :

$$E = E_0 \ (1 + B \ Nr^3)^{-1} \qquad\qquad (33)$$

$$\nu = \nu_0 \ (1 + C \ Nr^3)^{-1} \qquad\qquad (34)$$

where B and C are numerical functions of ν so that to a first approximation E/E_0 is independent of ν:

$$E/E_0 \approx (1 + 16 \ Nr^3 /9)^{-1} \qquad\qquad (35)$$

* For strength, two limit cases exist:
   For dilute concentration of cracks, with no interactions, strength depends on the crack size but not on the crack density. This behaviour is described by equation (25) for penny-shaped cracks where σ_u is a function of $l^{-1/2}$.
   For a high crack density, crack interactions must be taken into account. The case of interactions between rows of coplanar cracks has been studied and shows that either a decrease or an increase of σ_u can be foreseen, depending on the crack-network geometry. Actually, experimental results generally give a decrease of strength which seems to be due to a linking of contiguous cracks, therefore leading to an increase of the crack severity. However, in case of the toughness microcracking can induce a R-curve effect, as already discussed in the lecture on the mechanical properties of ceramics. Last but not least, the interaction and the coalescence of cracks can induce a blunting of the crack tip which reduces its severity for a given mean length.
   * Thermal expansion is an intrinsic property which does not depend on microstructure by itself. For example a chemically homogeneous glass has the same expansion coefficient in a cracked or in a non-cracked state. However, heterogeneous and anisotropic materials have an average expansion which depends on the elastic properties of each component, and thus also on microcracking. It can be shown that such a microcracking generally reduces the overall coefficient of thermal expansion, $\bar{\alpha}$, down to the value of the component with the lowest expansion. Consequently, microcracking can be considered as decreasing thermal expansion.
   The effects of microcracking on k, E, v, σ_u, and α leads to an effect on the thermal shock parameters, which can be seen, for example, for the parameters R and R′, respectively.

$$R = \sigma_u \ (1-v)/\alpha E \ ; \ R' = kR \qquad\qquad (36)$$

For randomly oriented cracks with crack-interaction effects such that the strength remains invariant, the variations of E and V vs. Nb³ lead to an increase of R, whereas R' can increase or decrease depending on the change of the (k/E) ratio vs. microcracking. However, E is generally more affected by microcracking than k, therefore R' is expected to be greater in microcracked materials.

It must be noticed that the invariance of σ_u vs. crack-interaction is observed in materials with rather coarse microstructure but not in dense and fine-grain ceramics, where a greater crack density increases the size of the largest flaws which decreases the strength.

Porosity is a microstructural characteristic which can be compared with microcracking. Strength and elastic moduli generally decrease according to an exponential law when the porosity (P) increases:

$$\sigma_u \approx \sigma_{u0} \exp(-k1\ P) \tag{37}$$

$$E \approx Eo\ \exp(-k2\ P) \tag{38}$$

k1 and k2 are constants close to 5 and 2.5, respectively. The faster decrease of σ_u in comparison with E shows that R – which is proportional to σ_u/E – should decrease if P increases: this has been observed in most cases.

REMARK. The assumption that all the microcracks are of the same size and of the same shape and that they propagate simultaneously is not realistic. Generally, the propagation of one crack releases the stress field around the others, therefore limiting their further extension.

This result leads to a different behaviour for shocks at decreasing or at increasing temperatures, which can be best understood in the case of a spherical sample:

For a decrease in temperature the surface temperature is lower than the average temperature, and surface TENSILE stresses result. Besides, the core temperature is higher than the average temperature and COMPRESSIVE stresses result. Since ceramics are much weaker in tension than in compression fracture initiates at the surface : the stressed volume is large and many cracks develop but to a small extent only. There is a spalling of the surface which does not always lead to severe damage.

For an increase of temperature the stresses are reversed, and cracks initiate at the core. The stressed volume is small and few cracks develop but they will be large: that means the fracture process is generally catastrophic.

7. THERMAL FATIGUE

Brittle materials are sensitive to sub-critical crack growth due to stress-corrosion at low temperatures and diffusional mechanisms at high temperatures. This provokes static fatigue and dynamic fatigue under the influence of "mechanical" stresses and also THERMAL FATIGUE when thermal stresses are concerned.

The rate of sub-critical crack growth, da/dt, is supposed to be expressed by the same expression as for isothermal conditions:

$$da/dt = A \, K^n \, \exp(-Q/RT) \tag{39}$$

where A is a constant, n is the stress-intensity exponent, K is the stress-intensity factor, Q is the activation energy, R and T are the gas constant and the absolute temperature.

For a given value of the thermal stresses, failure will occur for a critical crack size, ac, for which K = Kc. The thermal fatigue life is defined by the number of thermal cycles required to propagate the crack from its initial length, ao, to ac. This number of cycle, Nc, is:

$$N_c = B \int_o^{K_c} \frac{K \, dK}{da/dt \; \sigma^2 (t) \; Y^2} \tag{40}$$

where B is a constant, and $\sigma(t)$ is the transient stress.

Various simplifications can be made, in order to obtain a convenient analytical expression:

* Any temperature in the thermal cycle can be expressed as the fractional value of the maximum temperature Tmax at the beginning of the cycle. Therefore, the thermal activation of the crack growth can be expressed in terms of a factor $\exp(Q/RT)$.

* Similarly, any value of the thermal stresses can be expressed as the fractional value of the maximum stress, σmax.

* The number of cycles-to-failure is assumed to be defined in terms of an inverse function of the maximum stress intensity factor (Ki) encountered during the first thermal cycle, and is assumed to be an inverse function of the time over which thermal stresses act for each cycle.

This leads to:

$$N_c \approx \frac{C \, \exp(Q/RT_{max})}{A \, \sigma_{max}^2 \, Y^2 \, (n-2) \, K_i^{n-2}} \tag{41}$$

In order to establish the nature of a thermal fatigue curve for a given material, it is usual to determine the ratio of $(Nc)1$ to $(Nc)2$ which corresponds to two conditions of thermal fatigue. There are three basic ways to operate:

(1) dT varies by varying Tmax and keeping Tmin constant: this leads to:

$$(Nc)1/(Nc)2 \approx (dT1/dT2)^{-n} \exp(Q/R(1/T1-1/T2)) \qquad (42)$$

---> The two quantities n and Q are included in this expression.

(2) dT varies by varying Tmin and keeping Tmax constant: this leads to:

$$(Nc)1/(Nc)2 \approx (dT1/dT2)^{-n} \qquad (43)$$

---> Only n is contained in this expression.

(3) dT is constant; Tmax and Tmin vary by equal amounts. This leads to:

$$(Nc)1/(Nc)2 \approx \exp(Q/R(1/T1-1/T2)) \qquad (44)$$

---> Only Q is included in this expression.

These results show that the common practice measuring the thermal fatigue behaviour by keeping Tmin constant and varying Tmax and dT by the same amount (i.e. the first case of equation (42)) cannot be applied to obtain a quantitative value for n, thereby using the value of the slope of Ln(Nc) vs. Ln(dT) because the activation energy term has also be taken into account.
    Qualitatively, high thermal fatigue resistance requires high values of the thermal diffusivity and thermal conductivity and low values of Young's modulus and the coefficient of thermal expansion, i.e. the same requirements as for high thermal shock resistance). In addition to that, it also requires high values of the stress intensity exponent n and the activation energy Q, respectively. Typical values are n = 6 and Q = 700 kJ/mole for hot-pressed silicon nitride and n = 16 and Q = 100 kJ/mole for soda-lime-silica glass.

REMARK. For low values of Biot's number a very pronounced size effect exists. Major improvements in thermal fatigue life can be achieved by minor design changes in the shape or by a slight reduction in the component size.

8. CONCLUSION

The understanding of the factors which affect the thermal stress failure of ceramic specimens should permit the improvement of their thermal shock (fatigue) resistance:
    * For a given design and given thermal conditions, the best choice of a material can be supported by the various

figures-of-merit relevant to each case (R, R′, R′′′′...)

    * For a given material, improvements can be reached by modifying dimensions, geometry, or thermal conditions.

    * Other alternatives are to separate the "mechanical function" from the "thermal function", as it is the case for thermal barrier coatings, or to incorporate compressive residual stresses, as it is the case for tempered wind-screeens.

    The optimization of the thermal shock resistance of ceramic items remains a difficult task which requires a team approach with expertise in materials science and engineering, heat transfer, thermoelasticity, and fracture mechanics.

9. REFERENCES

For the last two decades most of the studies on the thermal shock resistance of ceramics have been made by Hasselman and co-workers. Among the main references :

(1) D.P.H Hasselman, Unified theory of thermal shock fracture initiation and crack propagation of brittle ceramics, J. Am. Ceram. Soc. 52 (11) (1969), 600-604

(2) D.P.H. Hasselman, Thermal stress resistance parameters of brittle refractory ceramics: a compendium, Am. Ceram. Soc. Bull. 49 (12) (1970) 1933-1937

(3) D.P.H. Hasselman and J.P. Singh, Analysis of thermal stress resistance of microcracked brittle ceramics, Am. Ceram. Soc. Bull. 58 (9) (1979) 856-860

(4) J.P. Singh, K. Niihara, and D.P.H. Hasselman, Analysis of thermal fatigue behaviour of brittle structural materials, J. Mat. Sc. 16 (1981) 2789-2797

(5) D.P.H.Hasselman, Thermal stress resistance of engineering ceramics, Mat. Sc. Engineering 71 (1985) 251-264

(6) W.D.Kingery, Chap. 12 and 16 in Introduction to ceramics, J.Wiley, New-York (1976)

TRANSFORMATION TOUGHENING OF CERAMICS

Nils Claussen
Technische Universität Hamburg-Harburg
P.O. Box 901403
2100 Hamburg 90
Federal Republic of Germany

ABSTRACT. Ceramic materials can be considerably toughened by utilizing the phase transformation of ZrO_2 particles. The transformation is nucleation-controlled and invariably stress-assisted. Three main toughening mechanisms are operative: stress-induced transformation, microcracking and crack deflection. Some toughened ceramics with low critical transformational stress exhibit transformation plasticity and memory effects analogous to martensitic metal alloys. Microstructural features of the three ZrO_2 toughened ceramic (ZTC) groups are presented: Partially - stabilized ZrO_2 (PSZ), tetragonal ZrO_2 polycrystals (TZP) and dispersion -toughened ceramics, e.g. $ZT-Al_2O_3$, ZT-mullite, etc. The mechanical properties of some ZTC are compared with predicted values. Since ZTC generally exhibit a disappointing high-temperature behavior, some strategies are outlined to overcome the characteristic deficiencies.

I. INTRODUCTION

The mechanical properties of brittle ceramics can be improved by exploiting the tetragonal (t) → monoclinic (m) phase tranformation of discrete zirconia (ZrO_2) particles, precipitates or grains dispersed within ceramic matrices. The toughening originates essentially from the crack shielding associated with the volume and shape change of the martensitic transformation, which reduces the stress intensity at the crack tip. This type of energy dissipation is analogous to that associated with crack-tip plasticity in ductile metals.

The micromechanics of transformation toughening is understood in principle, but specific details in the several different types of transformation-toughened ceramic systems now known are still being debated. The least ambiguous toughening mechanism is that associated with the direct crack shielding provided by the stress-induced martensitic transformation in a zone ahead of propagating cracks (stress-induced transformation toughening). Another crack-shielding mechanism, which may lead to improved mechanical properties, is the nucleation and extension of matrix microcracks, themselves caused by

K. P. Herrmann and L. H. Larsson (eds.), Fracture of Non-Metallic Materials, 137–156.
© *1987 by ECSC, EEC, EAEC, Brussels and Luxembourg.*

the stress fields around transformed m-ZrO$_2$ particles. In this case, the transformation is induced thermally during cooling following fabrication (microcrack toughening).

A further toughening effect, which, other than crack-shielding, reduces the driving force for the crack propagation, involves crack deflection. The cracks can be deflected by localized residual stresses associated with transformed m-ZrO$_2$ particles or directly by t-ZrO$_2$ particles, e.g., at elevated temperatures (crack-deflection toughening). A further important aspect of stress-induced transformation is the generation of desirable residual compressive surface stresses from transformation induced e.g. by grinding or machining; this is analogous to surface strengthening of glass and can result in considerable strength increases (surface transformation strengthening).

The large-scale yielding of some highly toughened ZrO$_2$ ceramics has opened a completely new and technologically important field in ceramic materials. Large process zones and extended shear band formation make these ceramics flaw-insensitive, i.e., their strength is, as with metals, controlled by the yield stress. Furthermore, the similar shape-memory effect as exhibited in austenitic metal alloys can be utilized in technological operation.

Three basically different classes of ZrO$_2$-toughened ceramics (ZTP) have been identified: a) Partially-stabilized ZrO$_2$ (PSZ), in which t-ZrO$_2$ particles are coherently precipitated within a cubic (c) stabilized ZrO$_2$ matrix. b) Tetragonal ZrO$_2$ polycrystals (TZP) which consist predominantly of fine ($< 1\mu$m) t-ZrO$_2$ matrix grains. These ceramics have bend strength > 2000 MPa, and thus represent the strongest class of ceramic materials made to date; and c) Dispersion-toughened ceramics, where t- or m-ZrO$_2$ particles are dispersed in ceramic materials such as ZrO$_2$-toughened Al$_2$O$_3$ (ZTA), ZrO$_2$-toughened mullite (ZTM), ZrO$_2$-toughened spinel (ZTS), etc.

Throughout this article "ZrO$_2$-toughened ceramics" (ZTC) is used synonymously with "transformation-toughened ceramics", as virtually all work reported to date has focussed on ZrO$_2$ (or HfO$_2$-alloyed ZrO$_2$) as the "toughening" agent. It is possible that phase transformations in other materials may also be suitable for toughening ceramics.

A collection of literature on ZrO$_2$ systems and ZrO$_2$ toughening is contained in two conference proceedings /1,2/.

2. PHASE TRANSFORMATION

Pure ZrO$_2$ exists in three polymorphic forms. The high-temperature cubic phase with the fluorite structure is stable from the melting point, 2953 to $\sim$ 2640 K. Between 2640 and 1440 K, a tetragonal distorted-fluorite structure is stable; below 1440 K, a further distortion to monoclinic symmetry occurs. It is generally accepted that the t $\to$ m transformation is martensitic in nature; the shape and volume increase (4.7% at room temperature) associated with this transformation invariably causes cracking in bulk ZrO$_2$. The start or M$_s$ temperature of the bulk martensitic t $\to$ m transformation is $\leq$ 1220 K.

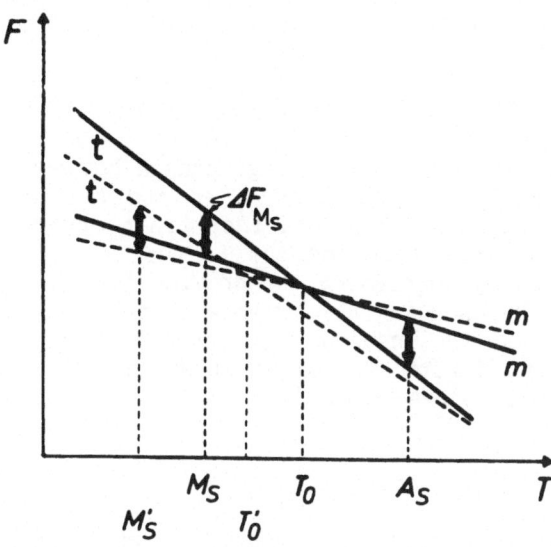

Fig. 1. Free energy (F) vs temperature (T) diagrams for martensitic reactions. t = parent phase, m = martensite phase. The solid and dashed curves are for pure ZrO_2 and ZrO_2 containing a stabilizer, respectively. T_0 and T_0' are the temperatures where parent and martensitic phases possess the same free energy. ΔF_{Ms} is required for surmounting the nucleation barrier. M_s and M_s' are the martensite start temperatures, and A_s the temperature for the reverse transformation /3/.

Fig. 1 further explains the energetics of the transformation. The negative effects of this uncontrolled t → m transformation in bulk materials can sometimes be avoided if the c → t transformation can be suppressed. This can be accomplished by "stabilizing" the c structure by alloying additions such as MgO, CaO, Y_2O_3 or rare-earth oxides.

Extensive reviews of the transformation behavior, especially the nucleation of martensite (m-ZrO_2) in confined partials are given in references /3-6/.

2.1 Transformation in Confined ZrO_2 Particles

The situation is different for fine t-ZrO_2 particles, whether present as discrete grains or precipitates or particles included in a ceramic matrix. Firstly, M_s is lowered, often to below room temperature, and in some cases, to below 0 K. Secondly, the martensitic transformation within confined particles is invariably stress-induced; for toughening ZrO_2-containing ceramics, the stress fields associated with propagating cracks or with surface grinding or machining are most important.

The transformation of a confined particle is governed by two factors /5/: a) the free energy barrier involved in nucleation (ΔF^*); and b) the change of the total free energy of the system (particle plus matrix):

$$\Delta F = - \Delta F_0 + \Delta U_T - \Delta U_I \tag{1}$$

where ΔF_0 is the chemical free energy change between parent (t-ZrO_2) and product (m-ZrO_2), which constitutes the driving force for the transformation, ΔU_T includes both the strain energy change and the changes in interface energy and ΔU_I is the interaction energy with the local stress. By increasing the driving force, e.g., by lowering temperature, or by changing ΔU_I by an externally applied stress, the transformation can be induced. For the reaction to take place, thermodynamic equilibrium has to be attained (ΔF 0) and the nucleation barrier must be overcome.

2.2 Nucleation of Martensite

The nucleation of "martensite" (m-ZrO_2), and hence the M_s temperature, is controlled by the size and shape of the particle, the chemistry of the system, the structure of the interface between matrix and confined particle, and the thermal expansion mismatch and elastic anisotropy of the system. The nucleation is invariably stress-assisted and heterogeneous, the stresses arising from thermal expansion mismatch between particle and matrix and any superimposed external stresses. For all particles other than perfect ellipsoids, stress concentrations are associated with shape inhomogeneities, e.g. grain facets, and scale with a characteristic particle dimension; this causes a particle-size dependent M_s temperature, and, for a given stress level, a particle-size-dependent propensity for transformation.

Fig. 2 further elucidates the size dependency of the transformation /7/. The left-hand side shows the free energies of pure ZrO_2 (m and t, cf. Fig. 1) and those of a small (F_1) and a larger (F_2) constraint t particle as a function of temperature. On the right-hand side, the respective free energies at room temperature are plotted versus the reaction coordinate. F_2 is at a higher energy level because the stresses controlling the elastic energy scale with the particle dimensions /8/. This energy term (see Eq. (1)) is

$$\Delta U_I = \varepsilon^T_{ij} \sigma_{ij} \tag{2}$$

where ε^T_{ij} is the unconstrained transformation strain and σ_{ij} are the stresses due to a thermal expansion mismatch between particle and matrix or between differently oriented grains (TZP). Since $\Delta F^* \propto 1/\Delta F^2$, both, an increase in the driving force (ΔF_0) and/or an increase in internal (or external) stresses will promote the propensity for the transformation.

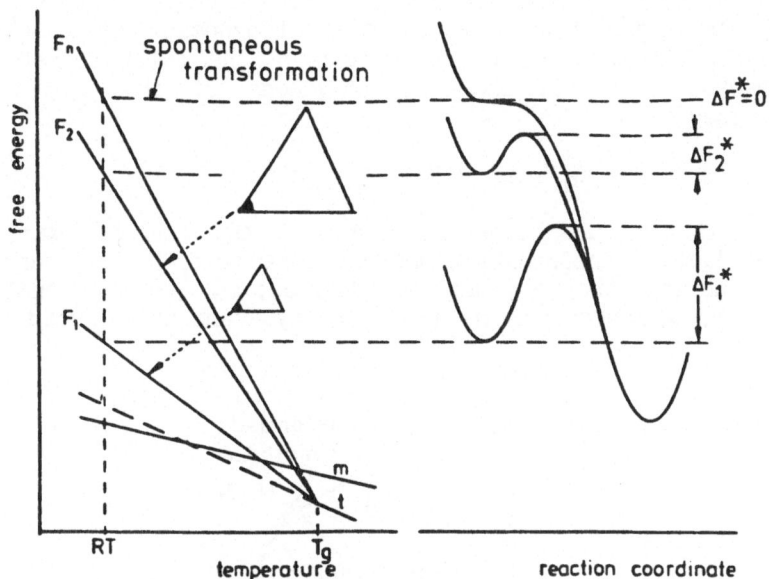

Fig. 2. Free energy vs temperature and vs reaction coordinate of a
small (F_1) and a larger (F_2) constrained t-ZrO_2 particle.
Since thermal expansion mismatch stresses scale with particle facett length F_2 is at a higher level than F_1. T_g is
the temperature (~ 800°C) below which thermal stresses no
longer relax.

3. TOUGHENING MECHANISMS

Toughening in ZrO_2-containing alloys may arise from several
mechanisms associated with the transformation. The principal crack-
shielding mechanisms are stress-induced transformation and stress-
induced microcracking; these phenomena change the stress field around
a crack tip such that the stress intensity factor is reduced. Thus,
increased far field (external) stresses are necessary to propagate
the crack, and the fracture toughness is increased. Crack deflection
reduces the driving force for crack propagation, i.e., directly
increase the critical crack-tip stress intensity factor.

Fig. 3 illustrates the crack shielding by a fully transformed or
microcracked zone. The crack-tip stress ($K^{TIP}/\sqrt{2\pi r}$) is lowered by
the transformation (or microcracking) provided the associated strain
has the same sign as the crack-tip strain, i.e., the process zone
volume must increase. The stress intensity in the near-tip region
(K^{TIP}) is lower than the stress intensity due to the applied field,
K^∞. The reduced stress intensity

$$\Delta K = K^\infty - K^{TIP} \qquad (3)$$

corresponds directly with an increased toughness. The crack propagates when K^{TIP} attains the toughness of the fully transformed or microcracked material, K_O. The toughness of the ZrO_2-toughened ceramic then becomes ($K^\infty = K_C$):

$$K_C = \Delta K_C + K_O \tag{4}$$

In the following the toughness increment ΔK_C will be examined for stress-induced transformation and microcracking. K_O is changed both by crack deflection and microcrack degradation. Detailed analyses of the toughening mechanisms are presented by Evans et.al. /6,8-10/.

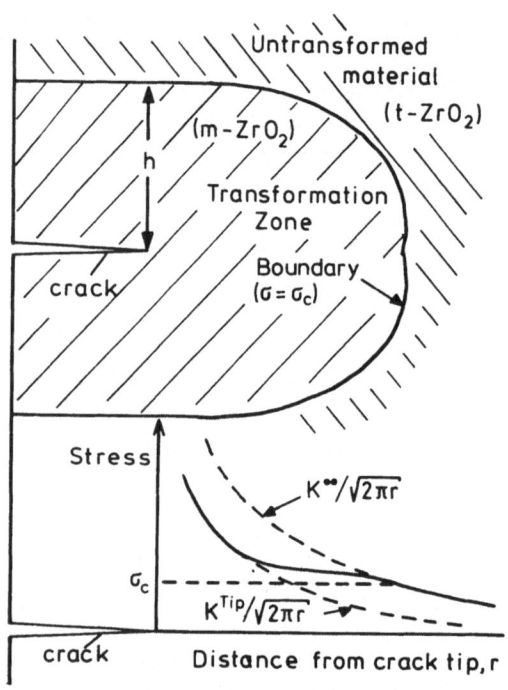

Fig. 3. a) Fully transformed or microcracked frontal process zone.
b) Modified stress field ahead of crack tip.

3.1 Stress-Induced Transformation

Stress-induced transformation is the dominant toughening mechanism in (PSZ) and (TZP) and less important in ZrO_2-containing dispersion-toughened ceramics. The prerequisite is the presence of t-ZrO_2 at the service temperature; as this is usually room temperature, M_s must be below room temperature. The toughening may be illustrated schematically as shown in Fig. 4.

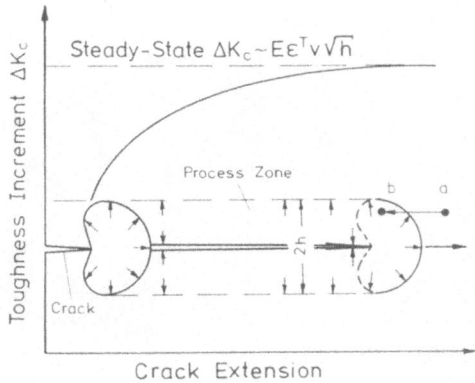

Fig. 4. Schematic diagram showing the development of the process zone with crack advance and the resultant toughness increment, ΔK_c; h is generally in the range 0.2-2 μm. Positions a and b of a volume element relate to the micrographs in Fig. 5.

On far-field loading of the crack, introduced in the untransformed material, the stress concentration around the crack tip will eventually reach a critical value at which the $t\text{-}ZrO_2$ particles within a "process" zone of width 2 h will transform. In the case of a dilational zone profile, ΔK determined by integrating the tractions along the transformation zone boundary, reveals that ΔK is zero, i.e., there is no toughening /10/ (This situation changes if a shear-band zone profile is considered).

On further loading, the crack has to propagate into the (compressive) process zone encountering enhanced crack resistance, i.e., ΔK increases, and consequently an R-curve is exhibited. After an extension of about 5 h, a steady-state toughness increment is attained, which, after integration of the traction over the transformation zone, is

$$\Delta K_c = c_{ij} \, v \, \epsilon^T \sqrt{h} \qquad (5)$$

where c_{ij} is the elastic constant tensor and v the volume fraction of transformable $t\text{-}ZrO_2$. For purely dilational transformation $c_{ij} = 0.22E/(1-\nu)$; this magnitude is more or less the same for reversible and irreversible transformation. If the transformation occurs in shear bands, $c_{ij} = 0.16E/(1-\nu)$ at the start of the propagating crack and $0.38E/(1-\nu)$ at steady-state /10,11/.

A volume element passing into the process zone (from position a to b in Fig. 4) interacts with the crack as shown experimentally in Fig. 5 /12/.

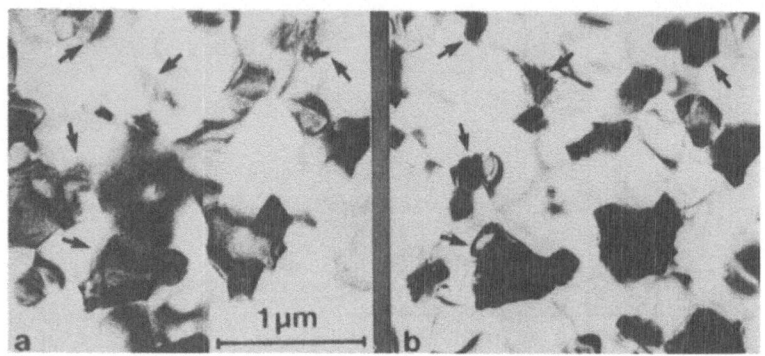

Fig. 5.   High-voltage transmission electron micrographs of Al$_2$O$_3$
           containing 15 vol. % ZrO$_2$ particles before (a) and after (b)
           loading by means of an in-situ straining stage. In the
           as-sintered composite, the ZrO$_2$ particles (arrowed) have t
           symmetry; after loading, most of them have transformed and
           undergone a characteristic twinning /12/.

           This is a sample of a sintered ZrO$_2$-toughened Al$_2$O$_3$ containing
15 vol.% ZrO$_2$. Most ZrO$_2$ particles (arrowed) have t symmetry (Fig.
5a). After a critical stress is reached in the vicinity of the
advancing crack, the particles transform (Fig. 5b). The transformed
m-ZrO$_2$ particles are invariably twinned, the twinning occurring to
minimize the shape strains accompanying transformation.
           The process zone height is for the case of irreversible trans-
formation:

$$h = \frac{\sqrt{3}}{12\pi}\ (1+\nu)^2\ (K_c/\sigma_c)^2 \tag{6}$$

where σ_c is the critical mean stress to initiate transformation.
           The preceeding paragraph is based on a stress-intensity analy-
sis, however, the same result can be derived applying thermodynamics
of crack advance /6/. In the Griffith approach, the net energy
increase associated with a crack increment occurs solely behind the
crack tip and is expressed as a change in the interaction energy (cf.
Eq. (2)). In front of the crack, a balance always exists between the
decrease in interaction energy and the energy dissipated by the
transforming particles.

The increment in the critical energy release rate is /6/

$$\Delta G_C = v \int_{-h}^{h} \Delta U_I^c \, dy \equiv 2 \, v \varepsilon^T \sigma_c h \qquad (7)$$

which has been demonstrated to be identical to ΔK_c of Eq.(5) /11/. At transformation conditions it follows from Eq.(1):

$$\Delta U_I^c = - \Delta F_0 + \Delta U_T \qquad (8)$$

For a small transformation zone, $\Delta U_T \sim 0$, because the crack surface relaxes the transformation strain, then ΔU_I^c is directly related to the driving force which is experimentally observed when measuring ΔK_C or ΔG_C as a function of temperature (Fig. 6).

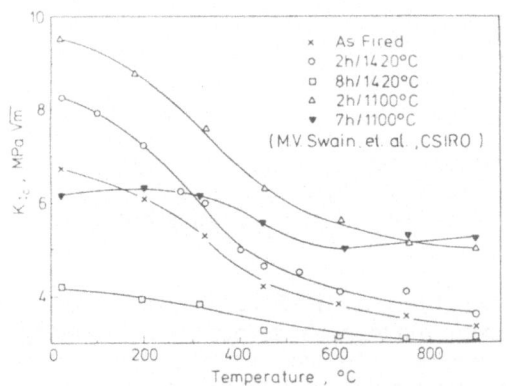

Fig. 6. Fracture toughness of various PSZ ceramics, heat-treated under different conditions, vs temperature /29/.

Another thermodynamic derivation of ΔG_C has been provided by Budiansky et.al./11/ by examining energy balance integrals. Again, no toughening is achieved as long as the crack does not extend into the transformed zone. However, when the zone is fully developed, the toughness increment derives as /6/

$$\Delta G_C = 2 \int_{0}^{h} U(y) \, dy \qquad (9)$$

where U (y) is the residual energy density in the wake. U (y) can be evaluated using Fig. 7.

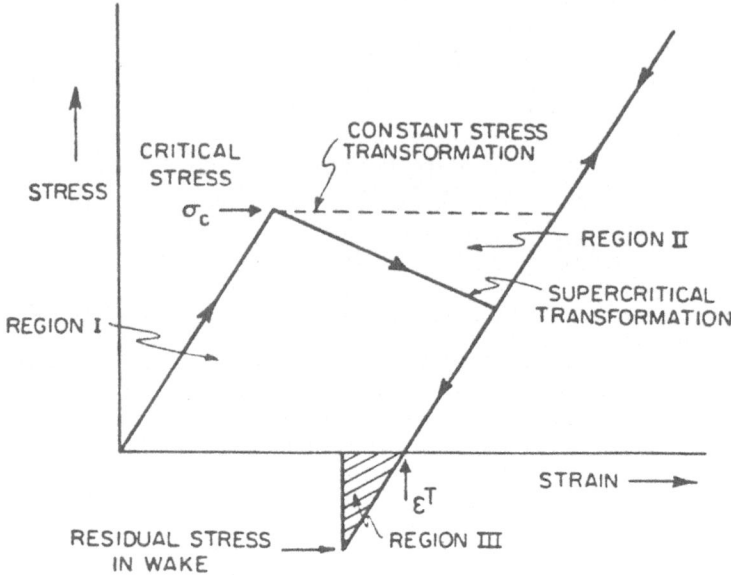

Fig. 7. Schematic stress-strain curve for supercritical martensitic transformation /6/. Hysteresis loop for a volume element (e.g. a in Fig. 4) passing through the process zone into the wake.

When passing from position a in Fig. 4 via b and beyond into the wake, a volume element undergoes a complete loading and unloading cycle. I.e., each volume element in the wake is subject to the residual stress work contained in the hysteresis loop in Fig. 7. The residual energy density is then represented by the areas of regions I, II and III. At supercritical transformation conditions (the stress drops below σ_c) the areas of II and III cancel. Then Eq. (9) is identical to Eq. (7).

3.2 Microcracking

Microcracking has been observed in all three ZTC types, i.e., PSZ, TZP and dispersion - toughened ceramics /13-17/, although it is more typical for the latter type where ZrO_2 particles are intergranularly dispersed.

Microcracks nucleate in regions of residual tension, hence, adjacent to transformed $m-ZrO_2$ particles, and due to thermal expansion mismatch. Relief of these stresses by extension, triggered by external loading, results in residual opening and associated dilatation. This dilatation is the main source of microcrack toughening /9,18/.

It can be shown that the volumetric strain induced by non-inter-acting microcracks is /18/:

$$\epsilon^M = \frac{16\ (1-\nu^2)}{3} \ \frac{\sigma_R f}{E}$$ (10)

where f is the microcrack density (correlates to the volume fraction of m-ZrO_2 particles if each particle is associated with one matrix microcrack), E is the modulus of the uncracked material and σ_R is the residual stress adjacent to ZrO_2 particles:

$$\sigma_R = (2/9)\ E\ \epsilon^T\ /\ (1-\nu)$$ (11)

The shielding and hence the toughness increment, ΔK_c can again be analyzed using the energy-balance integrals described in section 3.1 Fig. 8 shows the hysteresis loop for a volume element translating through the microcrack process zone (cf. Fig. 4). At the critical stress, σ_c, microcracks extend between ZrO_2 particles causing the dilatation, ϵ^M. On further loading and unloading the microcracked material exhibits a lower modulus than the uncracked material. The area under the loop corresponds to the energy density (cf. Eq. (9)).

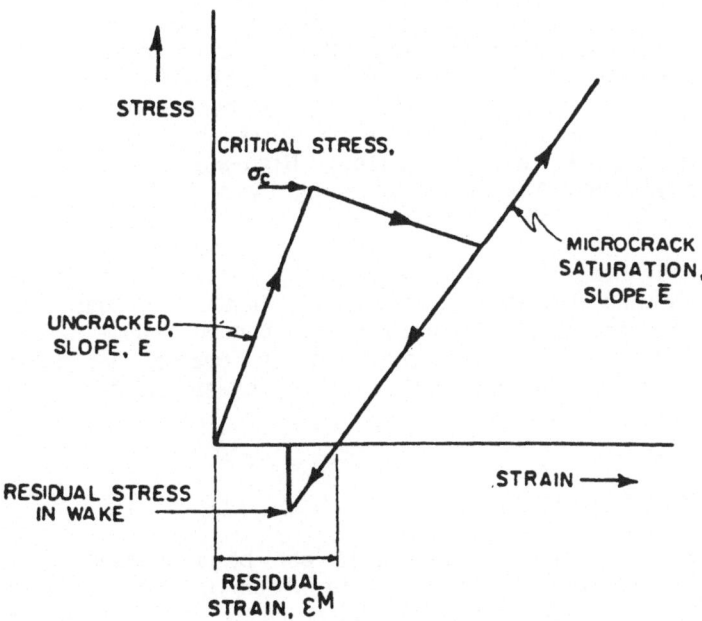

Fig. 8.   Same schematic as in Fig. 7 for supercritical stress-induced microcracking. Note that the modulus of the microcracked material is reduced /6/.

In the case of pretransformed but uncracked material /6/,

$$\Delta K_C = \frac{0.07}{(1-\nu)} E \, \varepsilon^T \, \sqrt{h} \, (1.2 \, f^{1/3} - f) \tag{12}$$

and for (less probable) simultaneous transformation and microcracking (Fig. 8)

$$\Delta K_C = \frac{0.14}{(1-0)} E \, \varepsilon^T \, \sqrt{h} \, (f + 0.6 \, f^{1/3}) \tag{13}$$

Comparison with stress-induced transformation (Eq. (5)) shows that for equivalent h simultaneous transformation and microcracking is more effective than transformation alone. However, it must be noted that K_O in Eq. (4) is reduced by the microcracking by a factor (1-f) /6/.

3.3 Crack Deflection

Cracks can be deflected either by the residual stress fields asso-ciated with transformed m-ZrO_2 particle or by the fracture resistant t or m particles. The toughening results from the reduced driving force on the deflected portion of the crack /18,20/. The driving force can be estimated by separatly evaluating the mode I, II and III strain energy release rates /20/. The toughening magnitude depends on volume fraction and shape of the particles as well as the intensity of the associated stress fields. High-aspect ratio rods induce maximum toughness. Maximum toughening by a factor of 1.7 have been observed in ZrO-m-ZrO_2 alloys /21/.

The advantage of this mechanism is that it is independent of temperature and dictates K_O (Eq. 4)), hence, it adds to the ΔK_C of other mechanism. E.g., in whisker-reinforced ZrO_2 - toughened cera-mics, K_O is considerably enhanced by cracks deflecting at the whiskers, while ΔK_C originates from stress-induced transformation /22/.

3.4 Surface Strengthening

The strength of ceramic materials can be increased by introducing compressive surface stresses, as is state of the art in glass tech-nology. In ceramics containing ZrO_2, these stresses are produced by transforming particles in the surface layer to m symmetry and main-taining those in the bulk with t symmetry. This localized volume expansion can be stress-induced, i.e. by grinding, by sand blasting, or by a low-temperature treatment /23/.

Another method of producing these surface stresses is by chemically destabilizing t or c particles near the surface and allowing them to transform. Transient compressive stresses are generated on cooling composites containing m-ZrO_2 to room temperature. When composites are thermally shocked from temperatures LT > A_f, these transient compressive stresses counteract the thermal stresses and hence may completely prevent failure. E.g., Si_3N_4-ZrO_2 composites have been quenched from 1000°C into cold water without any measurable degradation. Some techniques for the introduction of compressive stresses are listed in Table 1.

Table 1. Possible technique for introduction of residual compressive stresses /23/.

Technique	Transformation induced	Example
Grinding, sand blasting, quenching	By stresses	Al_2O_3-ZrO_2, PSZ
Heat treatment in HfO_2 powder in air in vacuum	chemically	Al_2O_3-ZrO_2 Si_3N_4-ZrO_2 Mg-PSZ,Mg-CSZ,TZP
Cooling in liquid He or N_2 for short periods	Low-temp. quenching (T_c<M_s)	Al_2O_3-ZrO_2
Coating with (a) higher vol fraction, or (b) larger particle size	M_s of coating > T M_s of bulk < T, T: applic. temp.	Al_2O_3-ZrO_2

Although grinding-induced transformation is comparable to the transformation in the crack-tip stress field, the exact mechanisms have not been fully investigated. There are a number of possible mechanisms which may be simultaneously active: (a) Transformation is triggered by the high shear strains caused by the grinding process; (b) a high density of parallel flaws, induced by the abrasive media, creates similar stress-field transformations to those prevailing at single sharp crack tips; (c) the dislocations produced in the process pile up at particles and consequently cause their transformation. From various experiments with ZTA and ZTM ceramics it becomes obvious that surface grinding results in a much larger transformed depth than that due to crack propagation /25/.

Obviously, techniques that avoid the damaging of the surface have a greater potential for enhancing the strength /24/. However, grinding-induced surface strengthening is technologically more important since grinding is usually a required machining operation.

The residual compressive stresses at depth x (x : distance from the surface) can be estimated for a flat plate by /26/:

$$\sigma^c (x) = \varepsilon^T v_o e^{-bx} \cdot E/ (1-\nu) \qquad (14)$$

assuming that the transformational depth is small with respect to the thickness of the plate, that the Young's moduli of matrix and particles are similar, and that the volume fraction of transformable particles is modest. v_o is the volume fraction of transformed $m-ZrO_2$ at the surface (x=0) and e^{-bx} is the transformation profile and b a constant. Surface stresses as high as 1 GPa have been measured in $Al_2O_3-ZrO_2$ (Y_2O_3) alloys /24/.

The strengthening magnitude depends on the residual compressive stress component of the stress intensity factor, K^R, the depth of the critical surface flaw with respect to the stress profile and the toughness of the material, K_c. At critical condition, the toughness of the surface - strengthened material is given by /27/

$$K_c^{ST} = K_c + K_c^R \qquad (15)$$

A detailed analysis of the strengthening by compressive surface stresses is presented in reference /24/.

3.5 Combined Toughening Mechanisms

Various toughening mechanisms can operate simultaneously. In some cases, they are additive, in others, counteracting. Due to the complexity it is rather difficult to account for the exact contribution of each specific mechanism. Generally, all shielding mechanisms can be expressed as /6/

$$K_c = \Sigma \sqrt{h} + K_o \qquad (16)$$

where Σ depends of the hysteresis or total crack closure forces. K_o is influenced by deflection and microcrack degradation. An alternative expression indicates a multiplicativity of shielding and reduction of crack-tip driving force /6/.

$$K_c = \left[\frac{0.046 (1+\nu) E v \varepsilon^T}{(1-\nu) (\sigma_c - \sigma_r)} + 1 \right] K_o \qquad (17)$$

where σ_r is the residual stress in the zone. E.g., crack deflection requires an increase in applied load in order to bring K^{TIP} to the critical level (cf. Eq. (3) and (4)). Consequently, h increases which leads to additional stress induced transformation or microcracking.

In dispersion-toughened ceramics for instance, it has been shown that both microcrack and transformation shielding is operating, the respective extend depending on chemistry and size of the ZrO_2 particles /17/. In ZTA, the increases in toughness due to microcracking

are comparable to those achieved by the stress-induced t→m transformation /16,17/. The microcracking clearly occurs to relieve
stresses introduced by the prior spontaneous t→m transformation, and
at a given distance from a crack, is more likely to occur adjacent to
the larger particles. Contrary to some previous theories, microcracking does not seem to accompany the stress-induced t→m transformation, i.e., a given particle can either transform in the stress
field of a propagating crack, or if already transformed, can cause
microcracking. If transformation occurs in a given stress field, the
stress reduction so engendered does not permit microcrack nucleation
/16/.

4. TRANSFORMATION PLASTICITY AND SHAPE-MEMORY EFFECT

It has recently been demonstrated that maximum strength of highly
toughened ZrO_2 ceramics does not correlate with maximum toughness
/37,28/, i.e., the basic fracture mechanical relation between
strength, σ, and toughness

$$\sigma = Y \, K_c \, / \, \sqrt{C_f} \qquad (18)$$

no longer obtains. Y is a constant and C_f the fracture controlling
flaw size. E.g., in a series of heat-treated Mg-PSZ ceramics, the
peak strength invariably occurred at values between 50 and 75 % of
the maximum toughnes /29/. Another example of this phenomenon was
demonstrated with Y-TZP where the strength was considerably increased
(to > 2000 MPa!) by the addition of Al_2O_3 while, at the same time,
the toughness was reduced from 10 to 5 MPa $\sqrt{m}$ /30/.

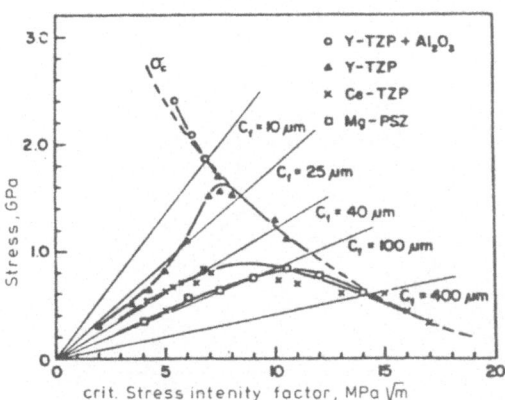

Fig. 9.   Fracture toughness of several PSZ and TZP ceramics. The
          toughness of a specific alloy was changed by varying the
          heat-treatment. The plot indicates that the strength is
          limited by the critical stress to initiate the transfor-
          mation, σ_c /28/.

Fig. 9 shows a collection of strength - toughness data for various PSZ and TZP ceramics /29/. It nicely demonstrates that the strength is limited by σ_c beyond a certain K_c value and becomes essentially insensitive to the preexisting flaw size. This is analogous to the mechanical behavior of steels in the ductile to brittle transition, where the inhomogeneity of the slip process causes microcracks to initiate at the yield strength. The microcracks rapidly become unstable and induce brittle failure /6/. In ZrO_2 alloys, the yield limited strength (yield limit equivalent to the critical stress to initiate transformation, σ_c) is probably controlled by the nucleation of microcracks which occur at shear band terminations /31/. The plastic strain at failure, manifested as a pronounced deviation from linearity in the stress/strain curves of Mg-PSZ /27/, may attain values of > 0.5 %, in certain TZP alloys even > 100 % when deformation is carried out at T > 1200°C /32/.

The prospect of yield-limited strength, as opposed to flaw-limited strength, implies an important change in the design philosophy of engineering parts, i.e., Weibull-statistics type of reliability considerations can be omitted /33/.

An interesting technological aspect of these "superplastic" ceramics is the utilization of large-scale plasticity as shape-memory parts /34/.

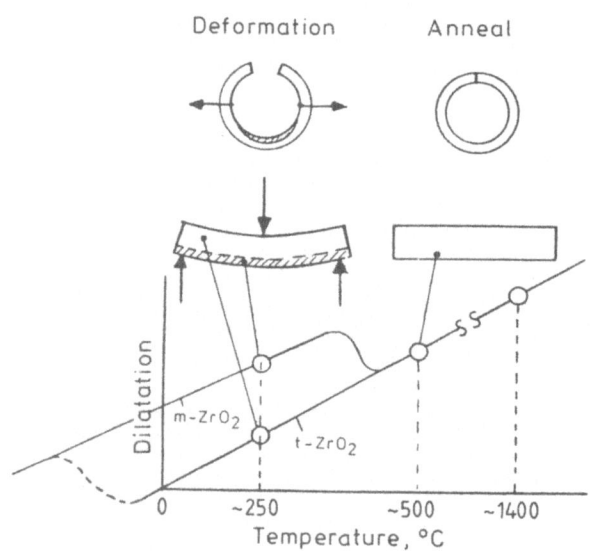

Fig. 10. Shape-memory effect in highly transformable Y-TZP ceramics. The stress and humidity induced transformation rate is a maximum at ca. 250°C.

In this case, Y-TZP is plastically deformed at low temperatures, i.e., the regions under high tensile and shear stresses are almost completely transformed to m symmetry (see example in Fig. 10).

The deformation process is usually carried out in the 250 to 300°C temperature range where the t → m transformation is especially enhanced in a humid atmosphere /14/. After heating the deformed part to $T > A_f$, the m part of the component is retransformed into the t form and consequently, the original shape is reasumed. This effect can be utilized in sensor devices, pipe fittings, high-temperature springs, etc.

5. TYPES OF TRANSFORMATION-TOUGHENED CERAMICS

Three basically different types of tough ZrO_2-containing ceramics have been developed: partially stabilized ZrO_2 (PSZ), tetragonal ZrO_2 polycrystals (TZP), and dispersion toughened ceramics /25/.

5.1 Partially Stabilized ZrO_2

PSZ ceramics contain insufficient solute (typically MgO, CaO or Y_2O_3) to prevent formation of t-ZrO_2; they are usually sintered in the c solid solution phase field, i.e., at relatively high temperatures ranging between ~1900 and 2100K, and consist of large (50-100 μm) c solid solution grains. During cooling, coherent t-ZrO_2 precipitates form and are usually coarsened or otherwise modified by aging at temperatures between 1370 and 1770K; this aging increases their propensity to transform under stress.

PSZ ceramics (mostly Mg-PSZ) have been items of commerce for some time, and have been optimized for high toughness and wear resistance. They find application as metal forming tools, dies, bearings, etc. and are being tested as automative components, such as came follower inserts, rocker faces, valve guides and seats, cylinder liners, piston caps, hot plates, etc. /35/.

5.2 Tetragonal ZrO_2 Polycrystals

TZP are fine-grained (0.1-1μm), predominantly single-phase fully dense t-ZrO_2 ceramics containing Y_2O_3 or rare-earth oxides. Sintering is optimally carried out in the t single-phase field; for example, 1.5 to 3.5 mole % Y_2O_3 alloys are sintered at temperatures between ~ 1600 and 1800K. This type of material is presently the toughest and strongest of all ZrO_2-containing ceramics, indeed probably of all polycrystalline ceramics made to date, with strengths > 2000 MPa and toughnesses > 10 MPa $\sqrt{m}$ having been achieved. The superior properties reflect the fine grain size and full density (probably achieved by the presence of a continuous glassy grain boundary phase), the lack of any intrinsic strength-limiting flaws and a very high volume fraction of the transformable phase. As for PSZ ceramics, they are presently being tested as components in engines and as metal forming tools; furthermore, many new applications have been developed, such as scissors, knives, textile cutters, golf club inserts, etc..

5.3 Dispersion-Toughened Ceramics

The dispersion-toughened ceramics encompass a variety of two-phase materials containing mainly intergranular ZrO_2 particles. The critical ZrO_2 particle size for which M_s is < 300K varies with different matrices but is usually between 0.5 and 1.2 μm. Small amounts of Y_2O_3 are usually added to further stabilize the t phase.

The strongest representative of this class of material is ZrO_2-toughened alumina (ZTA), with strengths > 1000 MPa. Weaker matrices than Al_2O_3 - mullite, spinel, etc. - have been appreciably strengthened (> 500 MPa) by ZrO_2 additions. Commercial applications are cutting tool tips, high-performance seals, wear parts in heat engines, etc. /25/.

6. HIGH-TEMPERATURE PROBLEMS

Compared with the two currently available main competitors for high performance engineering ceramcis (SiC and Si_3N_4), ZrO_2-toughened ceramics have a clear advantage in the high applied stress regime, while the two covalent representatives exhibit a much greater high temperature potential. There are two principal reasons why the present ZrO_2-toughened ceramic (ZTC) types must be considered to be low temperature ceramics; firstly, the creep rates in the usually oxide-based ZrO_2-toughened ceramics (especially PSZ and TZP) are high relative to those of SiC and Si_3N_4 and, secondly, the most important toughening mechanism, which results from the stress-induced phase transformation, decreases drastically towards the equilibrium transformation temperature T_O (which is equal to about 1000°C for bulk ZrO_2, cf. Fig. 1).

This disappointing characteristic feature limits the load-bearing application of ZTC to low or, at the most, medium temperatures. Therefore, it represents a materials science challenge to search for wayse of improving the high temperature mechanical properties of these materials. The typical low temperature toughness of ZTC remains an important property even when they are used at higher temperatures, e.g. under thermal up shock and down shock conditions the maximum stresses usually occur at rather low temperatures. However, a better strength performance up to 1000-1200°C would considerably widen their technical use.

In a recent paper several strategies for strengthening ZTC for high-temperature application have been discussed /36/. The conclusion is that a number of possibilities exist to develop microstructures for ZTC in order to use this class of high performance ceramics at high temperatures. However, for further clarification of microstructural design criteria, it is important to investigate the high temperature thermomechanical behavior of the present ZTC types to a greater extent, i.e. too little is known about the failure mechanisms, especially under static and dynamic stress conditions at elevated temperatures.

From the various strategies discussed the following measures seem to offer the best prospects for strength improvements at temperatures in the tetragonal stability range: (a) prevention or control of amorphous intergranular phases, i.e. by using high purity processing, by crystallizing the glass or by preventing the wetting of the grain boundaries; (b) addition of hard and high modulus second phases; (c) whisker (fiber) reinforcement, especially for TZP, ZrO_2-toughened mullite and ZrO_2-toughened cordierite.

REFERENCES

/ 1/ A.H. Heuer and L.W. Hobbs (eds.), Science and Technology of Zirconia, in Advances in Ceramis, Vol. 3, American Ceramic Society, Columbus, Oh, 1981.

/ 2/ N. Claussen, M. Rühle and A.H. Heuer (eds.), Science and Technology of Zirconia II, in Advances in Ceramics, Vol. 12, American Ceramic Society, Columbus, OH, 1984.

/ 3/ M. Rühle and A.H. Heuer, p. 14 ff in ref. 2

/ 4/ A.H. Heuer, N. Claussen, W.M. Kriven and M. Rühle, J. Am. Ceram. Soc., 65 (1982) 642.

/ 5/ A.H. Heuer and M. Rühle, to be published in Acta Met.

/ 6/ A.G. Evans and R.M. Cannon, to be published in Acta Met.

/ 7/ S. Schmander and H. Schubert, submitted to J. Am. Ceram. Soc.

/ 8/ A.G. Evans, Acta Met., 36 (1978) 1845.

/ 9/ A.G. Evans, p. 193 ff in ref. 2.

/10/ R.M. McMeeking and A.G. Evans, J. Am. Ceram. Soc., 65 (1982) 242.

/11/ B. Budiansky, J.W. Hutchinson and J. Lambropoulos, Int. J. Sol. Structures, 19 (1983) 337.

/12/ M. Rühle, A. Strecker, D. Waidelich and B. Krans, p. 256 ff in ref. 2.

/13/ A. King and P.J. Yarorsky, J. Am. Ceram. Soc., 51 (1968) 38.

/14/ M. Rühle, N. Claussen and A.H. Heuer, p. 352 ff in Ref. 2.

/15/ N. Claussen, J. Am. Ceram. Soc. 59 (1976) 49.

/16/ M. Rühle, N. Claussen and A.H. Heuer, to be published in J. Am. Ceram. Soc..

/17/ T. Kosmac, M.V. Swain and N. Claussen, Mat. Sci. Eng., 71 (1985) 57.

/18/ K.T. Faber, p. 293 ff in ref. 2.

/19/ J.W. Hutchinson, ref. 59 in ref. 6.

/20/ K.T. Faber and A.G. Evans, Acta Met., 31 (1983) 565.

/21/ H. Ruf and A.G. Evans, J. Am. Ceram. Soc., 66 (1983) 328.

/22/ N. Claussen, to be published in Fracture Mechanics of Ceramics, vol. 7/8, 1986.

/23/ N. Claussen and M. Rühle, p. 137 ff. in ref. 1

/24/ D.J. Green, F.F. Lange and M.R. James, p. 240 ff in ref. 2

/25/ N. Claussen, p. 325 ff in ref. 2

/26/ O. Richmond, W.C. Leslie and H.A. Wriendt, ASTM Trans. Q. 57 (1964) 294.
/27/ B.R. Lawn and D.B. Marshall, Phys. Chem. Classes, 18 (1977) 7.
/28/ M.V. Swain, J. Am. Ceram. Soc., 68 (1985) C 97.
/29/ M.V. Swain, to be published in J. Am. Ceram. Soc.
/30/ K. Tsukuma, K. Ueda and M. Shimada, J. Am. Ceram. Soc., 68 (1985) C 4.
/31/ Y. Fu, A.G. Evans and W.M. Kriven, J. Am. Ceram. Soc. 67 (1984) 626.
/32/ F. Wakai, GIRIN, personal communication.
/33/ I.W. Chen, to be published in J. Am. Ceram. Soc.
/34/ T. Soma and M. Matsui, Japanese Patent Appl. JP P131818-1983.
/35/ R.C. Garvie, p. 465 ff in ref. 2.
/36/ N. Claussen, Mat. Sci. Eng. 71 (1985) 23
/37/ M.V. Swain, to be published in Acta Met.

MECHANICAL BEHAVIOUR OF GLASSES

G. Fantozzi, G. Orange
I.N.S.A., G.E.M.P.P.M., UA 341
Bât. 502
69621 Villeurbanne Cédex   FRANCE

ABSTRACT. In the case of glasses, there is a significant subcritical
crack growth which occurs for values of the stress intensity factor
smaller than the critical value.
Practical applications of glasses require the knowledge of the slow
crack growth responsible for the degragation of materials.
Firstly, the methods of determination of the (K_I, v) diagram are described : DCB and DT techniques. Then, experimental results are presented :
the effect of glass composition, atmosphere, temperature ... on slow
crack growth is studied.
Secondly, the mechanisms of slow crack propagation are presented : chemical reaction, transport-controlled rupture and a comparison with experimental results is carried out.
Finally, the time dependence of strength, the strength-probability-time
diagram, the dynamical fatigue and proof testing are studied.

1.   INTRODUCTION

For brittle materials, catastrophic fracture occurs when the
stress intensity factor K or the strain energy release rate G reaches a
critical value : then, there is a very quick propagation of the crack.
However, for values of the stress intensity factor smaller than
the critical value ($K_I < K_{IC}$), the crack can grow slowly (with a velocity of the order of 1 to 10^{-6} m/s) up to the sudden fracture when K_I
reaches the K_{IC} value. This phenomenon of delayed fracture or static
fatigue or time-dependent failure is very significant in glasses. This
strength degradation with time is of great significance for structural
applications and must be taken into account for the prediction of safe
lifetimes of structures. Thus, practical applications of glasses require
the knowledge of the slow crack growth responsible for the degradation.
The use of the subcritical crack growth allows to predict the lifetimes
of stressed structures. Furthermore, the knowledge of mechanisms controlling the slow crack growth is very useful in order to improve the
mechanical properties of glasses.

K. P. Herrmann and L. H. Larsson (eds.), Fracture of Non-Metallic Materials, 157–180.
© *1987 by ECSC, EEC, EAEC, Brussels and Luxembourg.*

2. SLOW CRACK GROWTH

2.1 (K_I, v) diagrams

2.1.1 Principle : A typical stress intensity factor-crack velocity diagram is shown in Fig. 1. This diagram was already discussed in the lectures by Sommer (for glass) and Davidge (for ceramics) in this seminar.

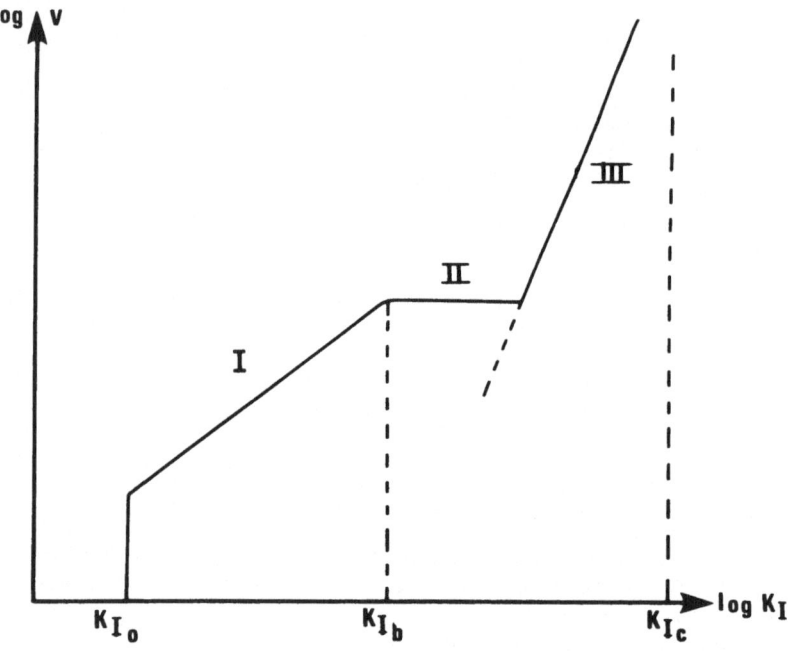

Figure 1. Stress intensity factor K_I-crack velocity v diagram

Four crack growth stages can be distinguished for $K_I \lessgtr K_{IC}$, each stage corresponding to a different growth process.

There is often a threshold value K_{IO} below which no crack propagation takes place : thus, the existence of this threshold allows us to define a perfectly safe region.

In range I and II, the crack velocity can be described by the following expression :

$$v = AK_I^n \qquad (1)$$

where A and n are constants.

For region I, $10 \leq n \leq 20$ and for region II, n = 0. In the two regions I and II the crack growth is controlled by a stress-induced corrosion mechanism near the crack tip. The crack growth velocity in

region I is reaction-rate controlled whereas in region II it is controlled by the diffusion of the corrosive species towards the crack tip.
The stage III occurs for high values of stress intensity factor ($n \sim$
100). The mechanisms controlling the crack propagation in region III
are not unsterstood.

2.1.2 Measurement techniques

Two types of methods can be used to determine the crack growth
velocity : on one hand direct methods that allow (K_I, v) diagram deter-
minations, on the other hand, indirect methods which allow us to determine
the A and n constants of Eq. (1).
Firstly, we shall describe only direct methods, indirect methods
being presented in the next paragraph.

Double cantilever beam (DCB) technique

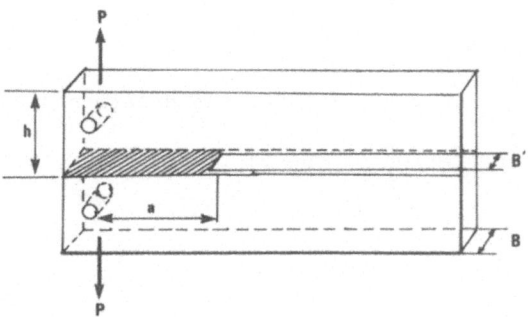

Figure 2.   DCB specimen

The DCB technique has been described by many authors [1, 2].
The DCB specimen configuration is shown in   Fig. 2. In order to confi-
ne the crack to the plane of symmetry, the specimen is grooved on each
face. Crack initiation may be performed either by thermal shock or with
a diamond saw. Following Wiederhorn [1], the crack length must be larger then 1.5 h. The expression for K_I is given by :

$$K_I = Pa \ (\frac{12}{BB' h^3})^{1/2} \left[1 + 1.32 \ (\frac{h}{a}) + 0.542 \ (\frac{h}{a})^2 \right]^{1/2} \tag{2}$$

The stress intensity factor depends on the crack length which
must be measured.
Two   methods can be used to obtain the (K_I, v) diagram.

-i- A constant load is applied and the crack propagation is observed optically during a time Δt. The crack velocity is given by :

$$v = \frac{\Delta a}{\Delta t}$$

for a mean value of K_I $(a + \frac{\Delta a}{2})$.

The crack propagation must be small in comparison with a. Velocities between 10^{-4} ms^{-1} and 10^{-11} ms^{-1} can be measured by this method.

-ii- Relaxation experiments can be performed |3|.

The specimen is loaded at a rapid crosshead speed to a load corresponding to about 90 % of K_{IC} and then the crosshead is stopped. There is a crack propagation and the load decreases. If the machine stiffness is high enough, the crack propagation takes place with decreasing velocity up to crack arrest. So, the (K_I, v) diagram can be obtained. The knowledge of the machine compliance allows to avoid the observation of the crack front.

Double Torsion (DT) technique

This technique is now very popular for the determination of slow crack growth in ceramics.
The schematic representation of the DT test configuration is given in Fig. 3.

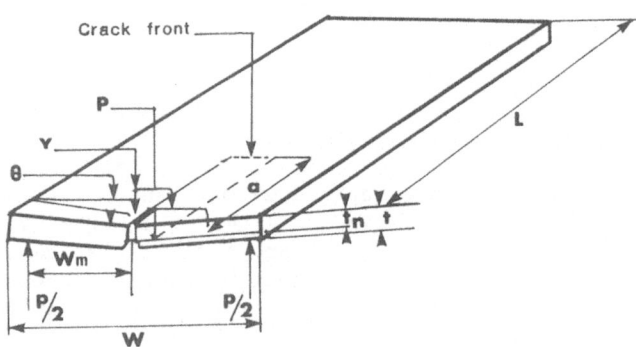

Figure 3. DT specimen configuration

The specimen (a plate which is partially cracked) is supported on two parallel rollers and the load is applied by a loading ball fixed to the machine crosshead. Each beam is loaded in torsion by four-point loading, causing the crack propagation. In order to obtain a crack growth along the specimen axis, the specimen is very often side grooved on one or both sides.

The concept of DT was introduced by Outwater and Gerry |4| and developed by many investigators |5-8|. The analysis of the double-torsion configuration has been reviewed by Fuller |5|.

The displacement of the load-point y for an elastic solid is related to the applied load by the relationship :

$$y = CP \qquad (3)$$

where C is the compliance which depends on the specimen geometry, the crack size and the material. Experimentally, we observe that C varies linearly with the crack size a :

$$C = \frac{y}{P} = Ba + D \qquad (4)$$

However, the linearity is not observed near the ends of the specimen (Fig. 4).

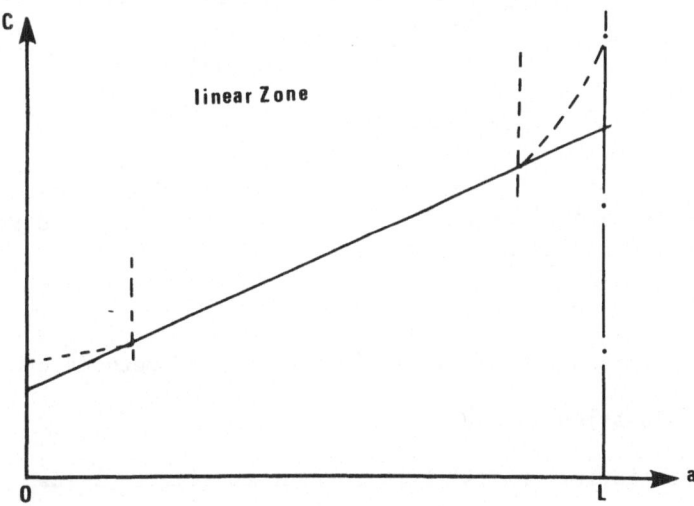

Figure 4. Schematic variation of compliance with the crack size.

So, the strain energy release rate G can be obtained and consequently the stress intensity factor K_I |5| :

$$K_I = P \left(\frac{EB}{2\,t_n}\right)^{1/2} \quad (\text{plane stress})$$

$$K_I = P \left(\frac{EB}{2(1-\nu^2)t_n} \right)^{1/2} \quad \text{(plane strain)} \tag{5}$$

where E is the Young's modulus and ν the Poisson's ratio We can note that K_I is independent of the crack size.

The preceding equations are valid only if the crack front profile is straight.

The constant B can be determined analytically. Fuller |5| has shown that B is given by :

$$B = \frac{3 W_m^2}{\mu W t^3 \phi(d)} \tag{6}$$

where $\phi(d)$ is a function of $d = \frac{2t}{W}$; a simplified expression is given by :

$$\phi(d) = 1 - 0.63202 \, d + 1.2 \, de^{-\pi/d} \tag{7}$$

So, the stress intensity factor is given by :

$$K_I = P W_m \left[\frac{3(1 + \nu)}{Wt^3 t_n \phi(d)} \right]^{1/2} \quad \text{plane stress} \tag{8}$$

$$K_I = P W_m \left[\frac{3}{Wt^3 t_n (1-\nu) \phi(d)} \right]^{1/2} \quad \text{plane strain}$$

The analytical solution is valid only between the end regions and so, the linear range is obtained for

$$0,5 \, W \leq a \leq L - W$$

For determining the (K_I, v) diagram several methods can be used.

-i- Relaxation method

Differentiating Eq.(4) with respect to time t yields

$$\frac{dy}{dt} = (Ba + D) \frac{dP}{dt} + PB \frac{da}{dt} \tag{9}$$

with $\frac{da}{dt}$ = v = crack growth rate.

During a load relaxation test, the precracked specimen is loaded up to a load P lower than the critical load leading to fast fracture at a rapid crosshead speed and then the crosshead is stopped (the displacement y is maintained constant). Therefore dy/dt = 0 and from Eq. (4) the velocity is obtained :

$$v = \frac{- Ba + D}{BP} (\frac{dP}{dt})_y = \frac{-y}{BP^2} (\frac{dP}{dt})_y \qquad (10)$$

$(\frac{dP}{dt})_y$ is the load relaxation rate.

Furthermore, y being constant, we have :

$$y = P(Ba + D) = P_{i,f} (Ba_{i,f} + D)$$

$P_{i,f}$ = initial or final load

$a_{i,f}$ = initial or final crack length

Introducing the result in Eq. (10) yields

$$v = - \frac{P_{i,f}}{P^2} \left[a_{i,f} + \frac{D}{B} \right] (\frac{dP}{dt})_y \qquad (11)$$

The crack growth velocity is obtained from the measurement of the instantaneous load, the loading relaxation rate and the initial or final value of a and P.

The load relaxation technique is very sensitive to small temperature fluctuations. The minimum crack velocity which can be measured is about $10^{-6} - 10^{-7}$ m s^{-1}. With a good control of temperature, it is possible to measure up to 10^{-9} m s^{-1}.

In addition, extraneous relaxations of the fixture and machine must be taken into account [7].

The crack front is not straight : the crack is longer on the tension side of the specimen than on the compression side (upper face) (Fig. 4).

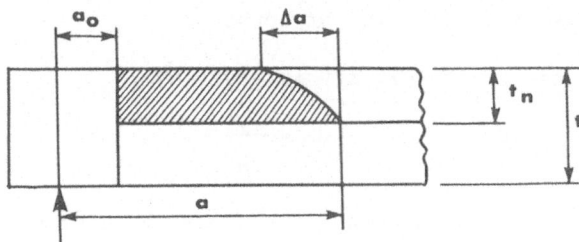

Figure 4.   Crack front of the DT specimen.

This geometry is independent of the crack length |8|.

Thus, the relationship (11) giving the crack velocity must be modified as

$$v = -\phi \frac{P_{i,f}}{P^2} \left(a_{i,f} + \frac{D}{B} \right) \left(\frac{dP}{dt}\right)_y \qquad (12)$$

$$\text{with} \quad \phi = t/\sqrt{\Delta a^2 + t^2}$$

other modifications have been proposed |6|.

Generally, D/B can be neglected in comparison with a.

-ii- Constant displacement rate method.

The specimen is loaded at constant displacement rate $\dot{y}$. A load plateau is observed |9| (Fig. 5).

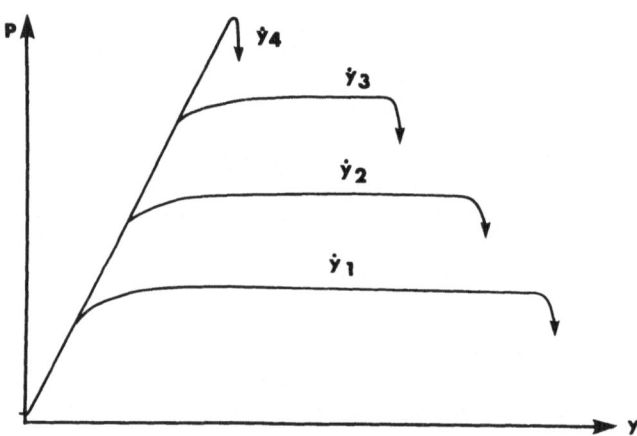

Figure 5. Load-displacement curve at constant $\dot{y}$

From Eq. 9, the crack velocity is given by :

$$\frac{dP}{dt} = 0 \qquad\qquad v = \frac{\dot{y}}{PB} \qquad (13)$$

To obtain the (K_I, v) diagram, tests must be carried out for different values of $\dot{y}$.

-iii- Constant load

A constant load is applied and the displacement rate is measured.

The crack velocity is given by Eq. (13). This method allows us to measure very low velocity.

The load relaxation method is the most used because it allows to obtain the (K_I, v) diagram with only one specimen. However, if a significant plasticity occurs during the crack growth, only the constant load method can be used and the crack propagation must be measured directly.

The experimental aspects of the DT have been reviewed in detail by Pletka et al. |7| and Mamoun |6|. A notch is cut with a diamond saw and a precracking is initiated by loading the specimen at very low crosshead speed or by thermal shock.

2.1.3 Experimental results

A lot of crack propagation data have been collected on glass |10, 11|. Wiederhorn |10| showed that the relative humidity has a significant effect on the subcritical crack growth in soda-lime glass, as can be seen in Fig. 6.

The three regions I, II, III are observed. In region I, crack growth depends on the stress intensity factor and the amount of water in the gas. In the plateau region II, crack growth does not vary with the stress intensity factor but depends significantly on the relative humidity. In region III, the crack velocity depends strongly on the stress intensity factor but is not dependent of the water content.

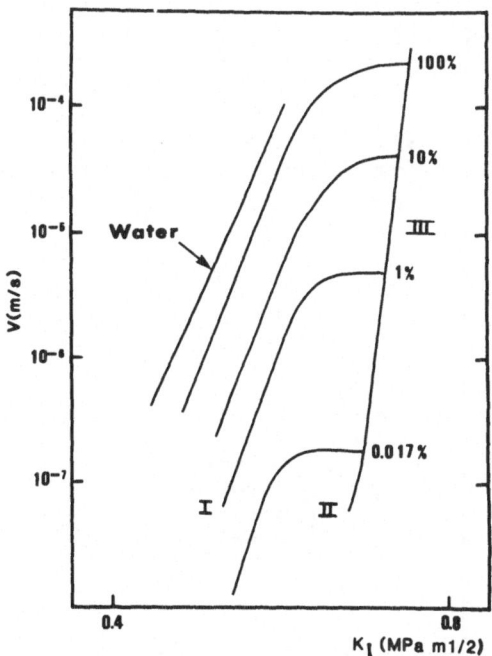

Figure 6. (K_I, v) diagram for soda-lime glass tested at room temperature in nitrogen gas : effect of relative humidity (After Wiederhorn |10|).

The slow crack growth depends on the concentration of H^+ and OH^- ions as can be seen in Fig. 7 and equally on the glass composition (Fig. 8). Thus, these results show that the external and internal chemical environment must be taken into account.

Frieman |11| studied the glass behaviour in water dissolved in alcohols of different chain length. He showed that, in region I and II, the relative humidity controls the crack growth whereas in region III, crack velocity is dependent of the alcohol chain length.

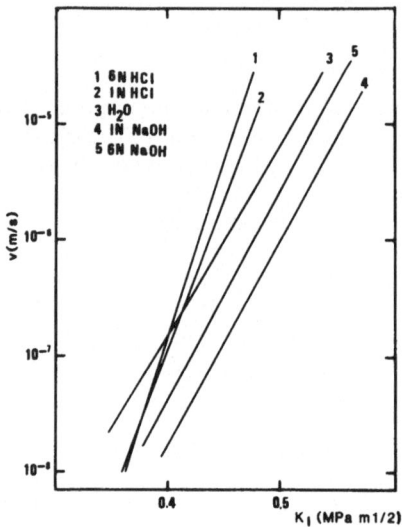

Figure 7. Effect of pH values on slow crack growth for silica glass tested in aqueous solutions (After Wiederhorn and Johnson |1|).

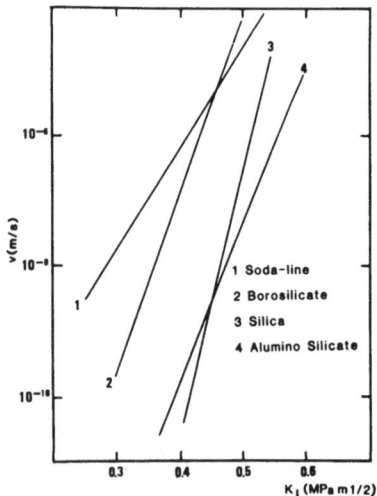

Figure 8. Effect of glass composition on (K_I,v) diagram obtained at room temperature in water (After Wiederhorn and Bolz |12|).

    Crack velocity is dependent on temperature as shown by Fig. 9 :
crack growth increases as a function of temperature.

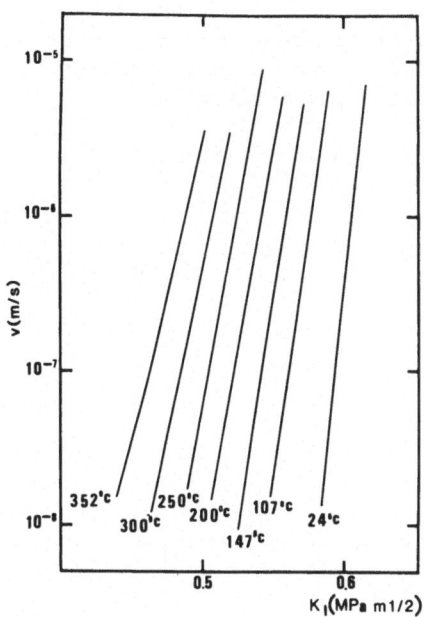

Figure 9. Effect of temperature on crack propagation, region III, in
61 % lead glass tested in vacuum of 10^{-5} Torr (After Wiederhorn et al.
|1|).

2.2   Mechanisms of slow crack growth

    Numerous theories have been proposed to explain slow crack growth
in glasses and a review is made by Wiederhorn |10|. Only essential fea-
tures of these theories are presented with an attempt to fit the experimental results.

2.2.1 Chemical reaction rate theory

    The model of Hillig and Charles |13| is the best known theory of
stress corrosion cracking.
    This theory assumes that crack growth results from a stress enhan-
ced chemical reaction. The rate of reaction depends on the curvature of
the crack surface.
    The expression for crack velocity is given by :

$$v_c = v_o \exp \left[(- \Delta E^* - V_m \gamma/\rho + \Delta V\, K_I/\sqrt{\pi\rho})/RT \right] \tag{14}$$

where ΔE^* is the activation energy of the chemical reaction without
stress, V_m the molar volume of the glass, γ the surface tension at

interface, ρ the radius of curvature of the crack tip, ΔV the activation volume, R the gas constant and T the absolute temperature.

Another approach of the chemical reaction rate theory is given by Lawn and Wilshaw |14|.

Let's consider a crack front with molecules A of the environment which react with molecules B of the material for breaking the bonds −B−B− (Fig. 10); following the reaction :

$$(A-A) + (-B-B-) \rightarrow (-B-A, A-B-) \qquad (15)$$

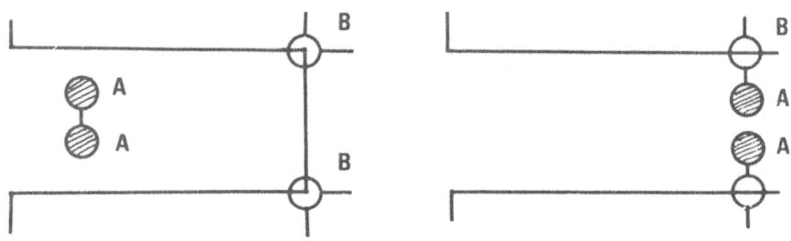

Figure 10. Rupture of bonds B−B near the crack-tip by chemical reaction with molecules A of environment.

In the case of glass, the reaction takes place with molecules of water present in the environment :

$$H-O-H + (-Si-O-Si) \rightarrow (-Si-OH \quad HO-Si-) \qquad (16)$$

The rupture of the bond requires the passage through an activated state :

$$\eta A + B \rightarrow B^* \quad \eta = \text{integer number} \qquad (17)$$

The total energy change of the system when the crack of unit width extends on da is the sum of mechanical and chemical terms :

$$\Delta U = -G_I da + (\mu_B^* - \mu_B - \eta \mu_A) N \, da \qquad (18)$$

where G_I is the strain energy release rate, μ_B, μ_B^* the chemical potentiel of B bonds before and after rupture, μ_A the chemical potential of A molecules in environment and N the surface density of B bonds.

The crack velocity is given by an Arrhenius equation :

$$v_c = \nu b \exp\left(-\frac{U_o}{kT}\right) \exp\frac{\alpha \, G_I}{kT} \exp\left(\beta\eta \, \mu_{A/kT}\right) \qquad (19)$$

where ν is the attempt frequency, b the lattice spacing, $\alpha = b^2/2$, $\beta = 1/2$, Uo the activation energy.

For an ideal gas, the chemical potential is given by :

$$\mu_A = \mu_A^o + kT \ln p_A \qquad (20)$$

where μ_A^o is a reference level and p_A is the gas pressure at the crack tip.

Inserting Eq. (20) into (19) gives :

$$v_c = C(T) \, p_A^{\beta\eta} \exp\left(-\frac{U_o}{kT}\right) \exp\left(\frac{\alpha \, G_I}{kT}\right) \qquad (21)$$

C(T) is a term which varies slowly with temperature.

2.2.2 Transport controlled rupture

When the species concentration is low or for high crack velocity, the crack growth is no more controlled by the bond rupture but by the transport of reactive species towards the crack tip. Let's consider that the chemical reaction is spontaneous. When a number dM of species A arrive at the crack tip, the number of broken B bonds is given by :

$$dN = dM/\eta$$

The crack velocity becomes :

$$v_t = \frac{da}{dt} = \frac{1}{N}\frac{dN}{dt} = \frac{1}{\eta N}\frac{dM}{dt} \qquad (22)$$

Different transport mechanisms are possible : gaseous diffusion along the crack interface, surface diffusion on the crack walls or bulk diffusion.

Consider the case of gaseous diffusion in an ideal gas. The number of molecules A which cross a unit area per unit time is given by :

$$\frac{1}{4} n_A \, \bar{v}_A \qquad (23)$$

where n_A is molecular density ($= p_A/kT$), $\bar{v}_A$ the mean molecular velocity ($= (8 \, kT/\pi m)^{1/2}$) and m the molecular mass.

If the reaction zone near the crack tip has a width equal to b, we have :

$$\frac{dM}{dt} = \frac{1}{4} b \, n_A \, \bar{v}_A \qquad (24)$$

From Eqs. (22), (23) and (24), we obtain :

$$v_t = \frac{bP_A}{\eta N} / (2\pi \, mkT)^{1/2} \qquad (25)$$

When the active species exist in a dilute solution, we replace the pressure by the concentration. Eq. (25) shows that chemical concentration plays a significant role as for the reaction controlled process (Eq. (21)) but temperature and load parameters have little or no effect.

However, the chemical reaction is not instantaneous and the molecular flow is determined by the pressure difference between source (P_A°) and tip (P_A). If the transport is slower than the reaction, the crack tip is depleted $(p_A \ll p_A^{\circ})$ whereas for the opposite, p_A is nearly p_A°. Eqs. (21) and (25) may be rewritten :

$$v = v_r \, (P_A/P_A^{\circ})^{\beta\eta}$$

where $\quad v_r = C(T) \, P_A^{\circ \beta\eta} \, \exp\left(-\frac{U_0}{kT}\right) \exp\left(\frac{\alpha \, G_I}{kT}\right) \qquad (26)$

$$v = v_t \, (1 - P_A/P_A^{\circ})$$

$$v_t = \frac{b.P_A^{\circ}}{\eta N} / (2\pi \, mkT)^{1/2} \qquad (27)$$

Eliminating P_A from Eqs (26) and (27), we obtain an implicit equation valid for the two processes (transport and reaction) :

$$\frac{v}{v_t} + \left(\frac{v}{v_r}\right)^{1/\beta\eta} = 1 \qquad (28)$$

For low G_I or K_I, v_r is much lower than v_t and v is nearly equal to v_r (the crack growth is reaction limited). For high G_I or K_I, v_r is much higher than v_t and $v \to v_t$ (the crack propagation is transport limited). Thus, when G_I or K_I increases, the crack velocity shows firstly a rapid increase (v_r increases exponentially with G_I Eq. (26)) then presents a saturation (v_t is independent of G_I or K_I, Eq.(27)).

2.2.3 Other mechanisms

Diffusional theories of slow crack growth have been proposed by Hasselman |15| and by Stevens and Dutton |16|. The crack growth occurs by diffusion of vacancies and atoms near the crack tip.

Hasselman assumes that crack growth occurs by elimination of plane of atoms and deposition on the crack surface by surface diffusion. The driving force for crack motion is the difference of energy of atoms situated near the crack tip and near the crack surface. For a stress higher than the fatigue limit, atoms near the crack tip tend to go towards the crack surface. In the model of Stevens and Dutton, the driving force is the difference in the chemical potential between the atoms situated near the crack tip and those in the bulk.

In both models, the diffusion of vacancies to the crack tip controls the crack propagation.

On the other hand, plastic flow mechanisms were suggested by many authors. These mechanisms are reviewed by Wiederhorn |10|. Such models must be taken into account but it will be necessary to show that they predict a correct description of the effect of many parameters on the crack growth.

Finally, Fuller and Thomson |17| have formulated an atomic model of fracture. Between the stress required for crack extension and the one required for crack healing, the crack is lattice-trapped and crack growth can occur, although no environmental effects are introduced.

2.2.4 Comparison with experimental results

The chemical reaction controlled crack growth can describe correctly the region I of the (K_I, v) diagram. The crack velocity is given by Eq. (21) which shows an exponential dependence of v with K_I or G_I.

Experimentally, Eq.(1) is very often used but it is impossible to distinguish between this equation and the exponential law. The reaction limited theory explains the dependence of crack growth with temperature and composition.

Eq. (21) is well verified as shown by the results of Fig. 11 obtained on soda-lime glass tested in liquid water. A value of U_o equal to about 10^5 J/mol is found which could correspond to the rupture of --Si-O-Si bonds in silica.

The effect of chemical concentration is also well described by Eq. (21). As shown in Fig. 6, the crack velocity increases with the concentration of the active species. From results of Fig. 6, we can plot the crack velocity as a function of relative humidity in nitrogen gas (Fig. 12). A straight line of slope $\beta\eta = 1/2$ is found, which corresponds to a value of $\eta = 1$. Thus, one molecule of H_2O interacts with one Si-O-Si bond following the reaction (16).

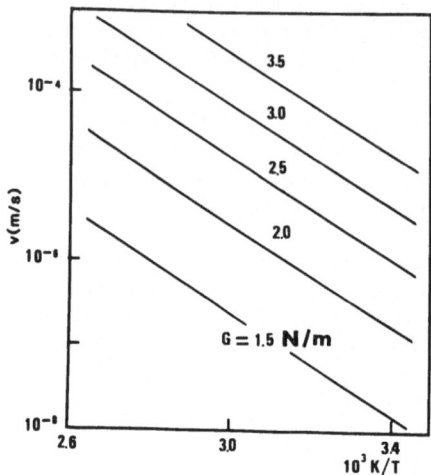

Figure 11. Crack velocity as function of temperature for a soda-lime glass tested in water, for different crack extension force (After Schönert et al. |14|).

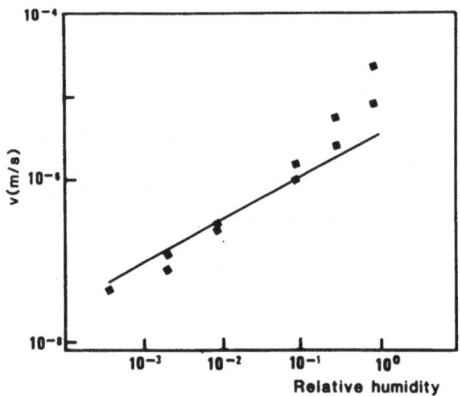

Figure 12. Crack velocity as function of relative humidity in nitrogen gas for a soda-lime glass (After Wiederhorn |14|).

However, a complete description of the interaction process at the crack tip must account for all experimental results, in particular the effect of chemical environment, pH, glass composition.

The activation energies U_o measured for different glasses

(Table I) show that the activation energy for crack motion in presence
of water is lower than activation energy for diffusion in glass (except
for alkali ions).

Table I - Activation Energies for crack motion in water or in
vacuum and for diffusion of elements in glass (After Wiederhorn
|10|).

Glass	U_o water $(10^5$ J/mol)	U_o vacuum $(10^5$ J/mol)	Diffusion	
			Element	Activation Energy $(10^5$ J/mol)
Silica	1.4		Na	1
Aluminosilicate (no alkali)	1.2	7	Ca	2
Aluminosilicate (high alkali)	1.2		O	3
Borosilicate	1.3	2.75	Si	3.7
Lead-alkali	1.		Al	3.3
Soda-lime silicate	1.1	6		

Crack propagation requires migration of the network formers (Si,
B, Al, O) and the activation energies for diffusion of these elements
are too high to explain the experimental results. So, the diffusional
theories cannot explain the crack motion in water.

The activation energies for crack growth in vacuum are higher
than in water and in this case, a diffusional process cannot be excluded. The crack growth behaviour in vacuum can be also explained by
the lattice trapping model.

The transport controlled crack growth can describe correctly
the region II of the (K_I, v) diagram. The crack velocity is independent of the stress intensity as shown by Eq. (25) but depends only on
the concentration as shown in Fig. 6. A plot of log v in region II
versus log of relative humidity for test in N_2 gas gives a straight
line of slope 1, as predicted by the theory.

The mechanisms responsible for the crack growth in region III
are not well undestood.

3. TIME DEPENDENCE OF STRENGTH, SPT DIAGRAM AND PROOF TESTING

3.1 Time dependence of strength : static fatigue and SPT diagram

The delayed fracture is due to the subcritical crack growth
which takes place for a constant applied stress lower than the critical stress leading to the catastrophic failure.

Failure occurs when the stress intensity factor reaches K_{IC} or
when the crack size reaches the critical value corresponding to the
applied stress σ_a.

The time to failure t_f is given by :

$$t_f = \int_{a_i}^{a_c} \frac{da}{v} \qquad (29)$$

a_i = initial crack size
a_c = critical crack size : $K_{IC} = Y \sigma_a \sqrt{\pi a_c}$
v = crack velocity

Substituting v by Eq. (1) and performing the integration yields :

$$t_f = \frac{2}{\sigma_a^2 \pi A Y^2 (n-2)} \left[K_{Ii}^{2-n} - K_{IC}^{2-n} \right] \qquad (30)$$

K_{Ii} = initial stress intensity factor for the largest flaw. For large n values, $K_{Ii}^{2-n} \gg K_{IC}^{2-n}$ and :

$$t_f = \frac{2 K_{Ii}^{2-n}}{\sigma_a^2 \pi A Y^2 (n-2)}$$

If σ_c is the strength of the glass without slow crack growth (in an inert environment) ($K_{IC} = \sigma_c Y \sqrt{\pi a_i}$), one can show that :

$$t_f = \frac{2 K_{IC}^{2-n}}{\pi Y^2 A(n-2) \sigma_c^{2-n}} \sigma_a^{-n} \qquad (31)$$

If statistical aspect is taken into account, one can show that

$$t_f = \sigma_a^{-n} f (Q_f) \qquad (32)$$

where $f(Q_f)$ is a function of the failure probability based on the Weibull analysis for σ_c.

Figure 13 shows results obtained for an usual glass tested in water at room temperature. The n value obtained from the slope of the straight line is 19.

Using Eq. (32), the probability of failure for the service stress can be determined for a desired lifetime.

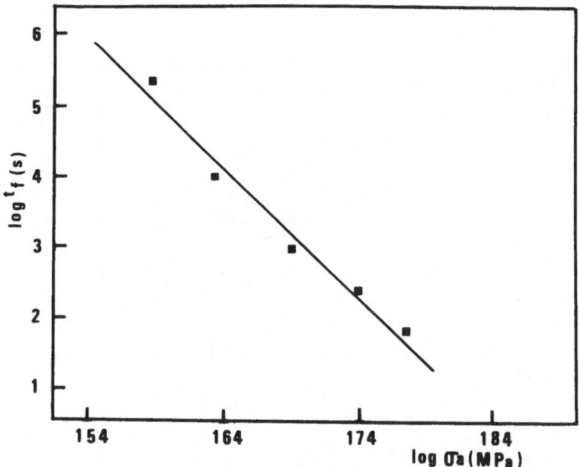

Figure 13. Stress dependence of time to failure of soda lime glass (After Jakus et al. |18|).

For a given failure probability, Eq. (32) shows that plots of log t_f versus log σ_a are a series of straigth lines of slope-n. Figure 14 shows a strength/probability/time diagram for glass at room temperature. This diagram gives the service stress for given probability of failure and time to failure.

The SPT diagrams are valid only for the conditions under which the specimens are tested.

3.2 Dynamic fatigue

3.2.1 Constant strain rate experiments

For a constant strain rate, the loading rate is proportional to the strain rate : $\dot{\sigma} = E \dot{\epsilon}$ and one can write :

$$\frac{d\sigma}{da} = \frac{\dot{\sigma}}{v} = \frac{\dot{\sigma}}{A K_I^n} = \frac{\dot{\sigma}}{A \sigma^n Y^n \pi^{n/2} a^{n/2}} \qquad (33)$$

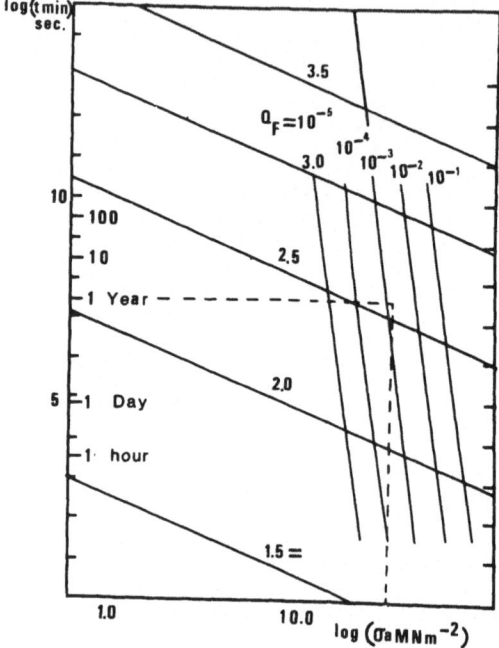

Figure 14. SPT diagram for a glass tested at room temperature (After Wiederhorn [19]).

Separating the variables and integrating, the fracture stress is obtained :

$$\int_0^\sigma \sigma^n \, d\sigma = \int_{a_i}^{a_c} \frac{\dot\sigma}{A \, Y^n \, \pi^{n/2}} \frac{da}{a^{n/2}}$$

$$\sigma^{n+1} = \frac{2(n+1) \, \dot\sigma}{(n-2)AY^2 \, \pi} K_{IC}^{2-n} \sigma_c^{n-2} \qquad (34)$$

The preceding equation is valid for mean fracture stress but the statistical aspect can be taken into account [20]. Eq. (34) shows that a logarithmic plot of fracture stress versus logarithm of stress rate $\dot\sigma$ gives a straight line with a slope of $1/(n+1)$.

Figure 15 shows such a plot in the case of a SiO_2 - TiO_2 glass.

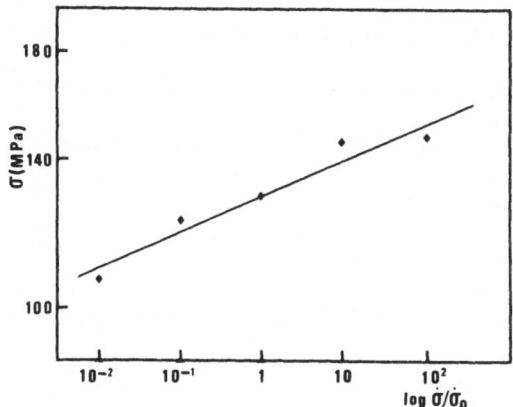

Figure 15. Loading rate dependence of strength for a SiO_2 - TiO_2 glass (After Wiederhorn |19|).

The failure time $t_{\dot{\varepsilon}}$ under constant strain rate condition up to the failure stress σ is higher than the lifetime t_f for a constant stress σ (Eq. (31)). These lifetimes are linked by the relation ;

$$t_{\dot{\varepsilon}} = (n+1) \, t_f \qquad (35)$$

This equation provides a connection between constant strain rate tests and delayed fracture tests. The knowledge of n allows us to determine the failure time at constant stress by using Eq. (31) : $t_f \, \sigma^n =$ Constant.

3.2.2 Cyclic fatigue

For a cyclic loading, with a mean stress σ_m and a variation amplitude $\sigma_o = \alpha \, \sigma_m$, the failure time t_c can be calculated from the failure time t_f under constant stress σ.
One can show that :

$$t_c = t_f \, g^{-1} (n, \, \alpha) \, (\frac{\sigma}{\sigma_m})^n \qquad (36)$$

So, a cyclic loading has no particular effect on parameters of subcritical crack growth.
Values of n obtained by different methods present large variations as shown by Freiman |11|. For soda-lime glass at room temperature, n varies between 12 and 32 depending on the type of test, surface conditions and environment.

3.3 Proof testing

Before use, a load higher than the service load is applied to the components in order to eliminate those which present a low strength. The proof test allows to guarantee a minimum lifetime for the survivors.

Proof test limits the largest size flaw present in the tested component.

If σ_p is the proof stress, components with flaws larger than a_p are eliminated :

$$\sigma_p \, Y \, \sqrt{\pi a_p} \; = \; K_{IC} \tag{37}$$

At the beginning of service, the maximum stress intensity factor is given by :

$$K_{Ii} \leq \sigma_a \, Y \, \sqrt{\pi a_p} \; = \; \frac{\sigma_a}{\sigma_p} \, K_{IC} \tag{38}$$

Substituting this relation into Eq. (30), the minimum time to failure under constant service stress after proof testing is obtained :

$$t_{min} \; = \; \frac{2 \, (\sigma_p/\sigma_a)^{n-2}}{\sigma_a 2\pi \, Y^2 (n-2) \, K_{IC}^{n-2}} \tag{39}$$

The minimum lifetime is a function of applied stress and of the ratio σ_p/σ_a.

$$t_{min} \; = \; \sigma_a^{-2} f \, (\sigma_p/\sigma_a) \tag{40}$$

If $\log t_{min}$ is plotted versus $\log \sigma_a$, a straight line with a slope of -2 is obtained for a given proof test ratio. An example for glass is shown in Fig. 14. Superimposed is the SPT diagram.

For low stresses, the proof test is not very useful but it is necessary for high stresses. Suppose for instance a survival time of 1 year at a stress of 40 MPa is required; a proof test ratio of 2.5 is necessary. Specimens passing the proof testing will survive the minimum lifetime with zero probability of earlier failure. In the absence of the proof test, the failure probability is about $2 \; 10^{-3}$.

However, some precautions must be taken when proof testing results are used, in particular the effect of slow crack growth during proof test must be taken into account |19|.

References

|1| Wiederhorn, S.M., "Subcritical crack growth in ceramics", pp.
613-46 in Fracture Mechanics of Ceramics, Wol. 2, ed. by R.C.
Bradt, D.P.H. Hasselman and F.F. Lange, Plenum Press, New York
(1974).

|2| Champomier, F.P., "Crack propagation measurements on glass : a
comparison between double torsion and double cantilever beam
specimens" pp. 60-72 in Fracture Mechanics Applied to Brittle
Materials, Part. II, ed. by S.W. Freiman, ASTM, STP 678, Philadelphia (1979).

|3| Virkar, A.V. and Gordon, R.S., "Application of load relaxation
techniques to study subcritical crack growth in brittle materials",
J. Am. Ceram. Soc., 59, 68-71 (1976).

|4| Outwater, J.O. and Gerry, D.J., "On the fracture energy, rehealing
velocity and refracture energy of cast epoxy resin, "J. of Adhesion, 1, 290-298 (1969).

|5| Fuller, E.R., "An evaluation of double-torsion testing-Analysis"
pp. 3-18 in Fracture Mechanics Applied to Brittle Materials,
Part II, ed. by S.W. Frieman, ASTM, STP 678, Philadelphia (1979).

|6| Mamoun, A. "Mise au point d'un dispositif de double torsion :
propagation sous-critique des fissures dans les céramiques"
Thesis, INSA, Lyon (1983).

|7| Pletka, B.J., Fuller, E.R., Jr and Koepke, B.G., "An evaluation
of double torsion testing-Experimental", pp. 19 37 in Fracture
Mechanics Applied to Brittle Materials, Part II, ed. by S.W.
Frieman, ASTM, STP 678, Philadelphia (1979).

|8| Evans, A.G., "A method for evaluating the time dependent failure
characteristics of brittle materials and its application to polycristalline alumina", J. Mat. Science, 7, 1137-1146 (1972).

|9| Evans, A.G., and Wiederhorn, S.M., "Crack propagation and failure
in predictions in silicon nitride at elevated temperature",
J. Mat. Science, 9, 270-278 (1974).

|10| Wiederhorn, S.M., "Mechanisms of subcritical crack growth in
glass", pp. 549-80 in Fracture Mechanics of Ceramics, Vol. 4 ed.
by R.C. Bradt, D.P.H. Hasselman and F.F. Lange, Plenum Press,
New-York (1978).

|11| Frieman, S.W., "Fracture mechanics of glass", pp. 21-78 in Glass
science and technology, Vol. 5, ed. by D.R. Uhlmann and N.J.
Kreidl, Academic Press, New-York (1980).

|12| Wiederhorn, S.M. and Bolz, L.H., "Stress corrosion and static
fatigue of glass", J. Am. Ceram. Soc., 10, 543-48 (1970).

|13| Hillig, W.B. and Charles, R.J., "Surfaces, stress dependent surfa-
ce reactions and strength", pp. 682-705 in High Strength Materials,
ed. by V.F. Zackey, John Wiley and Sons Inc., New-York (1965).

|14| Lawn, B.R. and Wilshaw, T.R., "Fracture of Brittle Solids" Cam-
bridge Solid State Science Series, Cambridge University Press,
Cambridge (1975).

|15| Hasselman, D.P.H., "Proposed theory for the static fatigue beha-
viour of brittle ceramics", pp. 297-315 in Ultrafine Grain Cera-

mics, ed. by J.J. Burke, N.L. Reed and V. Weiss, Syracuse, New-York (1970).

|16| Stevens, R.N. and Dutton, R., "The propagation of Griffith cracks at high temperature by mass transport processes", Mater. Sci. Eng., 8, 220-34 (1971).

|17| Fuller, E.R., Jr. and Thomson, R.M., "Lattice theories of fracture", pp. 507-48 in Fracture Mechanics of Ceramics, Vol. 4, ed. by R.C. Bradt, D.P.H. Hasselman and F.F. Lange, Plenum Press, New-York (1978).

|18| Jakus, K., Coyne, D.C. and Ritter, J.E., "Analysis of fatigue data for lifetime predictions for ceramic materials", J. Mat. Science, 13, 2071-80 (1978).

|19| Wiederhorn, S.M., "Reliability, life prediction and proof testing of ceramics", pp. 635-64 in Ceramics for high performance applications, ed. by J.J. Burke, A.E. Gorum and N. Katz, Brook Hill Publishing Company, Massachusetts (1974).

|20| Ritter, J.E., Jr., "Engineering design and fatigue failure of brittle materials", pp. 667-86 in Fracture Mechanics of Ceramics, Vol. 4, ed. by R.C. Bradt, D.P.H. Hasselman and F.F. Lange, Plenum Press, New-York (1978).

THERMAL CRACK GROWTH IN SELF-STRESSED GLASSY COMPOUNDS

K. P. Herrmann
Laboratory for Technical Mechanics
University of Paderborn
Pohlweg 47-49, D-4790 Paderborn
Federal Republic of Germany

ABSTRACT. Curved thermal cracks are considered running along special
principal stress trajectories of thermal stress fields originated in
different shaped glassy two-phase solids by a steady cooling process.
The resulting boundary-value problems of the stationary plane thermoelasticity are solved by means of the method of complex analysis as
well as of the finite element method. Moreover, using appropriate directional criteria established for crack path prediction, the further
extension of a thermal crack starting at the external surface of a
glassy two-phase compound with a circular cross section has been determined. Furthermore, the corresponding fracture mechanical data like
crack edge displacements, strain energy release rates and stress intensity factors, respectively, have been calculated by additional consideration of the influence of inner stress concentrators onto the paths of
quasistatic extending thermal cracks. Finally, the theoretical investigations are compared with calculations concerning the quasistatic
growth of interface cracks in thermally loaded glassy multiphase solids
as well as with the results of cooling experiments. Thereby the latter
concentrate on the determination of experimental principal stress trajectories in stable uncracked bimaterial specimens as well as on the
evaluation of fracture mechanical data governing the propagation behaviour of curved thermal cracks by means of the shadow optical method
of caustics.

1. INTRODUCTION

Composite structures subjected to nonstationary temperature fields will
fail due to thermal fracture if the arising thermal stresses reach the
ultimate strength of the material. Besides, the appearance of curved
crack paths has been observed very often in thermally loaded homogeneous
and nonhomogeneous materials, respectively, fracturing due to the extending thermal cracks. Therefore, an interesting problem in today's
fracture mechanics research represents the crack path prediction of
growing thermal cracks as function of the geometrical configuration of
a self-stressed body as well as of the applied thermal load distribution.

K. P. Herrmann and L. H. Larsson (eds.), Fracture of Non-Metallic Materials, 181–205.
© *1987 by ECSC, EEC, EAEC, Brussels and Luxembourg.*

Several authors |1-3| already considered curved or kinked cracks either
as interface cracks along circular inclusions or in connection with the
assessment of existing crack propagation criteria |4-7|. A comprehensive
survey about fracture behaviour of glasses has been given in a book by
Kerkhof |8|. Moreover, curved thermal cracks in glasses have been in-
vestigated in the past by Hieke |9|, Hieke and Loges |10|, Blauel |11|,
Karihaloo and Nemat-Nasser |12|. Finally, the crack path prediction of
extending thermal cracks in self-stressed glassy compounds as function
of the geometrical configuration as well as of the applied thermal load
distribution has been studied by Grebner and Herrmann |13|, Grebner |14|
and Herrmann and Grebner |15-19|.

In this paper, an overview about slow thermal crack growth in different shaped glassy two-phase compounds with circular, quadratic and
hexagonal cross sections will be given which is based on the correspond-
ing results in the references |13-19|. The two-phase solids consist of
homogeneous, isotropic and linearly elastic materials with differing
thermoelastic properties varying discontinuously at the straight inter-
face Γ from the values E_I, ν_I, α_I of the region I to the values $E_{II}, \nu_{II}, \alpha_{II}$
of the region II (E - Young's modulus, ν - Poisson's ratio, α - linear
coefficient of thermal expansion). Moreover, perfect contact of the two
materials at the material interface Γ is presumed. Figures 1 - 3 show the
corresponding cross sections of the cylindrical composite structures
which have been submitted to a constant temperature distribution
$T = T_I = T_{II} \neq T_O$ where T_O represents the temperature of the unstressed
initial state. The resulting thermal stress problems can be treated as
plane strain states by assuming temperature-independent thermoelastic
properties of the compound materials.

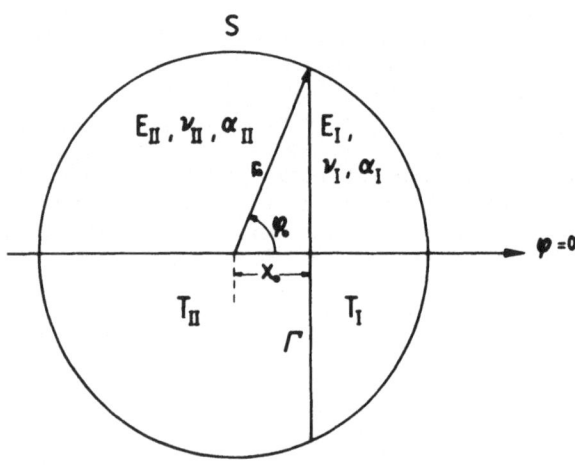

Fig. 1 Circular cross section

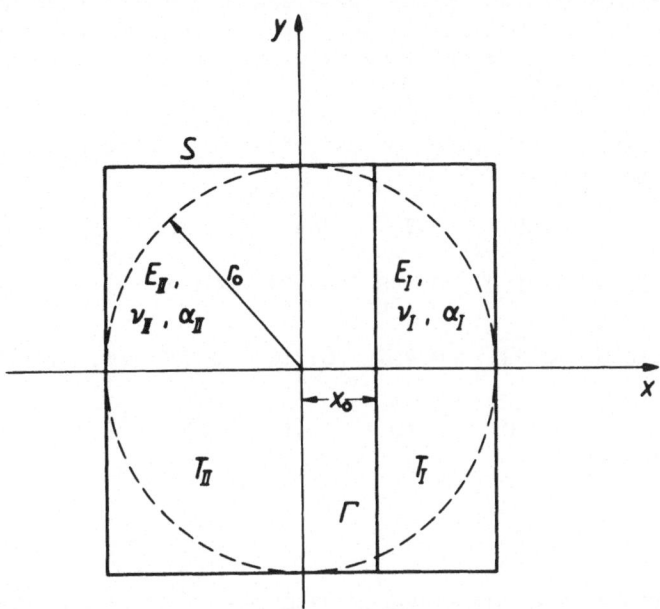

Fig. 2   Quadratic cross section

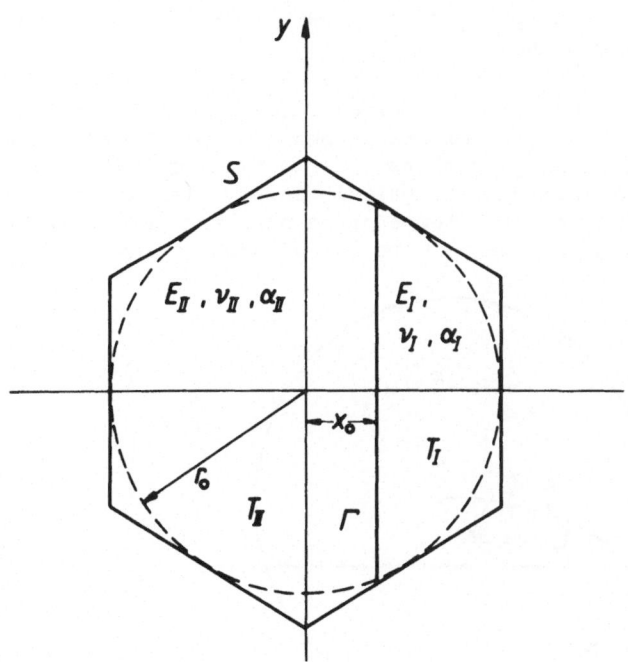

Fig. 3   Hexagonal cross section

Figures 1 - 3.   Cross sections of different shaped uncracked two-phase glassy compounds.

The material properties of the two-phase composite structures ZK5/BK7, BK4/BK12, F2/ZKN7 (optical glasses) and the corresponding three-phase composite structures, respectively, with a glass seal as gluing layer, are given in Table I.

Table I Material properties of two-phase and three-phase composite structures.

Notation	ZK5	BK7	BK4	BK12	F2	ZKN7	Glass seal
Young's modulus E[GPa]	68.180	80.148	69.474	84.036	57.389	71.024	40.000
Poisson's ratio ν	0.241	0.211	0.210	0.206	0.221	0.214	0.200
Linear coefficient of thermal expansion $\alpha(10^{-6}\mathrm{K}^{-1})$	8.70	7.10	7.14	6.75	8.20	5.40	8.20

2. FORMULATION OF BOUNDARY-VALUE PROBLEMS OF THE PLANE THERMO-ELASTICITY FOR NONHOMOGENEOUS TWO-PHASE SOLIDS

2.1 Fields of principal stress trajectories in self-stressed stable glassy compounds

Cooling experiments |14| performed for a variety of glassy compounds showed that curved thermal cracks began to propagate in disk-like specimens (geometrical parameters: $r_0 = 16.5\,\mathrm{mm}$, $\varphi_0 = 70^\circ$ and 90°, $t = 3\,\mathrm{mm}$) if special combinations of optical glasses with differing properties were glued together at a temperature near the transformation point of the glasses and afterwards were cooled down to lower temperatures.

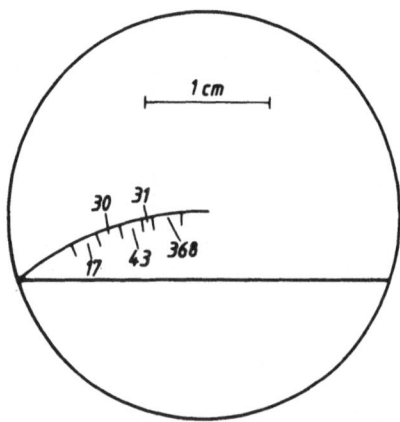

Figure 4. Slow thermal crack growth in a two-phase compound with a circular cross section (material combination: optical glasses BK4/BK12).

Further, these experiments give crack paths following in a reasonable agreement a special principal stress trajectory belonging to the existing thermal stress field in the associate uncracked two-phase specimen. Figure 4 shows the temporal development of thermal crack growth in a two-phase compound with a circular cross section. Thereby the numbers between the markings show the minutes needed for further crack extension. Moreover, the Figs. 5 - 8 give in addition the results of cooling experiments in different shaped bimaterial specimens (material combination: BK7/ZK5) containing inner stress concentrators. It can be seen that depending upon the position of such an inner stress concentrator (hole) the dominating curved thermal crack is passing the hole or will be catched by the hole. Besides, the experiments demonstrate the existence of a secondary crack originating after a long time in the opposite intersection point of the interface Γ with the external surface S which is creeping with very low velocity into the larger segment of the two-phase composite structure up to its arresting point.

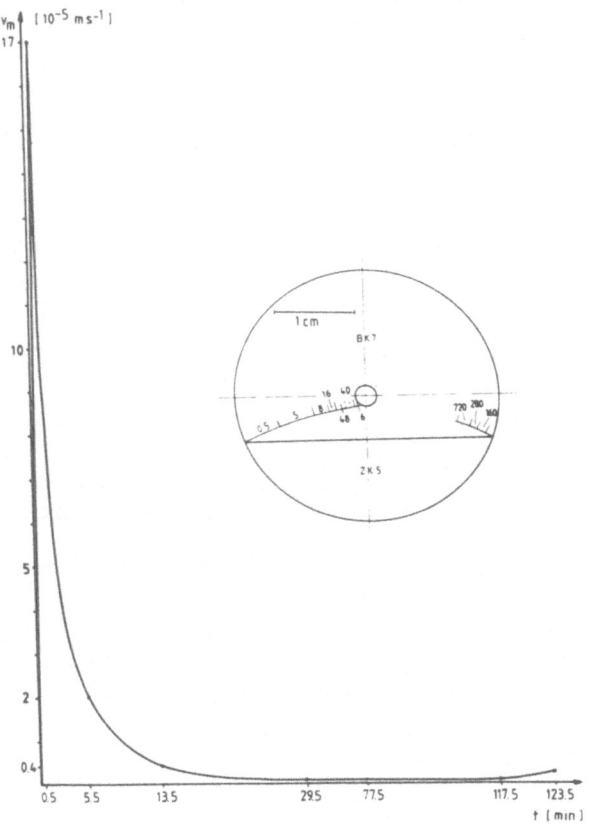

Fig. 5 Circular cross section with a concentric hole

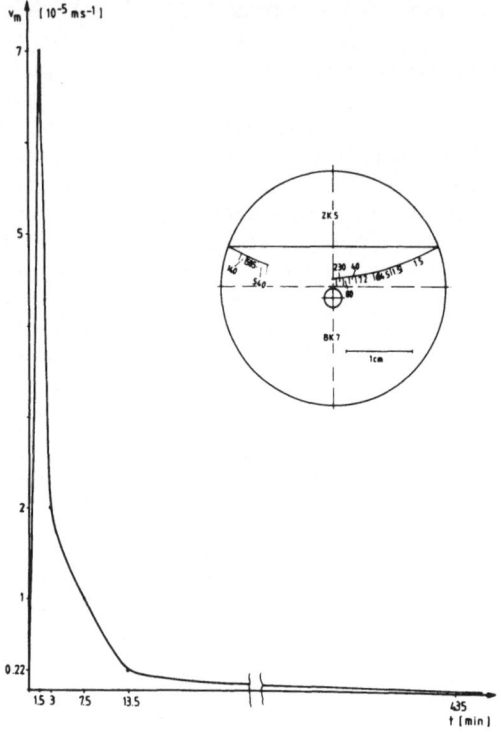

Fig. 6 Circular cross section with an eccentric hole

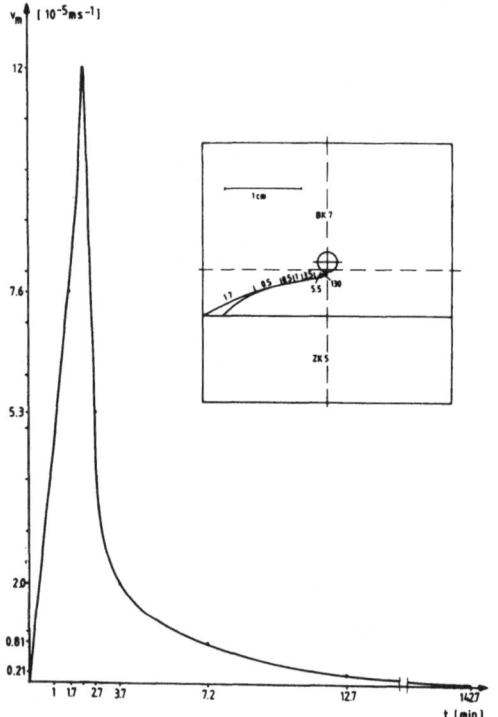

Fig. 7 Quadratic cross section with a circular hole

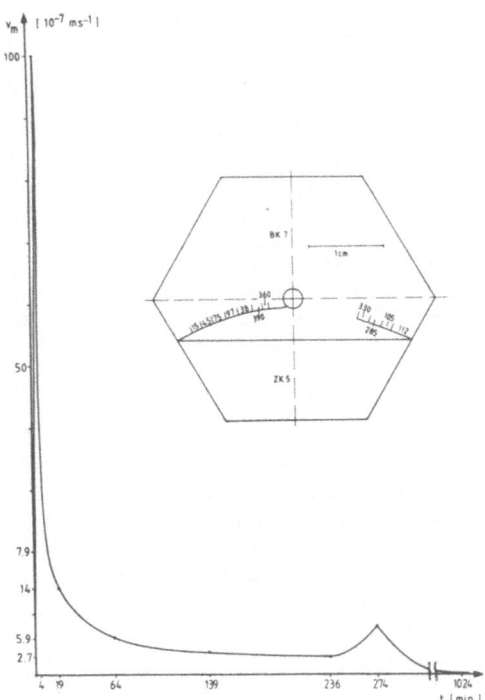

Fig. 8   Hexagonal cross section with a circular hole

Figures 5 - 8.   Temporal development of slow thermal crack growth in
different shaped two-phase composite structures (material combination:
optical glasses ZK5/BK7) with inner stress concentrators obtained from
cooling experiments |18|.

Because these crack paths shown in the Figs. 4 - 8 follow smoothly (especially in the initial stage) special principal stress trajectories of
the associated thermal stress fields existing in the corresponding uncracked self-stressed two-phase solids the latter should be considered
first. Thereby, by assuming the existence of a plane strain state in
those uncracked two-phase compounds the following boundary-value problem
of the plane thermoelasticity has to be solved:

$$\sigma_{ij} = 2\mu \left\{ \varepsilon_{ij} + \frac{\nu}{1-2\nu} \varepsilon_{kk} \delta_{ij} \right\} \tag{1}$$

$$\sigma_{ij,j} = 0 \tag{2}$$

$$\varepsilon_{ijk} \varepsilon_{\ell mn} \varepsilon_{km,jn} = -\alpha \varepsilon_{ijk} \varepsilon_{\ell mn} T_{,jn} \delta_{km} \tag{3}$$

with the boundary conditions

$$\sigma_{ij}n_j = 0 \quad ; \quad (i,j = x,y) \tag{4}$$

where n is the unit normal vector with respect to the stress-free boundary S. Furthermore, the continuity conditions

$$[\sigma_{xx}(x,y)]_{x=x_o} = [\sigma_{xy}(x,y)]_{x=x_o} = 0 \tag{5}$$

$$[u_x(x,y)]_{x=x_o} = [u_y(x,y)]_{x=x_o} = 0 \tag{6}$$

have to be fulfilled at the material interface Γ with the following definition of the jump relations

$$[\omega(x,y)]_{x=x_o} = \omega^I(x_o,y) - \omega^{II}(x_o,y) \tag{7}$$

By introducing Airy's stress function F, the boundary-value problem (1)-(6) can be transformed into a boundary-value problem of the bipotential theory. Thereby in case of a two-phase compound cylinder with a circular cross section a closed form solution of this boundary-value problem has been given for plane strain conditions by Hieke and Herrmann [20]. By applying the expression

$$F_j(r,\varphi) = V_j(r,\varphi) - W_j(r,\varphi) \quad ; \quad (j=1,2) \tag{8}$$

the boundary-value problem in that case reads as follows:

$$\nabla^2 W_j(r,\varphi) = \frac{E_i \alpha_i}{1-\nu_i} T_j(r,\varphi) \tag{9}$$

$$; \quad (\begin{matrix} j=1,2 \\ i=I,II \end{matrix})$$

$$\nabla^4 V_j(r,\varphi) = 0 \tag{10}$$

$$V_j(r_o,\varphi) = W_j(r_o,\varphi) \tag{11}$$

$$\{\frac{\partial V_j(r,\varphi)}{\partial r}\}_{r=r_o} = \{\frac{\partial W_j(r,\varphi)}{\partial r}\}_{r=r_o} \tag{12}$$

where in addition the continuity conditions (5)-(6) have to be added. Further, by using Dirichlet's discontinuous function the temperature distribution in the circular cross section of the two-phase solid has the following shape

$$T(r,\varphi) = T_1(r,\varphi) + T_2(r,\varphi)$$

$$= \frac{1}{2\pi i}\{T_I \oint \frac{e^{[r\cos\varphi - r_o\cos\varphi_o]z}}{z} dz + T_{II} \oint \frac{e^{[r_o\cos\varphi_o - r\cos\varphi]z}}{z} dz \} \tag{13}$$

Finally, the closed form solution of the boundary-value problem, the Airy stress function $F(r,\varphi)$, reads as follows:

$$F(r,\varphi) = \sum_{j=1}^{2} F_j(r,\varphi) = r_0^2 \text{Re}[\bar{z}\psi(z) + \chi(z)] - \frac{1}{2}(r\cos\varphi - r_0\cos\varphi_0)^2 [M_I T_I C_1 + M_{II} T_{II} C_2]$$

(14)

with the definitions

$$\psi(z) = \frac{M_I T_I}{8\pi}[2i(z^{-1} - 2\tilde{x}_0 + z)\ln\frac{z - e^{-i\varphi_0}}{z - e^{i\varphi_0}} - 2\sin\varphi_0(2 - \tilde{x}_0 z) -$$

$$-2\varphi_0(2z^{-1} - 4\tilde{x}_0 + z)] + \frac{M_{II} T_{II}}{8\pi}[2i(z^{-1} - 2\tilde{x}_0 + z)\ln\frac{z - e^{i\varphi_0}}{z - e^{-i\varphi_0}} +$$

$$+2\sin\varphi_0(2 - \tilde{x}_0 z) + 2(\varphi_0 - \pi)(2z^{-1} - 4\tilde{x}_0 + z)]$$

(15)

$$\chi(z) = \frac{M_I T_I}{8\pi}[i(-z^{-2} + 4\tilde{x}_0^2 - 4\tilde{x}_0 z + z^2)\ln\frac{z - e^{-i\varphi_0}}{z - e^{i\varphi_0}} +$$

$$+2\sin\varphi_0(z^{-1} - \tilde{x}_0 + z) + 2\varphi_0(z^{-2} - 2\tilde{x}_0^2)] +$$

$$+\frac{M_{II} T_{II}}{8\pi}[i(-z^{-2} + 4\tilde{x}_0^2 - 4\tilde{x}_0 z + z^2)\ln\frac{z - e^{i\varphi_0}}{z - e^{-i\varphi_0}} -$$

$$-2\sin\varphi_0(z^{-1} - \tilde{x}_0 + z) - 2(\varphi_0 - \pi)(z^{-2} - 2\tilde{x}_0^2)]$$

(16)

and the following meaning of the applied abbreviations:

$$M_i = \frac{E_i \alpha_i}{1 - \nu_i} \quad ; \quad (i = I, II) \quad ; \quad C_1 = \begin{cases} 0 \text{ for } x < x_0 \\ 1 \text{ for } x > x_0 \end{cases} \quad ; \quad C_2 = \begin{cases} 0 \text{ for } x > x_0 \\ 1 \text{ for } x < x_0 \end{cases}$$

(17)

Further, the quantities $\tilde{x} = x/r_0$, $\tilde{y} = y/r_0$, $\tilde{r} = r/r_0$, $z = \tilde{r}e^{i\varphi}$ mean dimensionless variables and the complex potentials $\psi(z)$ and $\chi(z)$, respectively, are defined in the region $\tilde{r} < 1$. The corresponding stress- and displacement states can now be calculated by using the theory of Kolosov-Muskhelishvili [21]:

$$\sigma_{xx}^{(F_j)}(x,y) = \text{Re}\{2\psi_j'(z) - \bar{z}\psi_j''(z) - \chi_j''(z)\}$$

(18)

$$\sigma_{yy}^{(F_j)}(x,y) = -M_i T_i C_j + \text{Re}\{2\psi_j'(z) + \bar{z}\psi_j''(z) + \chi_j''(z)\}$$

(19)

$$\sigma_{xy}^{(F_j)}(x,y) = \text{Im}\{\bar{z}\psi_j''(z) + \chi_j''(z)\}$$

(20)

$$2G_i\{u_x^{(F_j)}(x,y) + iu_y^{(F_j)}(x,y)\}$$

$$= r_0\{(3 - 4\nu_i)\psi_j(z) - z\bar{\psi}_j'(\bar{z}) - \bar{\chi}_j'(\bar{z})\} + \frac{dW_j(x)}{dx}$$

(21)

with the definition

$$W_j(x) = \frac{1}{2} M_i T_i [x-x_o]^2 c_j \quad ; \quad \left(\begin{array}{l} j=1,2 \\ i=I,II \end{array}\right)$$ (22)

Finally, the total stress- and displacement states in the circular cross section of a thermally loaded two-phase solid can be obtained by superposition

$$\sigma_{xx}^{(F)}(x,y) = \sigma_{xx}^{(F_1)}(x,y) + \sigma_{xx}^{(F_2)}(x,y)$$ (23)

$$\sigma_{yy}^{(F)}(x,y) = \sigma_{yy}^{(F_1)}(x,y) + \sigma_{yy}^{(F_2)}(x,y)$$ (24)

$$\sigma_{xy}^{(F)}(x,y) = \sigma_{xy}^{(F_1)}(x,y) + \sigma_{xy}^{(F_2)}(x,y)$$ (25)

$$u_x^{(F)}(x,y) = u_x^{(F_1)}(x,y) + u_x^{(F_2)}(x,y)$$ (26)

$$u_y^{(F)}(x,y) = u_y^{(F_1)}(x,y) + u_y^{(F_2)}(x,y)$$ (27)

Figures 9 - 10 show the fields of principal stress trajectories for two glassy compounds with circular cross sections consisting of two optical glasses BK4 (region I) and BK12 (region II) and with the geometrical parameters $r_o = 16.5$ mm, $\varphi_o = 70°$ and $90°$, respectively. Further, the applied constant temperature distribution in these cross sections read: $\Delta T = T_I = T_{II} = -560$ deg C.

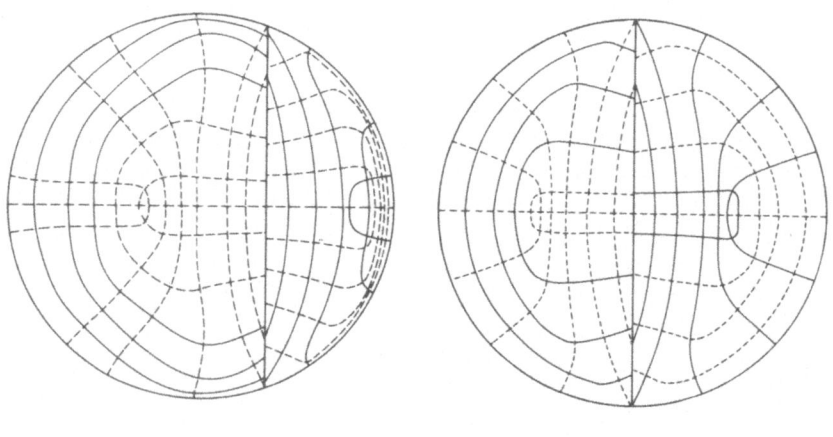

Fig. 9   Fig. 10

Figures 9 - 10. Principal stress trajectories in two uncracked thermally loaded bimaterial specimens with circular cross sections (solid lines: tension stresses, dotted lines: pressure stresses, $\Delta T = -560$ deg C, material combination: optical glasses BK4/BK12, $\varphi_o = 70°$ and $90°$, respectively) |14,16|.

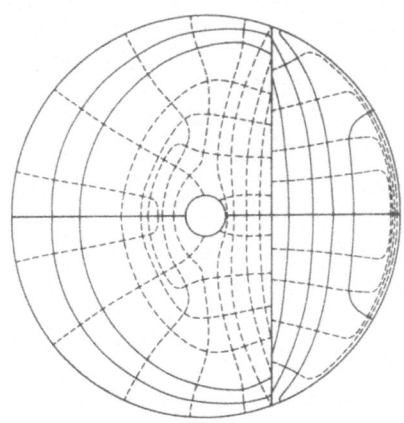

Figure 11. Principal stress trajectories in the circular cross section of a doubly connected thermally loaded uncracked two-phase composite structure (solid lines: tension stresses, dotted lines: pressure stresses, $\Delta T = 560$ deg C, material combination: optical glasses BK7/ZK5, $\varphi_O = 70^O$) |18|.

Moreover, Fig. 11 gives the thermal stress state for a doubly connected self-stressed two-phase composite structure (material combination: BK7 (region I) / ZK5 (region II)) containing an inner stress concentrator (hole). In this case the geometrical parameters were chosen to $r_1=16.5$ mm, $r_2=r_1/10$, $\varphi_O=70^O$ and the applied temperature distribution reads $\Delta T = T_I = T_{II} = 560$ deg C. Furthermore, Figs. 9 - 11 show the existence of two orthogonal sets of principal stress trajectories which embrace two singular points on the symmetry line of each cross section. Thereby the positions of these singular points of the plane thermal stress fields can be calculated from the equations

$$\sigma_{xx} - \sigma_{yy} = 0 \qquad , \qquad \sigma_{xy} = 0 \tag{28}$$

which are valid in those points.

Finally, further results for thermally loaded glassy compounds with a quadratic and a hexagonal cross section, respectively, have been obtained and are given in the Figs. 12 and 13. A comparison of these theoretically determined self-stress states with corresponding fields of experimental principal stress trajectories in stable uncracked bimaterial specimens |14,17| obtained by using the method of photoelasticity shows a very good coincidence. Further, the theoretically as well as the experimentally gained fields of principal stress trajectories show very clearly that for the cases $2\varphi_O \neq \pi$ only one principal stress trajectory of the two existing sets runs from one intersection point of the interface Γ with the external boundary S or S_O, respectively, to the opposite intersection point. Moreover, cooling experiments performed for different shaped glassy bimaterial specimens have indicated that for

so-called unstable material combinations a curved thermal crack starts
with a special initial velocity from one of the two intersection points
and is first of all running with decreasing velocity rather smoothly
along this special principal stress trajectory located entirely in one
of both glass segments (for the cases $2\varphi_o \neq \pi$ in the larger segment) to
the opposite intersection point.

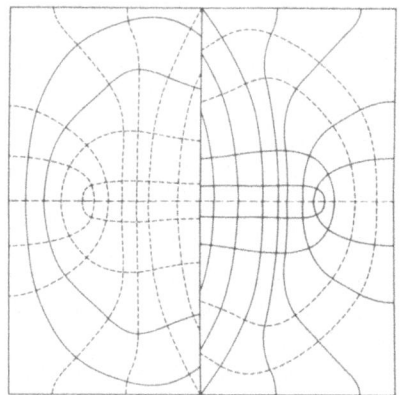

Figure 12. Principal stress trajectories in the quadratic cross
section of a self-stressed uncracked two-phase glassy compound
(solid lines: tension stresses, dotted lines: pressure stresses,
$\Delta T = -560$ deg C, material combination: optical glasses BK4/BK12, $\varphi_o = 90°$).

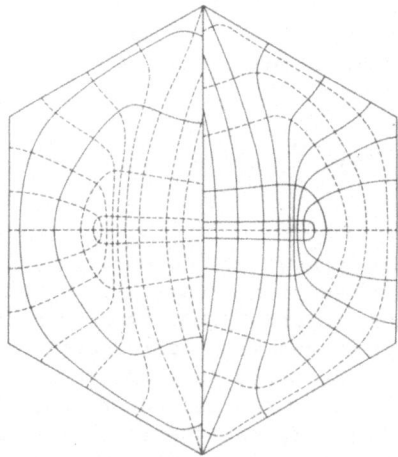

Figure 13. Principal stress trajectories in the hexagonal cross section
of a self-stressed uncracked two-phase solid (solid lines: tension
stresses, dotted lines: pressure stresses, $\Delta T = -560$ deg C, material
combination: optical glasses BK4/BK12, $\varphi_o = 90°$).

Besides, the experimental as well as the theoretical thermal stress fields show for the case $2\varphi_0 = \pi$ the existence of two principal stress trajectories (one in each semicircle) running into the two intersection points. But the cooling experiments mentioned above show that one crack only arises starting again from one of the two intersection points. Remarkable deviations from the slope of this special principal stress trajectory could be stated for thermal cracks extending very near to an inner stress concentrator (ref. Figs. 5 - 8). Further results concerning curved thermal crack growth in disk-like glassy compounds obtained by cooling experiments can be found in |17|. Finally, it should be mentioned that a closed form solution of the corresponding boundary-value problem Eq. (4) - Eq. (6) for a nonhomogeneous linear viscoelastic solid (Maxwell-body) could be obtained |22|. Thereby the thermoviscoelastic stress analysis has been performed by application of the correspondence principle of the linear viscoelasticity to the associated thermoelastic problem.

2.2 Crack path prediction in unstable bimaterial specimens

The cooling experiments show the existence of stable and unstable two-phase glassy compounds, respectively, depending upon the applied material combinations. Thereby the crack path prediction as well as the calculation of fracture mechanical data governing the quasistatic growth of a curved thermal crack have been performed by applying the concepts of linear elastic fracture mechanics. Further, a theoretical crack path prediction in thermally loaded two-phase composite structures requires the solution of different mixed boundary-value problems of the plane thermoelasticity which can be described by the basic equations Eq. (1) - Eq. (3) of continuum mechanics together with the continuity equations Eq. (5) - Eq. (6) at the straight material interfaces Γ and in addition by the boundary conditions Eqs. (4) where latter have to be numerically fulfilled at the stress-free boundary S (external surface) and at the two crack surfaces $S^+ \cup S^-$, respectively. Because of the complicated shape of the stress-free boundary of such a cracked bimaterial specimen a numerical calculation was performed by applying the finite element method in order to predict the curved crack path by means of appropriate directional criteria for crack propagation |16|.

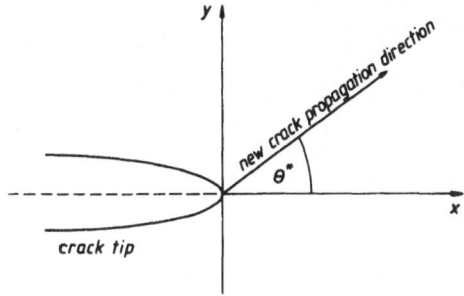

Figure 14. Crack tip coordinates and direction of further crack growth.

Thereby these criteria make use of the near-tip stress and displacement fields in the vicinity of the original crack characterized for a general plane loading situation by the stress intensity factors K_I and K_{II}, respectively. Moreover, the application of those criteria just mentioned to a cracked solid yields equations for the determination of the angle ϑ^* describing the direction of further crack growth (ref. Fig. 14). Investigations performed by Bergkvist and Guex |5| showed in case of small angles ϑ^* the existence of an approximate solution

$$\vartheta^* = -2K_{II}/K_I \tag{29}$$

valid for all projective criteria.

Based on this important result a finite element calculation concerning the theoretical prediction of a curved crack path observed experimentally in a two-phase glassy compound (material combination: BK4 (region I) / BK12 (region II); ref. Fig. 4) has been performed. Thereby the same geometrical parameters were used as for the calculation of the thermal stress state in the associate stable circular bimaterial specimen. Further, the temperature distribution in the cross section was again chosen to $\Delta T = T_I = T_{II} = -560$ deg C.

The approximate calculation procedure starts as already explained in |16| with the determination of the stress intensity factors K_I and K_{II} for the leading crack tip of the initial stage of an extending thermal crack. Moreover, this initial crack length was approached by a secant of that special principal stress trajectory connecting the surface points of the discontinuity area Γ of the original thermal stress field existing in the associate uncracked self-stressed bimaterial specimen. Then the new direction of the initial thermal crack can be determined by using Eq. (29). According to this calculated new direction the initial crack was lengthened by a small straight piece $\Delta a \approx 1.2$ mm.

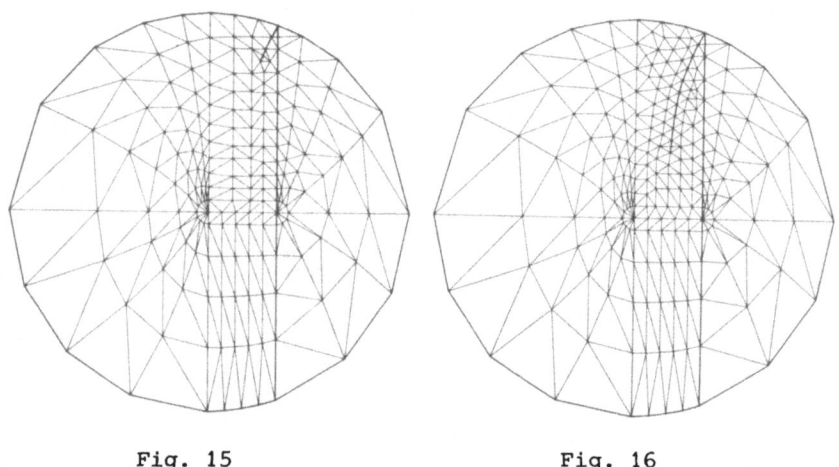

Fig. 15   Fig. 16

Figures 15 - 16. Finite element meshes for different crack lengths used for crack path prediction |16|.

Afterwards a new finite element mesh has to be generated. Finally, the repeated performance of this approximate procedure yields the wanted crack path. Thereby the numerical calculations were performed on the computer PRIME 550 at the University of Paderborn by the aid of the standard finite element program ASKA and by using the substructure technique as well as by applying triangular linear strain six-node elements. Figures 15 - 16 show two examples of typical finite element meshes used for crack path prediction. Further results concerning the associate crack edge displacements for several steps of the numerical crack path prediction of a quasistatic extending thermal crack are given in reference |16|. Finally, Fig. 17 shows a comparison of the numerically determined crack path with that special principal stress trajectory belonging to the associate self-stress field connecting the intersection points of the material interface Γ with the external sur-face of the uncracked two-phase glassy compound. A rather good coinci-dence of these two curves can be stated. Furthermore, Fig. 17 also contains experimental results obtained by cooling experiments performed with unstable disk-like glassy compounds consisting of two circular segments of different optical glasses. More details about these experi-ments can be found in reference |17|. However, experimental and theoretical results show a fairly good agreement as can be seen from Fig. 17. Moreover, it should be mentioned that the numerical as well as the experimental results confirm a more general stability consi-deration performed by Cotterell and Rice |4|.

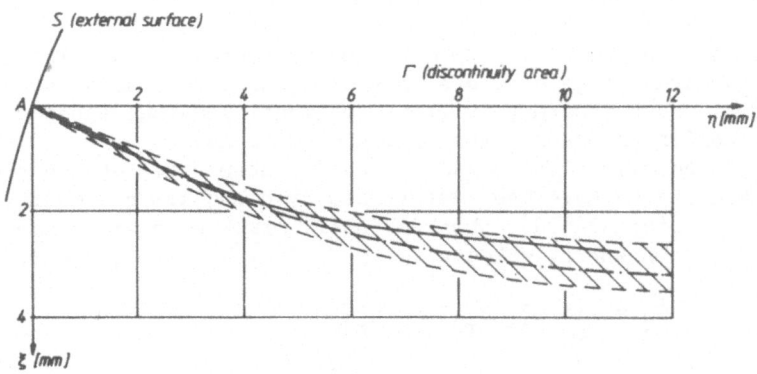

Figure 17. Comparison of a numerically determined crack path (solid line) with a principal stress trajectory (dot-dash line) of an associ-ated uncracked two-phase solid as well as with crack paths obtained by cooling experiments (hatched region) |16|.

3. DETERMINATION OF FRACTURE MECHANICAL DATA

By modelling the quasistatic extending curved thermal crack path in a self-stressed two-phase glassy specimen by the aid of that special principal stress trajectory of the original thermal stress field connecting the intersection points of the material interface Γ with the external boundary S the crack edge displacements for several stages of crack growth have been calculated. Figure 18 shows the finite element mesh for a hexagonal cross section of a two-phase solid containing 2200 nodal points.

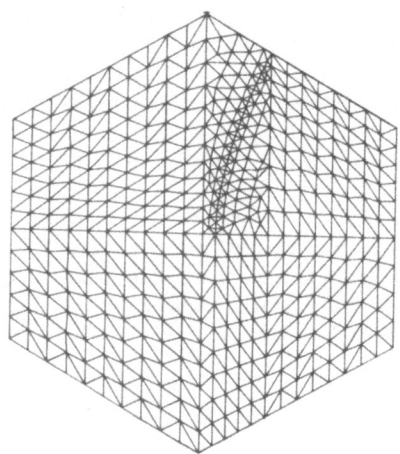

Figure 18. Finite element mesh for a two-phase compound with a hexagonal cross section.

Moreover, the Figs. 19 - 20 give the corresponding results for the calculated crack edge displacements of a quasistatically extending curved thermal crack. Thereby it should be mentioned that these calculations were performed under plane strain conditions.

In addition, the strain energy release rates at the tips of these thermal cracks have been determined by using Irwin's modified crack closure integral $|23|$. Then for small crack extensions $\Delta a \ll a$ the desired strain energy release rates G_j, $(j=I,II)$ are given by the formulae

$$G_I = \lim_{\Delta a \to o} \frac{1}{2\Delta a} F_{rc}(u_{rc} - u_{rd}) \tag{30}$$

$$G_{II} = \lim_{\Delta a \to o} \frac{1}{2\Delta a} F_{\vartheta c}(u_{\vartheta c} - u_{\vartheta d}) \tag{31}$$

where the nodal point forces F_{rc} and $F_{\vartheta c}$, respectively, are responsible for the coalescence of the nodes c and d (ref. Fig. 21 for comparison) in case of crack closure.

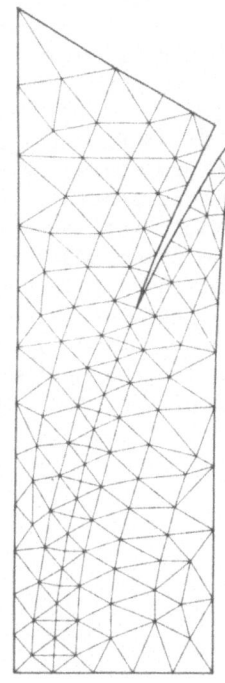
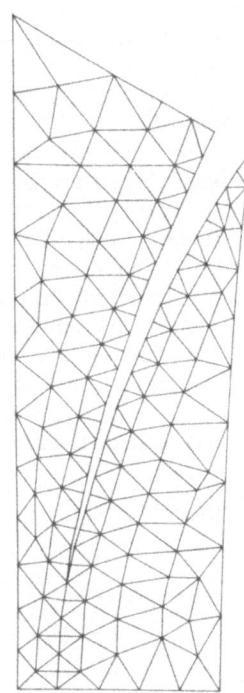

Fig. 19   Fig. 20

Figures 19 - 20. Crack edge displacements for different crack lengths
of a curved thermal crack in a two-phase compound with a hexagonal
cross section |16|.

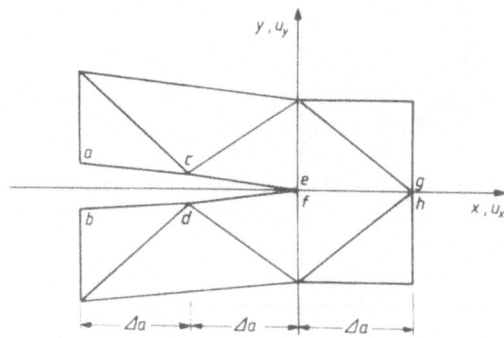

Figure 21. Finite element mesh at the crack tip used for the calculation of the strain energy release rates.

Figure 21 shows the finite element mesh at the crack tip used for the calculation of the strain energy release rates. Further, Fig. 22 gives the strain energy release rate G_I as a function of crack length for three different shaped bimaterial specimens (material combination: BK4 (region I) / BK12 (region II), $\Delta T = -560$ deg C) with circular, hexagonal and quadratic cross sections, respectively. Because of $G_{II} = 0$ the curved thermal crack is under mode I-loading only during its proposed growth along a special principal stress trajectory. Thus, it always follows $G = G_I$ at the leading crack tip of the quasistatic extending thermal crack. Further, Fig. 22 shows a remarkable influence of the shape of the external boundary of the two-phase compounds on the G_I-values. Besides, because of vanishing values of the strain energy release rate G_I for $a \rightarrow o$ it can be concluded that the curves belonging to the circular and the hexagonal cross section, respectively, should also have both a maximum value for very small crack lengths approaching the experimentally obtainable G_{IC}-value in the vicinity of the starting point of the curved thermal crack.

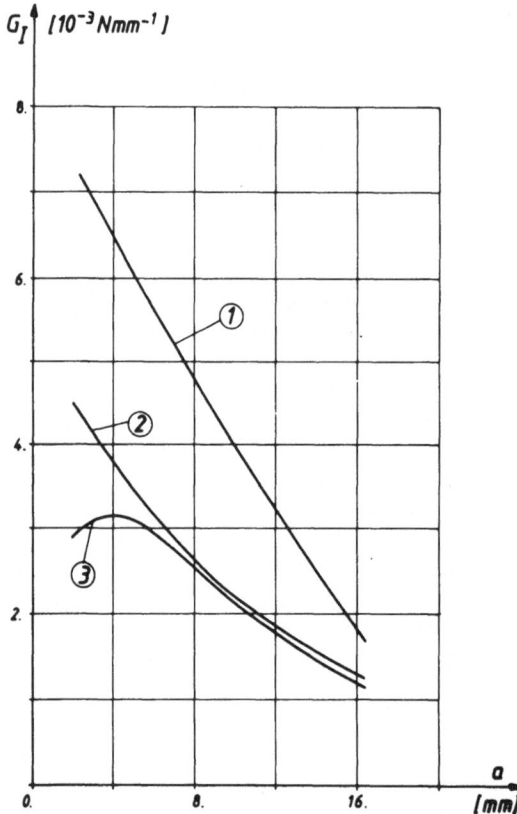

Figure 22. Strain energy release rate G_I in dependence on crack length for three two-phase glassy compounds with different cross sections. (1): circular, (2): hexagonal, (3): quadratic (material combination: optical glasses BK4/BK12, $\Delta T = -560$ deg C) |16|.

Further, Fig. 23 gives a comparison of experimentally determined stress intensity factors obtained by the optical method of caustics in disk-like glassy two-phase compounds and theoretically determined stress intensity factors gained by a finite element calculation. Experimental and theoretical results show a reasonable agreement. More detailed informations about experimental investigations concerning curved thermal crack growth in nonhomogeneous materials with different shaped external boundaries can be found in reference |17|. Besides, strain energy release rates G_j,(j=I,II) were calculated for curved thermal cracks extending quasistatically in glassy two-phase compounds containing inner stress concentrators |18|.

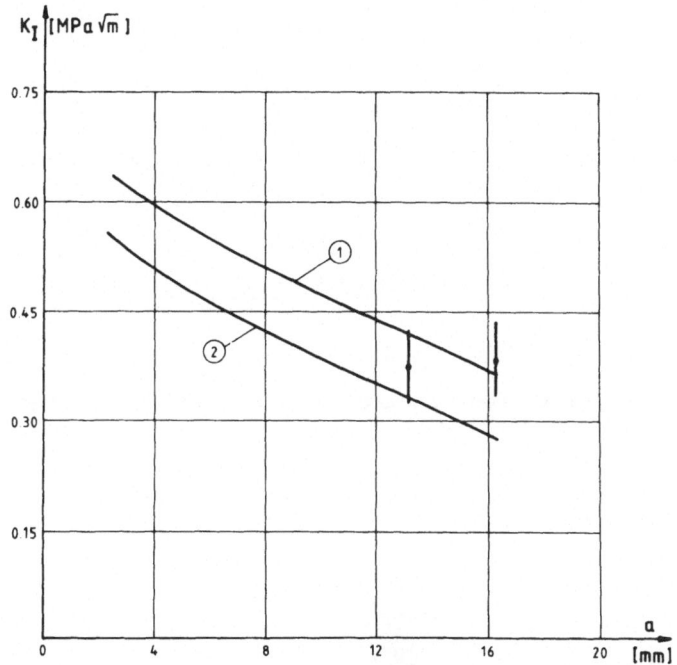

Figure 23. Comparison between theoretically and experimentally determined stress intensity factors for a cracked self-stressed two-phase solid (material combination: optical glasses F2/ZKN7, $\Delta T = -260$ deg C) with a hexagonal cross section ((1): plane strain, (2): plane stress).

4. STRAIGHT INTERFACE CRACKS IN THE MATERIAL INTERFACES OF THERMALLY LOADED MULTIPHASE COMPOUNDS

For the sake of comparison straight interface cracks in glassy multiphase compounds have been studied starting at a surface point of the discontinuity area Γ and extending quasistatically into the interior towards the symmetry line of a multiphase solid. In this case the crack is under mixed-mode loading and strain energy release rates G_I and G_{II} are obtained. Thereby a typical cross section of a three-phase compound cylinder containing an interface crack in one of the two interfaces is shown in Fig. 24.

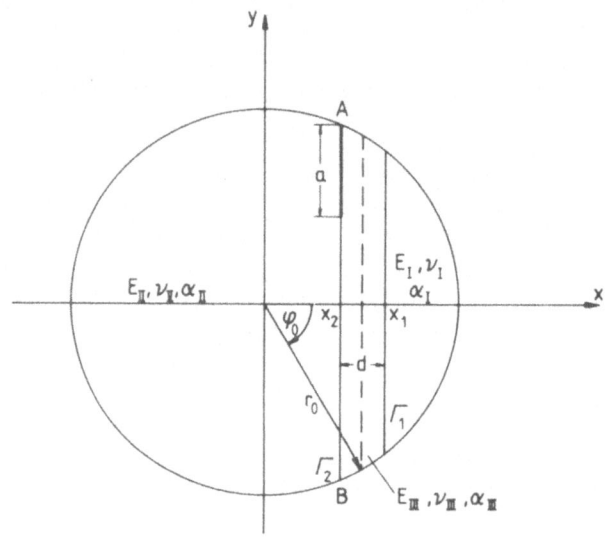

Figure 24.   Circular cross section of a cracked three-phase solid.

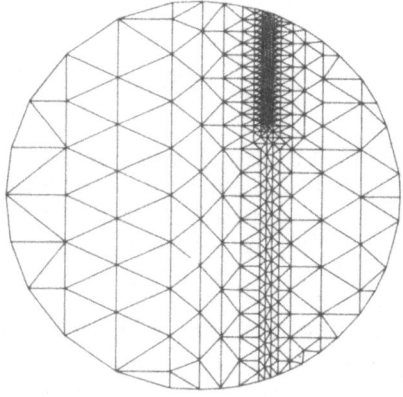

Figure 25.   Finite element discretization of a multiphase compound containing an interface crack |19|.

urther, this three-phase composite structure has been submitted to a
constant temperature distribution $T = T_I = T_{II} = T_{III} \neq T_O$ where T_O again
represents the temperature of the unstressed initial state. By assuming
plane strain conditions as well as temperature-independent thermoelastic
properties the resulting mixed boundary-value problem can be transformed
into a boundary-value problem of the bipotential theory $|19|$. For dif-
ferent crack configurations in multiphase solids with one or two material interfaces the corresponding boundary-value problems have been
solved again by means of the finite element method. Figure 25 shows the
finite element mesh applied for the modelling of an interface crack
starting at point A of the external boundary and extending quasistatically along the interface of a three-phase or a two-phase solid ($d = 0$),
respectively, into the interior of a composite structure. Thereby the
mesh consists of seven substructures with about 2000 nodal points. The
triangular elements (TRIMP6) with quadratic displacement functions are
focused especially in the neighborhood of the prospective crack path.
Moreover, in case of the cracked three-phase solid (ref. Fig. 24) a combination of the optical glasses ZK5 (segment I), BK7 (segment II) and a
glass seal (gluing layer III) has been used in the calculations. In
case of a cracked two-phase solid a combination of the optical glasses
BK4 (segment I) and BK12 (segment II) has been considered too. The material properties of these compounds are given in Table I. The geometrical data used in the calculations have been chosen to $r_O = 16.5$ mm,
$d = 0.5$ mm, $\varphi_O = 70^O$ and the temperature distribution in the cross section
reads $\Delta T = -560$ deg C.

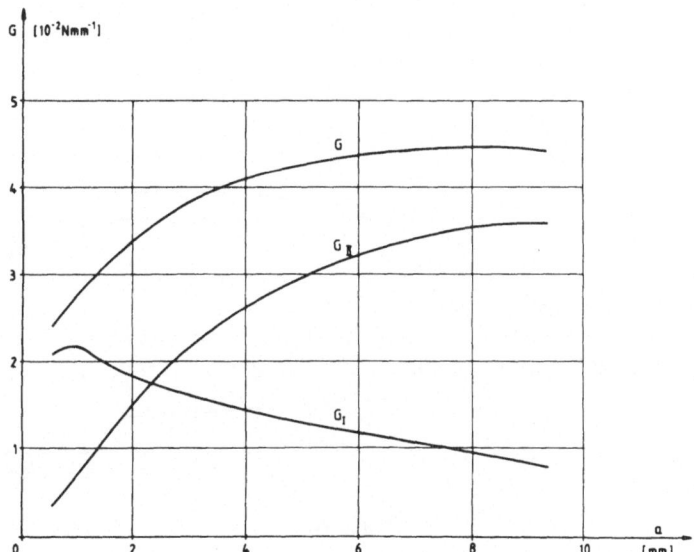

Figure 26. Strain energy release rates as functions of crack length
for an interface crack in a thermally loaded three-phase composite
structure (material combination: optical glasses ZK5/Glass seal/BK7,
$\Delta T = -560$ deg C).

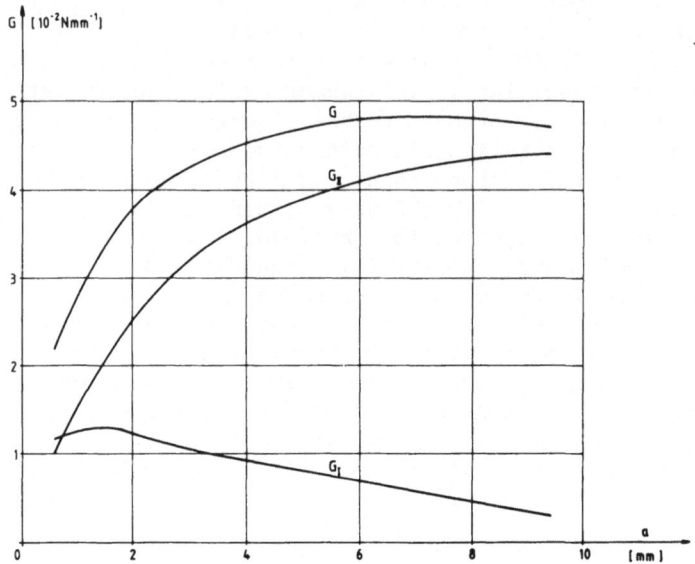

Fig. 27

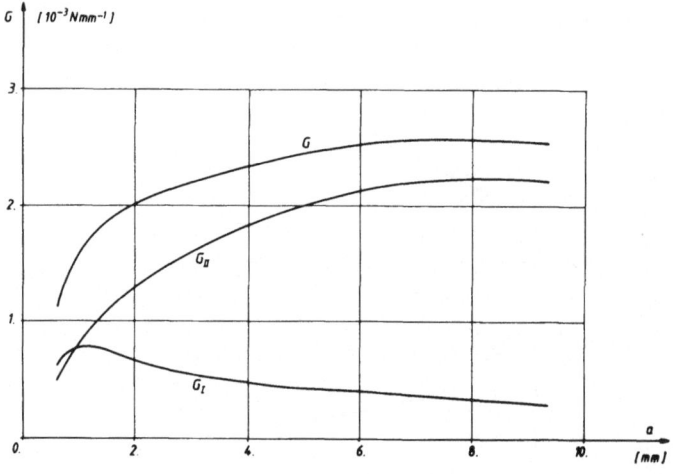

Fig. 28

Figures 27 - 28. Strain energy release rates as functions of crack
length for an interface crack in a thermally loaded two-phase composite structure (material combination: optical glasses ZK5/BK7 and
BK4/BK12, respectively, $\Delta T = -560$ deg C).

Due to the existing thermal stress states in the compound materials mixed-mode loading of the crack surfaces occurs. However, the interface crack lying in the material interface of a three-phase or a two-phase composite structure, respectively, is constrained to move in a straight line. Further, the calculation of strain energy release rates $G_j, (j=I,II)$ at the tips of quasistatically extending thermal interface cracks has been performed by using the method of reference $|23|$. Finally, the Figs. 26 - 28 show the results of finite element calculations concerning the strain energy release rate G for interface cracks extending quasistatically from the point A of the external boundary S into the interior of a thermally loaded three-phase or two-phase solid, respectively. As the graphs show there exists a mixed-mode loading at the crack tips. Further, it can be seen that the total values of the strain energy release rate G as a function of crack length are higher in case of a two-phase compound (material combination: ZK5/BK7) in comparison with those of a three-phase compound (ZK5/GS/BK7). Thus, the gluing layer leads in case of the considered material combination to a reduction of the crack inclination of the three-phase composite structure in comparison to the corresponding two-phase compound. Further results concerning the crack edge displacements of quasistatically extending interface cracks are given in reference $|19|$.

Furthermore, a comparison of the strain energy release rates determined for the two-phase compounds ZK5/BK7 and BK4/BK12, respectively, shows a difference between corresponding values of about one order of magnitude. In addition, the results belonging to the two-phase composite structure BK4/BK12 allow a further comparison with the results of finite element calculations given in Fig. 22. Thereby, the total strain energy release rate G at the tip of a curved thermal crack extending quasistatically along a principal stress trajectory is decreasing with increasing crack length in contrast to the behaviour of an interface crack. On the other hand, in case of a curved thermal crack the critical G_{IC}-value of about $8.47 \cdot 10^{-3}$ N/mm for the optical glass BK12 is reached near the initiation point of the crack at the external surface of the glassy compound whereas the corresponding values for the strain energy release rate at the tip of an interface crack never reach this critical value. Thus, the investigations concerning the interface crack confirm the experimental results obtained by a series of cooling experiments $|17|$ that only curved thermal cracks arise propagating in one of both segments of a thermally loaded glassy compound.

5. CONCLUSIONS

Thermal crack propagation in two-phase and three-phase glassy compounds, respectively, submitted to well-defined thermal stress fields has been studied experimentally by the aid of cooling experiments as well as theoretically by applying the finite element method. Thereby the fields of theoretical principal stress trajectories obtained for self-stressed uncracked two-phase solids show a very good coincidence with the corresponding experimentally gained thermal stress fields. Further, cooling experiments show a rather smooth coincidence of the experimentally obtained crack paths with special principal stress trajectories of the associated uncracked specimens. But inner stress concentrators (small circular holes) depending upon their locations in the bimaterial specimens have a remarkable influence on the shape of the crack path. Moreover, by a nearly exact modelling of the crack path there exists a difference in the propagation behaviour of curved thermal crack paths running in two-phase compounds with or without inner stress concentrators, respectively. In case of different shaped two-phase composite structures curved thermal crack propagation takes place under mode I-loading only and the corresponding G_I-values show a decrease with increasing crack length because of the decreasing self-stress energy in the cracked bimaterial specimens. Moreover, the fracture mechanical data governing the thermal crack growth in these two-phase compounds show a remarkable influence of the shape of the external boundary. On the contrary, in case of a composite structure with inner stress concentrators there exists a mixed-mode situation starting from a certain crack length. Thereby the G_I-values again decrease with increasing crack length whereas the G_{II}-values show the opposite behaviour. Finally, straight interface cracks extending quasistatically in the material interfaces of thermally loaded multiphase compounds have been investigated. In contrast to the slow propagation of curved thermal cracks there exists a mixed-mode loading situation for interface cracks from the beginning. Furtheron, a comparison of the corresponding finite element calculations with cooling experiments performed with brittle multiphase composite structures showed that curved thermal cracks only are realized in the experiments because the interface cracks do not reach the critical values of the strain energy release rate valid for the components of the glassy compounds.

REFERENCES

|1| England, A., 'An arc crack around a circular elastic inclusion', J. Appl. Mech. 33 (1966), 637-640.
|2| Toya, M., 'A crack along the interface of a circular inclusion embedded in an infinite solid', J. Mech. Phys. Solids 22 (1974), 325-348.
|3| Herrmann, K., 'Curved thermal crack growth in the interfaces of a unidirectional Carbon-Aluminum composite', Mechanics of Composite Materials, Recent Advances (Eds. Z. Hashin and C. T. Herakovich), Pergamon Press, New York (1983), 383-397.
|4| Cotterell, B. and Rice, J.R., 'Slightly curved or kinked cracks', Intl.J. of Fracture 16 (1980), 155-169.
|5| Bergkvist, H. and Guex, L., 'Curved crack propagation', Intl. J. of Fracture 15 (1979), 429-441.
|6| Nemat-Nasser, S., 'On stability of the growth of interacting cracks, and crack kinking and curving in brittle solids', Numerical Methods in Fracture Mechanics (Eds. D. R. J. Owen and A. R. Luxmoore), Pineridge Press, Swansea (1980), 687-706.
|7| Palaniswamy, K. and Knauss, W. G., 'On the problem óf crack extension in brittle solids under general loading', Mechanics Today 4

(Ed. S. Nemat-Nasser), Pergamon Press, New York (1978), 87-148.

|8| Kerkhof, F., Bruchvorgänge in Gläsern, Verlag der Deutschen Glastechnischen Gesellschaft Frankfurt (Main) (1970).

|9| Hieke, M., 'Die Rißentstehung in Glasscheiben unter Eigenspannungen', Z. Naturforsch. 15a (1960), 543-546.

|10| Hieke, M. and Loges, F., 'Das Zerreißen von Gläsern unter dem Einfluß definierter Eigenspannungsquellen', Z. Angew. Phys. 22 (1966), 14-19.

|11| Blauel, H., Ph. D. Dissertation, Freiburg University (1970).

|12| Karihaloo, B. L. and Nemat-Nasser, S., 'Thermally induced crack curving in brittle solids', Analytical and Experimental Fracture Mechanics (Eds. G. C. Sih and M. Mirabile), Sijthoff & Noordhoff, Alphen aan den Rijn (1981), 265-272.

|13| Grebner, H. and Herrmann, K., 'Quasistatische Ausbreitung eines gekrümmten Wärmespannungsrisses in einem Zweikomponentenmaterial'. DVM-Berichte, 12. Sitzung des AK Bruchvorgänge, Freiburg (1980), 207-214.

|14| Grebner, H., Ph. D. Dissertation, Paderborn University (1983).

|15| Herrmann, K. and Grebner, H., 'Curved thermal crack growth in a bounded brittle two-phase solid', Intl. J. of Fracture 19 (1982), R 69 - R 74.

|16| Herrmann, K. P. and Grebner, H., 'Curved thermal crack growth in nonhomogeneous materials with different shaped external boundaries', I. Theoretical results, Theoretical and Applied Fracture Mechanics 2 (1984), 133-146.

|17| Herrmann, K. P. and Grebner, H., 'Curved thermal crack growth in nonhomogeneous materials with different shaped external boundaries', II. Experimental results, Theoretical and Applied Fracture Mechanics 2 (1984), 147-155.

|18| Herrmann, K. P. and Grebner, H., 'Slow thermal crack growth in thermally loaded two-phase composite structures containing inner stress concentrators', Life Assessment of Dynamically Loaded Materials and Structures (Ed. L. Faria) 2, Laboratório Nacional de Engenharia e Tecnologia Industrial, Lissabon (1984), 691-700.

|19| Herrmann, K. P. and Grebner, H., 'Quasistatic thermal crack extension in the interfaces of bounded self-stressed multiphase compounds', Theoretical and Applied Fracture Mechanics 4 (1985), 127-135.

|20| Hieke, M. and Herrmann, K., 'Über ein ebenes inhomogenes Problem der Thermoelastizität', Acta Mechanica 6 (1968), 42-55.

|21| Muskhelishvili, N. I., Einige Grundaufgaben zur mathematischen Elastizitätstheorie, Fachbuchverlag Leipzig (1971).

|22| Herrmann, K. and Grebner, H., 'Über ein Randwertproblem der Thermoviskoelastizitätstheorie', ZAMM 60 (1980), T 120 - T 122.

|23| Rybicki, E. F. and Kanninen, M. F., 'A finite element calculation of stress intensity factors by a modified crack closure integral', Eng. Fracture Mechanics 9 (1979), 931-938.

FRACTURE MECHANICS APPLIED TO WOOD

Ph. JODIN
G. PLUVINAGE
LABORATOIRE DE FIABILITE MECANIQUE
FACULTE DES SCIENCES UNIVERSITE DE METZ
ILE DU SAULCY
F-57045 METZ CEDEX
FRANCE

ABSTRACT. The goal of this paper is to show the applicability of fracture mechanics concepts to a very anisotropic medium : wood. Because of its behavior, fracture of wood can be analysed with the elastic fracture mechanics principles. Then two important applications are shown such as chipping and gluing. With respect to the variable nature of wood, probabilistic fracture mechanics is a very powerful tool to analyse the safety of wooden structures.

1. INTRODUCTION.

Usually, fracture mechanics is associated with homogeneous and isotropic media. This is an idealized view of most man-made materials. In opposite, most of natural materials are non-homogeneous and non-isotropic. Meanwhile, they are very extensively used as building materials. This is really the case of wood which has been used in this way since times out of mind.

Wood has a very interesting ratio of mechanical properties to density with respect to other materials (Table I). It can be seen that wood has a "quality factor" of the same level as a building steel, but the energy associated with wood industry is ten times lower than for other material. This economic advantage is conforted by the fact that wood, as a raw material, will be always available (provided our forests are not destroyed by accident, pollution or unrationalized tapping).

All these advantages are compensated by some disadvantages, the most important of them being the high variability of the material. This results generally in very high safety factors which affect considerably the economic interest of wood products. Moreover, wood, as a natural material, contains many natural defects, such as knots for instance, which must be taken into account for the evaluation of mechanical properties.

From this point of view, fracture mechanics and probabilistic fracture mechanics can constitute useful tools for the estimation with a

K. P. Herrmann and L. H. Larsson (eds.), Fracture of Non-Metallic Materials, 207–226.
© *1987 by ECSC, EEC, EAEC, Brussels and Luxembourg.*

Matérial	:	Tensile strength MPa	:	Breaking length m
	:		:	
Steel wire, maximum	:	320	:	41.10^3
Iron wire, hard drawn	:	550 – 840	:	7 – 10.8
Steel, building	:	520 – 620	:	6.7 – 8
Copper wire, hard drawn	:	420 – 490	:	4.7 – 5.4
Flax, irish	:	600 – 1100	:	40 – 75
Rayon, acetate	:	1000	:	75
Silk	:	350	:	25
Cotton	:	280 – 800	:	18 – 53
Hemp	:	800 – 900	:	52 – 58
Coniferous woods	:	50 – 150	:	11 – 30
Broadleaved woods	:	20 – 260	:	7 – 30
Bamboo	:	100 – 230	:	10 – 35

Table I. Tensile strength and breaking length of some materials / 12 /.

better precision of the available engineering properties. The scope of this lecture is to show the applicability of fracture mechanics to wood and to give an exemple of an application of probabilistic fracture mechanic to a wooden structure.

2. STRUCTURE OF WOOD FROM A MECHANICAL POINT OF VIEW

The description of the structure of wood may depend on the unit volume considered by the observer. For mechanical engineering applications, a cubic unit volume including three to four growth rings, in the radial direction, seems to be an adequate scale. Of course, with the aid of ultramicroscopic techniques, it is now possible to observe objects of a macromolecular size, which corresponds to the smallest structural unit of wood.

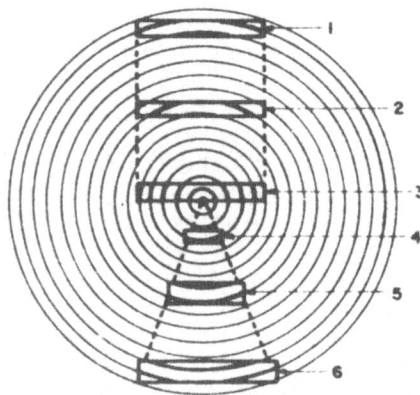

Fig.1 Influence of the position in the log for the growth rings.

Let us consider a large rectangular piece of wood (Fig.1). It can be considered as extracted from a cylindrical trunk of a tree. Of course, the orientation of the growth rings depends on the position in the log and on the orientation of the sample. But the principal directions of material symmetry can be already defined :
- the longitudinal direction (L) which coincides with the axis of the trunk, also called grain direction. This is the direction of the major length of wood cells.
- the radial direction (R) which coincides with the radius of the trunk. This is also the direction of wood rays.
- the tangential direction (T) which coincides with a tangent to annual growth rings.

In a board sawn in a log, there is one longitudinal direction and an infinity of radial and tangential directions. If the elementary volume is taken out at a large distance from the center, its width being small enough, the growth rings can be considered as planes and, therefore the material's symmetry is plane orthotropy. This stage of precision is enough to describe the wood behavior for engineering applications.

3. BASICS OF LEFM APPLIED TO ANISOTROPIC MEDIA.

If a piece of dry wood (that is to say the relative moisture content of which is about 10 to 15 per cent) is loaded slowly by a single force, the load displacement diagramm which is recorded is like that given in Fig.2. It can be seen that it exhibits a linear portion, corresponding to a linear elastic behavior. But the slope of the straigth portion depends highly on the orientation of the load with respect to the principal directions of material symmetry.

Let us at first recall what is exactly an anisotropic elastic medium. Consider the strain and stress tensors ε_{ij}, σ_{kl}. According

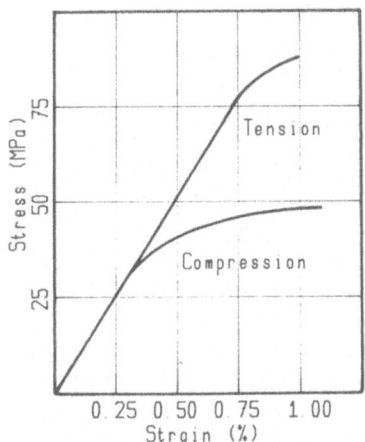

Fig.2 Stress-strain diagramm for wood / 12 /.

to Hooke's law, they are related together by a fourth order tensor denoted S_{ijkl} and called the compliance tensor. The 3-D relationship between ε_{ij} and σ_{kl} is :

$$\varepsilon_{ij} = S_{ijkl} \cdot \sigma_{kl} \tag{1}$$

with $i,j,k,l = 1,2,3$ where $1,2,3$ refer to the three axis of space reference. With a simplified notation (1) becomes :

$$\varepsilon_i = S_{ij} \cdot \sigma_j \tag{2}$$

with $i,j = 1$ to 6.

The successive introductions of materials symmetries produce zeroing of some terms of the compliance tensor, which, for plane orthotropy, is written as :

$$
\begin{vmatrix}
S_{11} & S_{12} & S_{13} & 0 & 0 & 0 \\
S_{21} & S_{22} & S_{23} & 0 & 0 & 0 \\
S_{31} & S_{32} & S_{33} & 0 & 0 & 0 \\
0 & 0 & 0 & S_{44} & 0 & 0 \\
0 & 0 & 0 & 0 & S_{55} & 0 \\
0 & 0 & 0 & 0 & 0 & S_{66}
\end{vmatrix}
\tag{3}
$$

In this table, the non-diagonal terms are symmetrical and the components can be expressed in an engineering manner as :

$$S_{11} = 1/E_1 \ , \ S_{22} = 1/E_2 \ , \ S_{33} = 1/E_3 \ ,$$

$$S_{44} = 1/G_{23} \ , \ S_{55} = 1/G_{13} \ , \ S_{66} = 1/G_{12} \ ,$$

$$S_{12} = S_{21} = -\upsilon_{21}/E_2 = -\upsilon_{12}/E_1 \ , \tag{4}$$

$$S_{13} = S_{31} = -\upsilon_{31}/E_3 = -\upsilon_{13}/E_1 \ ,$$

$$S_{23} = S_{32} = -\upsilon_{32}/E_3 = -\upsilon_{23}/E_2$$

where E_i are Young's modulus, G_{ij} shear modulus and υ_{ij} Poisson's ratio.

Of course, the measuring techniques of these nine components of the compliance matrix are not easy to use. An extensive study of this problem has just been published / 1 /.

Let us now consider a singular crack in an anisotropic body. The stress field around the crack tip will be highly dependent on the crack orientation and on the applied loads. For more simplicity, the crack will be assumed to lay in a plane of symmetry. Considering the crack front orientation and the plane of the crack, a notation system is generated which is described on Fig.3.

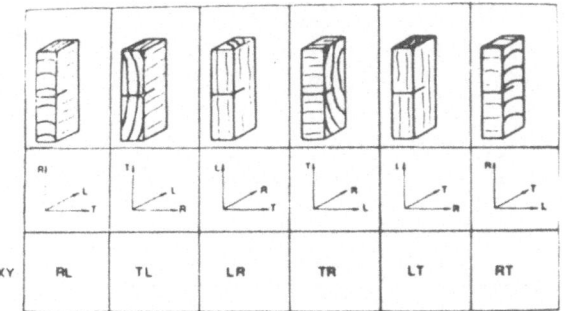

Fig.3 Systems of sampling orientation /3/

When the load is applied to the cracked body, if it is assumed that the strain energy is entirely recoverable, that is to say the non-recoverable strains are concentrated enough around the crack tip to be negligible, the critical strain energy release rate can be used as a fracture criterion, provided the crack orientation is specified. The measurements of G_{IC} can be achieved using the compliance method. Briefly, it can be estimated using the relationship /2/

$$G_{1c} = \frac{P_c^2}{2B} \frac{\partial C}{\partial a} \qquad (5)$$

where P_c is the load at fracture, B the specimen thickness, C the compliance and a the crack length. Strictly speaking, this relationship is only valid when the crack propagates in its plane. For some species /3/ in specified directions (TL or RL), the values of G_{IC} have been given.

When the crack lies in a plane perpendicular to the grain direction (LT or LR specimens), compliance experiments can be achieved and a "pseudo" strain energy release rate G_c can be derived. But, since the crack doesn't propagate in its original plane but in a perpendicular direction, the value of G_c must be taken very cautiously /3/ and is not a valid G_{IC} value.

Sih et al. /4/ have published a paper on the stress distribution around the tip of a singlular crack in an anisotropic body where the authors have established the expression of stresses in such a material. They also derived a relationship between K_I and G_I provided the crack propagates in its original plane. In plane stress conditions, this relationship takes the form :

$$G_{Ic} = S^* \cdot K_{Ic}^2 \qquad (6)$$

$$S^* = \left[\frac{S_{11} \cdot S_{22}}{2}\right]^{\frac{1}{2}} \cdot \left[\left(\frac{S_{22}}{S_{11}}\right)^{\frac{1}{2}} + \frac{2S_{12} + S_{66}}{2S_{12}}\right]^{\frac{1}{2}} = S_a^* \cdot S_b^* \qquad (7)$$

where 1 and 2 refer, respectively, to the direction of crack propagation and to normal direction. S_{ij} are taken from the compliance matrix given in (3).

Sampling	TL (m^2/N)	LT (m^2/N)
S_{11}	0.714×10^{-11}	8.62×10^{-11}
S_{22}	8.62	0.714
S_{12}	-0.370	-0.370
S_{66}	9.28	9.28
S^*	5.68	1.63

Table II. Values of compliance parameters for beech /3/.

This formula is established for mode I. For mode II and mode III there are equivalent relationships given in the same publication.
They are only valid if the crack propagates in its own plane, which can be the case for wood, if the crack lies in a TL or RL plane, instead of what happens in isotropic homogeneous materials.
The details of measurement's techniques can be found in /3/. Briefly, DCB specimens are cut off from boards, with a special attention to growth rings orientation. These specimens are notched with a saw cut sharpened with the aid of a razor blade. From each specimen, it is possible to derive a compliance versus crack length plot, which is fitted with an exponential function :

$$C = A.\exp(Ba) \qquad (8)$$

from which it is easy to derive (fig.4) :

$$dC/da = AB.\exp(Ba) \qquad (9)$$

For each crack length, it is then possible to compute a G_{Ic} value.

A mean value $\overline{G_{Ic}}$ and its standard deviation σ is derived from all the the experiments. It can be shown (Fig.5) that the critical strain energy release rate is independent on the crack length.
The conversion from G_{Ic} to K_{Ic} is easy with formula (6), provided

the S_{ij} are adequately measured (Table II).

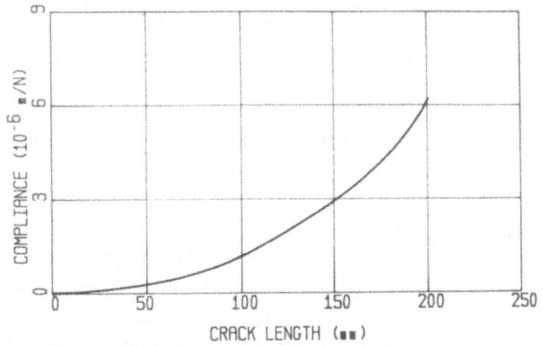

Fig.4 The compliance curve for a DCB specimen /3/.

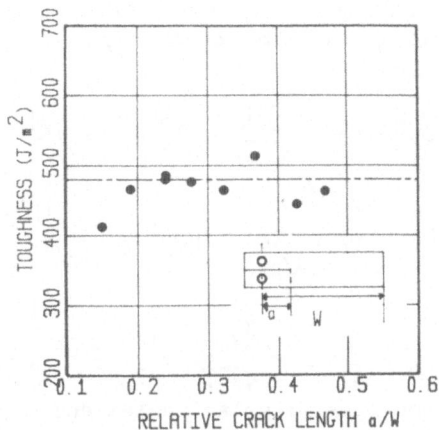

Fig.5 Toughness versus crack length /3/.

4. APPLICATION TO SOME PROBLEMS

As wood is very extensively used in structures and, additionally, as a natural material, exhibits many defects, it seems evident that fracture mechanics can be used for the analysis of a lot of fracture problems. In this range are found :

. problems of safety of structures depending on the noxiousness of some defects such as knots, cracks or notches.

. problems of fracture due to drying or to growth stresses relaxed during sawing or machining

. problems of the toughness of glued joints which are very widely used in wood industry.

. problems of machining including splitting, planing, slicing and cutting.

It is necessary to distinguish problems related to green wood from those related to dry wood. Indeed, green wood does not behave as a linear elastic material. In this work we just select problems treated by

LEFM and, consequently, the problems of wood drying or of growth stresses will not be adressed here. They need a non-linear viscoplastic analysis which is out of the scope of this lecture. In the remaining sections, we have selected to give an analysis of the chip production process, the toughness of a glued joint and the noxiousness of knots which will be presented in section 5.

4.1. Chipping

The production of a chip during a cutting operation is schematized in figure 6. A crack is formed in front of the tool and the different loads exerted by the faces of the tool on the rear part of the chip result in crack advance, i.e chip production. It is evident from figure 6 that the crack is submitted to mode I + mode II opening, the ratio of mode I over mode II depending on the geometry of the tool. By using the stress equations given by Sih et al. /4/, it is possible to derive the stress field around the crack tip, i.e at the point where the chip is formed :

$$\sigma_{xx} = \frac{K_I}{\sqrt{2\pi r}} \cdot Re_x^I \ \big[s_{ij}, \psi \big] + \frac{K_{II}}{\sqrt{2\pi r}} \ Re_x^{II} \ \big[s_{ij}, \psi \big] \tag{10}$$

$$\sigma_{yy} = \frac{K_I}{\sqrt{2\pi r}} \cdot Re_y^I \ \big[s_{ij}, \psi \big] + \frac{K_{II}}{\sqrt{2\pi r}} \ Re_y^{II} \ \big[s_{ij}, \psi \big] \tag{11}$$

$$\tau_{xy} = \frac{K_I}{\sqrt{2\pi r}} \cdot Re_{xy}^I \ \big[s_{ij}, \psi \big] + \frac{K_{II}}{\sqrt{2\pi r}} \ Re_{xy}^{II} \ \big[s_{ij}, \psi \big] \tag{12}$$

The Re ( ) terms are functions of polar angle ψ and of elastic constants. The estimation of stresses needs the evaluation of stress intensity factors K_I and K_{II}. They are related to the cutting loads and the

solution used is that proposed by Tada et al. /6/ :

$$K_I = C \ \frac{\sqrt{6}}{b \ t^{3/2}} \ (P_v \cdot a - P_H \cdot h) \tag{13}$$

and

$$K_{II} = C \ \frac{P_H}{\sqrt{2b^2 t}} \tag{14}$$

where C = 1 fore plane stress

$$C = \sqrt[4]{1 - \nu^2} \text{ for plane strain.}$$

b is the specimen thickness, t is the chip thickness and the other parameters are defined in figure 6.
K_{IC} is determined using DCB specimens and the compliance method descri-

bed previously. Since there are no recommanded specimens for K_{IIc} determination, this parameter can be estimated using a relationship given by Williams and Birch /2/. It derives from the computed G values for mixed modes.

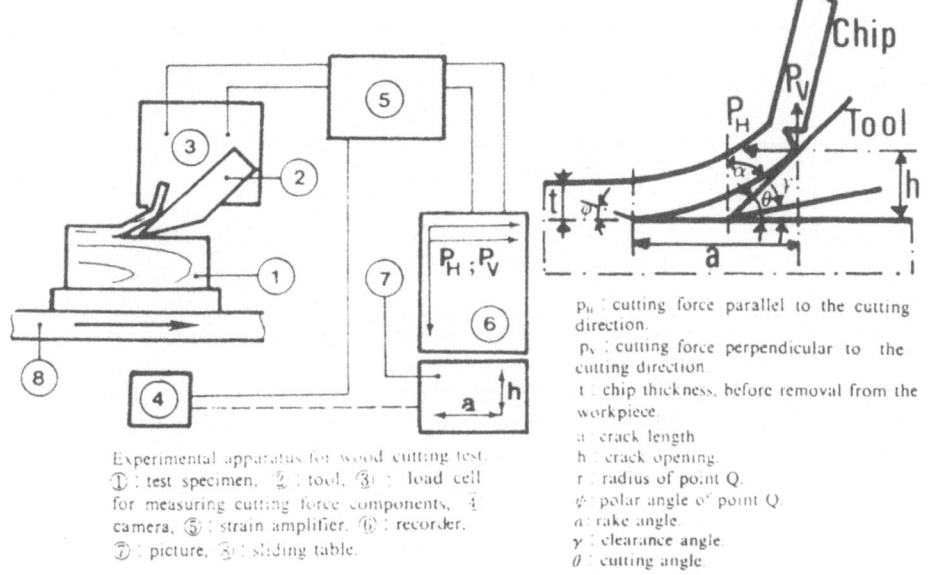

Experimental apparatus for wood cutting test.
① : test specimen. ② : tool. ③ : load cell for measuring cutting force components. ④ : camera. ⑤ : strain amplifier. ⑥ : recorder. ⑦ : picture. ⑧ : sliding table.

Fig.6 Experimental parameters for chipping experiments /5/.

$$G_I = K_I^2 . S^* \tag{15}$$

$$G_{II} = K_{II}^2 . \frac{S_{11}}{\sqrt{2}} . S_b^* \tag{16}$$

with $G = G_I + G_{II}$ (17)

This gives :

$$G = S^* \left[K_I^2 + \left(\frac{S_{11}}{S_{22}} \right)^{\frac{1}{2}} K_{II}^2 \right] \tag{18}$$

Since the crack propagates always in the TL plane, although there is a mixed mode loading system, it can be reasonably assumed that the fracture energy G is that required for the creation of a new TL surface, that is to say the G_{Ic} value related to the RL loading system.

Using eq.(6) and taking from table III for beech K_{Ic} = 1 MPa$\sqrt{m}$, one obtains from (18) a relationship between K_I and K_{II} :

$$K_I^2 + 0.328 \, K_{II}^2 = 1 \, (MPa)^2 .m \tag{19}$$

	Elastic constants (m^2/N)				Fracture Toughness MPa$\sqrt{m}$	
Specimen	$S_{11}=\dfrac{1}{E_x}$	$S_{22}=\dfrac{1}{E_y}$	$S_{12}=\dfrac{-\nu_{xy}}{E_x}$	$S_{66}=\dfrac{1}{G_{xy}}$	K_{Ic}	K_{IIc}
Spruce	5.58×10^{-11}	8.18×10^{-10}	-2.49×10^{-11}	1.37×10^{-9}	0.659	1.280
Beech	8.00×10^{-11}	7.40×10^{-10}	-3.20×10^{-11}	1.00×10^{-9}	1.000	1.740

Table III. Experimental parameters for chipping experiments /5/.

It is then possible to draw an intrinsic fracture curve for wood in mixed mode (Fig.7). The fig.8 gives the values of K_I and K_{II}

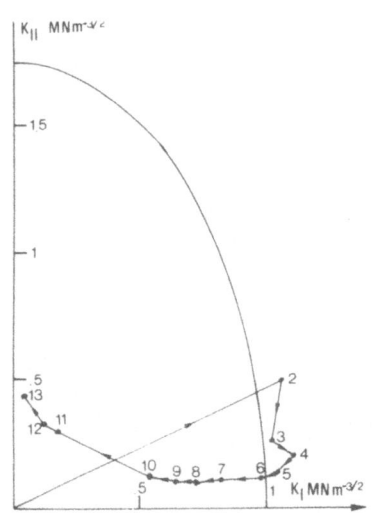

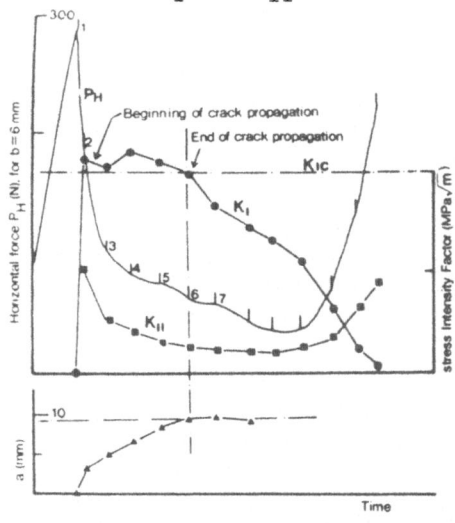

Fig. 7 Mixed mode fracture curve for chipping experiments for beech /5/.

Fig.8 Experimental values of K_I and K_{II} for beech from /5/

for the cutting experiments on beech with a cutting angle of 30 degrees. The loading pattern is given on Fig.7. It can be easily shown that propagation of the crack should occur between point 2 to 6, because they are on the rigth side of the intrinsic curve. This corresponds exactly to the experimental observations with the camera (Fig 8).

It is clear from the results that the chip is formed when the combination of stress intensity factors K_I and K_{II} is beyond an intrin-

sic fracture curve for mixed mode I + II. Practically, for a particular geometry of the tool, it is possible to derive the loads exerted on it, if the critical stress intensity factors are known. Moreover, a factor of machinability can be easily derived.

4.2. Gluing

Gluing is a popular process for jointing. In wood construction, it is used mainly in furniture, but now also in carpentry. The wellknown glue-laminated timbers works are an example of the structural use of gluing. The manufacturer of such structures has to avoid the presence of gluing defects due to the manufacturing itself such as lack of glue deposition, bubbles or due to unexpected cracking originated by shrinkage, for instance. The noxiousness of these defects located in the glue joint may be estimated with the aid of fracture mechanics. Moreover, the fitness of the glue to the species jointed can be also estimated with the aid of fracture mechanics /7/.

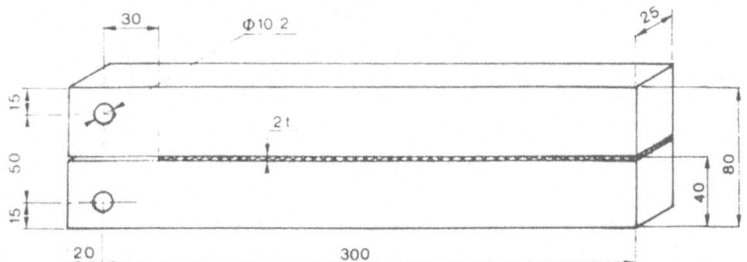

Fig.9 DCB specimen for the study of gluing /7/.

For solid wood DCB specimens, K_{Ic} is derived from the G_{Ic} value

experimentally measured with relation (6). This derivation is no more possible for glued specimens (Fig.9), because the equivalent Young's modulus E* taken into account in this relation is not definable in this case. Moreover, the stresses in the glue joint, which can be, in

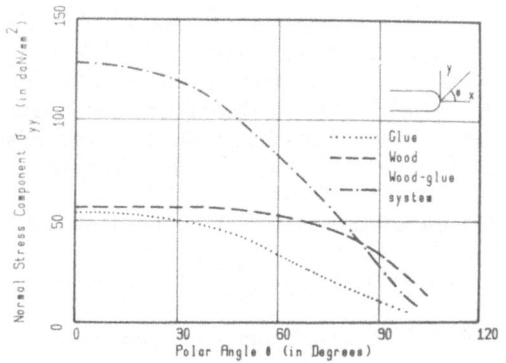

Fig.10 Stress distribution for the glue.

a first step, considered as an isotropic homogeneous material, may be widely influenced by the very near presence of an orthotropic substrate. If the stress field around the crack in the glue can be defined, then, the stress intensity factor can be derived. This is the process which is to be exposed in this paper.

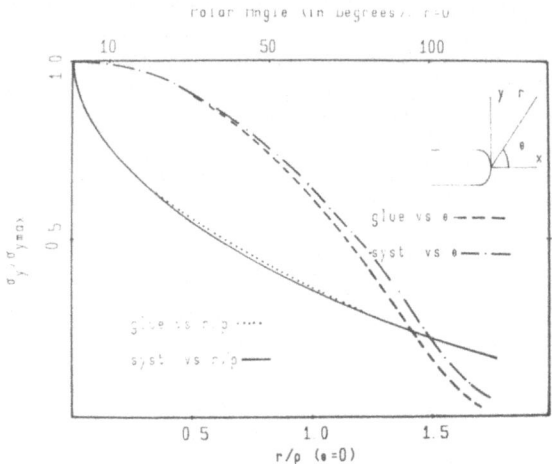

Fig.11 Normalized stress distribution for the glued system.

Let us first recall that the stress distribution around the crack in a DCB specimen made from solid wood can be calculated with the aid of relations (10) to (12). The same calculation can be made for an isotropic medium such as the glue, even with a blunted crack tip, using the relations given by Crager and Paris /8/. For a thin glued joint, between two large pieces of wood (Fig.9), the calculation of the stress distribution has been made by a finite element computation. The main results of these calculations are that the shapes of the stress

distributions (σ^{iso} for the pure glue and σ^{syst} for the glued system) are the same, the stress levels being different (Fig.10 and 11). Similarly, the stress distributions along a line perpendicular to the glue joint are equivalent to an isotropic distribution in the glue, exhibit a "jump" when they pass through the interface glue–wood, and, in the wood, are equivalent to an orthotropic stress distribution (Fig.12).

From these considerations, it can be postulated that :

$$K_I^{iso}(glue) = K_I^{ortho}(wood). \ \sigma_{max}^{syst}/\sigma_{max}^{iso} \qquad (20)$$

Thus, if it is assumed that crack propagation occurs in the glue :

$$K_{Ic}^{iso} = K_{Ic}^{a} \qquad (21)$$

where K_{Ic}^{a} is the critical stress intensity factor for the glue

Thus, $K_{Ic}^{ortho} = K_{Ic}^{eq}$ (22)

where K_{Ic}^{eq} is the apparent critical stress intensity factor for the system. Consequently, equation (20) can be rewritten as :

$$K_{Ic}^{eq} = K_{Ic}^{a} \cdot \sigma_{max}^{iso} / \sigma_{max}^{syst}$$ (23)

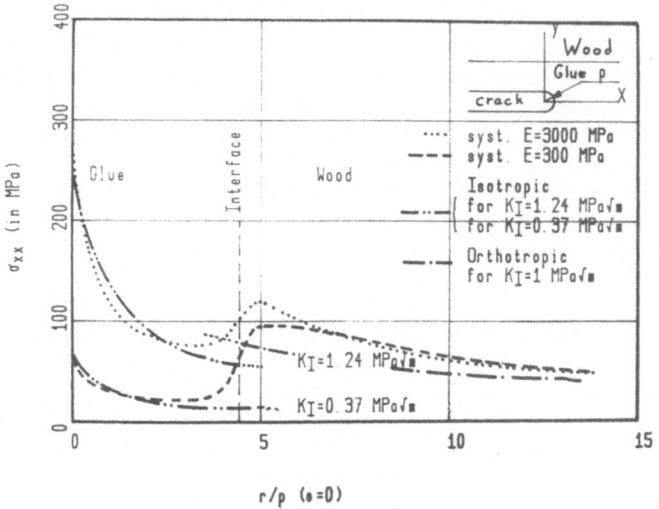

r/p (θ=0)

Fig.12 Stress distribution for the glued system with interface.

From figure 13 it is clear that the square of σ_{max}^{syst} is proportionnal

to the modulus of elasticity (MOE) of the glue. Thus, the ratio $(\sigma_{max}^{syst}/\sigma_{max}^{iso})^2$ is proportionnal to the ratio E/E^* of the MOE of the glue

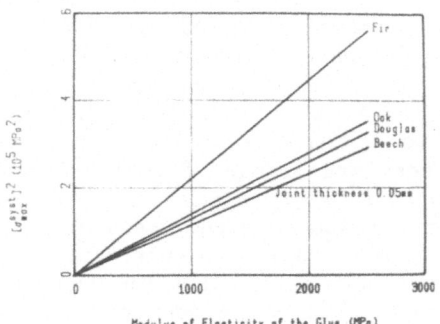

Modulus of Elasticity of the Glue (MPa)

Fig.13 Evolution of maximum stress with modulus of elasticity of the glue and of the wood.

over the equivalent modulus of elasticity of the wood. Finally :
$$K_{Ic}^{eq} = C \sqrt{E/E^*} \, K_{Ic}^a \tag{24}$$
where C is a coefficient depending on the geometry of the glue joint and, particularly, on its thickness.

This result is very interesting, because it shows the mechanical interaction between the adhesive and the substrate. It allows also to predict the thoughness of a glue joint from the toughness of the glue and from the mechanical properties of the materials jointed. This phenomenom is called "the rheological mutation of the glue".

It has been shown in this part how simple basic elements of fracture mechanics applied to orthotropic media can be used to modelize some popular manufacturing processes which imply the presence of geometrical defects. Of course, wood is a so variable material that very sharp numerical values are a nonsense and the laws derived from these studies must be considered cautiously. But it is clear that fracture mechanics may be a powerful tool to improve these processes and, finally, to produce a better material with a better yield.

5. PROBABILISTIC FRACTURE MECHANICS APPLIED TO A WOODEN STRUCTURE

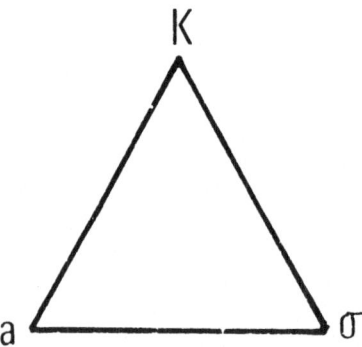

Fig. 14 Fundamental relationship of fracture mechanics.

The reliability of a structure may be schematized by the popular triangle defect-load-toughness (Fig.14). If two of these data are well defined then the third one is also accurately determined, say, for instance, for a given value of toughness, at a given load, the critical crack dimension is well defined. But, for real materials, and very especially for wood, these parameters are not easily defined and are given with a rather large range of scattering around a mean value. Let us treat an example /9/.

Consider the popular wooden structure which is a pallet (Fig.15). It is constituted by boards nailed on bearings. The load is constituted by the goods placed on the pallet and the support may be of various types such as the ground floor, the fork of a truck. The raw material constituting the floor of the pallet is generally a low grade wood. This means that there are many defects such as knots in the boards. Moreover, the mechanical properties vary considerably from a board to another. This means that for a given loading, say the nominal load, the

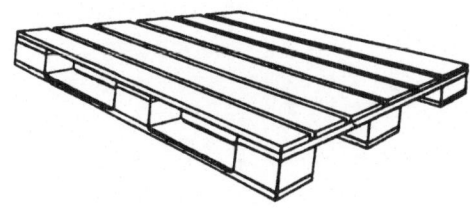

Fig.15 Pallet

failure of the structure is not sure. It must be expressed in terms of probability of failure.

The problem consists in observing the probability of occurrence of a defect of critical size, or greater, in any position in a board under a given load. This problem is rather complex because it implies the analysis of :

- the stress distribution in any part of the pallet,
- the probability of occurrence of a knot in any part of the pallet
- the probability of a knot, if any, to be of critical size.

It can be treated following two paths. In the first one, the problem is simplified by adding the following assumption :

- the fracture occurs at the most loaded point in the most loaded board.

The steps in the solution of the problem are thus :

1) Determine the influence of a knot on the strength of a board.

2) Determine the most loaded board and, in this board, the most loaded point.

3) Determine the probability of occurrence of a critical knot in this point.

4) Compute the probability of failure of the pallet.

This first point is an interesting application of fracture mechanics to wood. A non-adherent knot can be, at first, compared with a circular hole (Fig.16). Fracture always originates at the point quoted A. It can be shown (Fig.17) that the critical stress is inversely proportionnal to the square root of the relative diameter of the hole. This implies that interpretation in terms of fracture mechanics is possible.

According to Bowie /10/, the stress intensity factor of a small crack of length Xc, assumed to be a characteristic of the material, emanating from a hole of diameter Øa is :

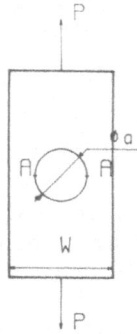

Fig. 16 Fracture mechanics specimens with knots

$$K_I = \sigma_0 . \sqrt{\pi X_c} \ . \ \sqrt{\Phi_a / 2X_c + 1/2}$$

(25)

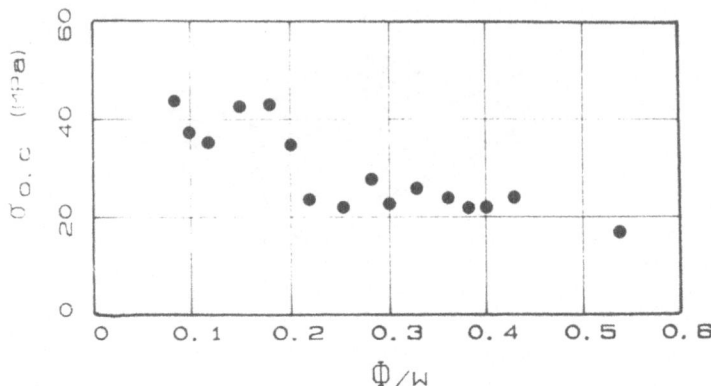

Fig.17 Fracture stress versus diameter of the hole /9/.

where σ_0 is the applied stress. In critical conditions, (25) becomes

$$K_{Ic} = \sigma_{0,c} \cdot \sqrt{\pi X_c} \cdot \sqrt{\phi_a/2X_c + 1/2} \qquad (26)$$

or $\quad K_{Ic}/\sqrt{\pi X_c} = \sigma_{0,c} \cdot \sqrt{\phi_a/2X_c + 1/2} \qquad (27)$

In this equation, the first member is equivalent to the fracture stress of an infinite plate containing a small defect of length Xc. Since this length is very small, it can be assumed that the fracture stress of the plate is equal to the ultimate stress of the material σ_u. Then (27) gives :

$$\sigma_u/\sigma_{0,c} = \sqrt{\phi_a/2X_c + 1/2} \qquad (28)$$

or $\quad \phi_a = 2X_c \left(\sigma_u/\sigma_{0,c} \right)^2 - X_c \qquad (29)$

If the ratio $\left(\sigma_u/\sigma_{0,c} \right)^2$ is not smaller than 5, then Xc in the second

member of (29) may be neglected with respect to the first term and (29) becomes :

$$\sqrt{\phi_a} \cong \sqrt{2X_c} \left(\sigma_u/\sigma_{0,c} \right) \qquad (30)$$

which shows that ϕ_a versus $\left(\sigma_u/\sigma_{0,c} \right)^2$ is a straigth line. This has

been verified experimentally (Fig.18). This curve is typical of a given material in given environnemental conditions and defines the allowable knot diameter ϕ_a for a given load σ_0.

The second point of the analysis, the determination of the most loaded point in the board, is rather simple. This can be done by experimental, analytical or numerical methods. In the example presented here, this has been done by a measurement of the strains in a border board when the pallet is supported on each corner and the result is shown

in Fig.19. The most loaded point is taken as shown. The level of stress
is computed using a simple Hooke's law (E = 15000 MPa). This stress $\sigma_{0.c}$
allows to determine via the characteristic curve defined before (Fig.18)
the critical diameter of a knot $\emptyset_c$ which in this case turns out to be
28 mm.

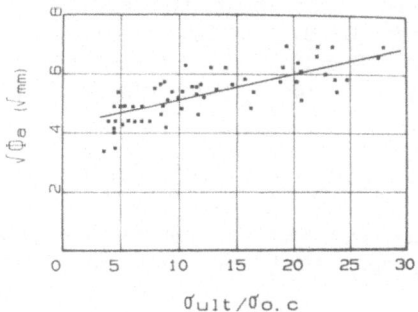

Fig.18 Characteristic curve of the influence of knot on the fracture
stress of a board /9/.

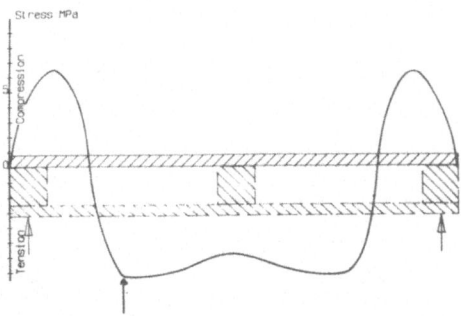

Fig.19 Stresses in a border board of a pallet /9/.

It is then necessary to estimate the probability of occurrence of a
knot in this place and the probability of this knot to be of critical
size (Third point). This is a manufacturing problem in the sense that
this estimation can be done only on the raw material used in the
factory. This has been done on a sample of 648 boards used for pallets
where the position and the diameter of the knots, if any, have been
reported on each board. The following distribution diagram (Fig.20) has
been obtained and is considered as a characteristic of the material
used in the factory considered. Apparently, there is no privileged
position and the probability of occurrence of a knot in a portion of
length of 40 mm is P=0.0694, wherever it may be in the board which
length is 1.20 meters. The next step, which is the probability of a
knot to have a diameter greater than the critical one, is determined
from the graph of the cumulated frequency versus diameter (Fig.21).
There, this probability is given by a curve fitting procedure on the
histogramm :

$$P\ (\emptyset \geqslant \emptyset_c) = (1 + A.\emptyset_c)\ exp\ (-A.\emptyset_c) \qquad (31)$$

Where A is a parameter describing the cumulated distribution of knots diameter and is here 0.08. It is obtained by fitting Eq.31 to the experimental results (Fig.21). One finds that, for the example here described, $P(\emptyset > \emptyset_c) = 0.345$. Then, the probability of failure of a pallet is simply :

$$P_F = P.P\ (\emptyset > \emptyset_c) = 0.024 = 2.4\ per\ cent$$

This is the simplest way to compute the probability of failure of a pallet. A more complicated one, but may be more accurate, can be explored.

The first step of the preceding analysis remains the same, but, after that, it is assumed that the critical knot can be found in any place of the board. If y is the abscissa along the board, after the determination of the stress along the board $\sigma_{o(y)}$ (Fig.19), the critical knot is represented by the function $\emptyset_c(y)$ since :
$$\emptyset_c = f(\sigma_u/\sigma_{o,c}(y))$$

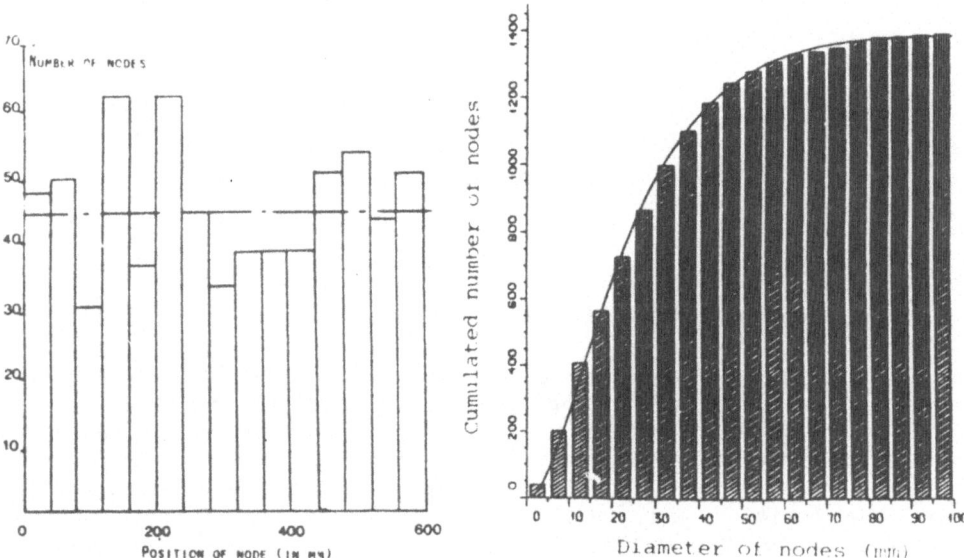

Fig. 20 Histogram of distribution of knots in 648 boards /9/.

Fig.21 Histogram of cumulated distribution of knot size /9/.

If the theory of Jayatilaka and Trustrum /11/ is used, the board's length is divided into successive "links", each link being associated with a knot. The probability of rupture of one link is equal to the probability of a knot to be greater than the critical one multiplied by the probability of occurrence in the link.

$$P_{R1} = P.P\ (\emptyset\ \ \emptyset_c(y)\ ) \tag{32}$$

Then the probability of survival of a board is the product of the probability of survival of each link. There are N links in the board, then :

$$P_{SN} = (1 - P_{R1})^N \tag{33}$$

Then, the probability of rupture of the board is :

$$P_{RN} = 1 - P_{SN} = 1 - (1 - P_{R1})^N \tag{34}$$

If N is large then (34) becomes :

$$P_{RN} = 1 - exp\ (-\ NP_{R1}) \tag{35}$$

This relation can be set in an integral form all over the board (Jayatilaka's theorem /11/), using eq. (31).

$$P_{RN} = 1 - exp[1 - P.\int_0^1 A^2 \Phi_c(y).e^{-A\Phi_c(y)}.dy] \tag{36}$$

where

$$\emptyset_c\ (y) = \left[\frac{7.584}{\sigma_o(y)} + 4.24\ /^2\right] \tag{37}$$

and $\sigma o(y)$ is a polynomial fit of the sixth degree of the curve given in fig. 19. P = 0.694.
In these conditions P_{RN} = 0.017 which is to be compared with the preceding value of 0.024.
It has been seen that, since wood is a very variable material, and especially when low grade pieces are used, probabilistic fracture mechanics can be used successfully. Moreover, the modelization of the behavior of a board containing a knot may be reasonably carried out with the aid of fracture mechanics, the knot being the most frequent defect in low grade boards. Of course, other defects like cracks are quite frequent, and can be evidently treated by fracture mechanics. Decay or inclined grain affect the strength of the material and are not treated by fracture mechanics. The scope of probabilistic fracture mechanics applied to wood is to introduce in manufacturing of wooden structures woods of low grade in view to lower the costs of these structures and to use as best the cut down trees.

6. CONCLUSION.

It has been shown that fracture mechanics, in its fundamental principles, rules the relations between a material characterized by its toughness, containing defects and an applied load. For brittle materials, this relation has a local character and is very sensitive to orientation of stresses with respect to crack planes and to anisotropy of material's properties.

Since wood is a very anisotropic material and exhibits a brittle behavior, techniques of toughness measurements have been applied successfully, provided the respective orientations of load and crack planes are well specified. It's also necessary to make some reserves in cases where the crack does'nt propagate in its original plane.

It has been shown that the principles and applications of fracture mechanics to wood find interesting developments in manufacturing applications such as machining or gluing. A better comprehension of the phenomena occurring near the crack tip could bring a helpful tool in the improvements of the processes.

It has been seen also that wood is a variable material, exhibiting many natural defects. The tendency is to use the best woods for structural applications and leave low grade ones for fire, paper pulp or particle boards. A statistical approach of fracture mechanics when low or medium grades are used in structures allows to give a failure probability which could be considered as acceptable when the safety of people is not under consideration.

REFERENCES

/1/ G. PLUVINAGE "Rapport de contrat : étude comparative de la mesure des constantes élastiques du bois" Rapport MRT 81.G.1058 (1985)

/2/ J.G. WILLIAMS and M.W. BIRCH "Mixed mode fracture in anisotropic media" Cracks and fracture, ASTM STP 601 (1976), 125-137.

/3/ P. TRIBOULOT, Ph. JODIN and G. PLUVINAGE "Validity of fracture mechanics concepts applied to wood by finite element calculation" Wood Science and Technology 18 (1984), 51-58.

/4/ G.C SIH, P.C PARIS and G.R IRWIN "On cracks in rectilinearly anisotropic bodies' Int. J. of Fracture 1 (1965) 3, 189-203.

/5/ P. TRIBOULOT, I. ASANO and M. OHTA "An application of fracture mechanics to the wood cutting process" Mokuzai Gakkaishi, 29 (1983), 2, 111-117.

/6/ H. TADA et al "The stress analysis of cracks handbook' Del Research Corporation, Hellertown, Pa. (1973).

/7/ G. DUCHANOIS, Ph. JODIN and G. PLUVINAGE "Mesure de la tenacité et étude du comportement des joints bois-colle" Rapport de contrat MIR N 83.G.0866 (1985).

/8/ M. GRAGER and P.C PARIS "Elastic field equations for blunt cracks with reference to stress corrosion cracking" Int. Fract. Mech. 3 (1961), 247-252.

/9/ D. BAJOLET "Application de la mécanique probabilistique de la rupture au problème des palettes en bois" Rapport de DEA. Université de Metz (1982).

/10/ O.L BOWIE "Analysis of an infinite plate containing radial cracks originating at the boundary of an internal circular hole' J. Math. and Phys. 35 (1956), 60.

/11/ A.S JAYATILAKA and K. TRUSTUM J. Mat. Sci., 13 (1978), 455.

/12/ F.F.P. KOLLMANN and W.A. COTE, Jr. "Principles of Wood Science and Technology" Springer Verlag, New York (1968).

FRACTURE MECHANICS OF POLYMERS AND ADHESIVES

J. G. Williams
Mechanical Engineering Department
Imperial College of Science and Technology
LONDON SW7 2BX
U.K.

ABSTRACT. A review of the use of Fracture Mechanics to characterise the toughness of polymers and adhesives is given. Examples are included of the determination of toughness at fracture initiation and the importance of stress state is emphasised. Size effects are discussed for both specimen dimensions and bond thickness in adhesives. Crack propagation is described for slow crack growth, including the effects of environments and also for high speeds. Some discussion of impact testing is also included.

1. INTRODUCTION

Polymers are now widely used as engineering materials and an important aspect of their behaviour in this role is their toughness and durability. Most adhesives are polymeric and their increasing use in engineering is largely influenced by the high toughness of new materials. Traditionally the fracture behaviour of polymers has been assessed via a battery of empirical tests such as Izod impact, falling weight and long term failure. These can be useful but they do not provide a logical framework in which to compare materials or to make long term predictions. Such a framework is provided by Fracture Mechanics and much effort has been devoted over the last fifteen years or so in establishing the basis on which it may be applied to polymers alone, as adhesives and as the matrices of composites. The two lectures outlined here will attempt to summarize the current situation as it appears to the author. A rather wide range of topics will be covered and the reader is referred to the recently published books listed at the end of these notes which give detailed discussion on all the topics concerned.

2. AREAS OF APPLICATION

Polymers themselves are used in a wide range of products and in many cases their resistance to fracture can be critical. Examples of this are in gas and water pipes where corrosion resistance and ease of installation gives a major impetus to the use of plastics but good resistance to the failure is very important, particularly for gas. In these cases, fracture mechanics has been used widely to characterise materials, as a method of predicting long term performace and sometimes, sadly, to explain field failures. It is worth noting, however, that many less critical areas make use of the method because

227

K. P. Herrmann and L. H. Larsson (eds.), Fracture of Non-Metallic Materials, 227–255.
© *1987 by ECSC, EEC, EAEC, Brussels and Luxembourg.*

although cracking or brittle failure may not be a calamity it is often costly and inconvenient. Thus, the use of the method to understand failures and to avoid them can be justified on economic and not just on safety grounds. Failure of a washing machine tub will probably not kill anyone but it can be costly. Adhesives have many critical applications, particularly in aerospace and failure can be either cohesive or adhesive within the joint. Both can be viewed as cracking and thus conveniently described in fracture mechanics terms.

In all these cases a critical role of fracture mechanics is <u>characterising</u> the toughness of the materials used and providing a well defined basis on which to compare them. It is this role that the method has been most thoroughly explored and much of the discussion here will be concerned with this aspect. Of course, the numbers can be used as a design tool or as the basis for inspection methods and some mention will be made of these here, but such uses are in their infancy for polymers.

3. SPECIAL FEATURES OF POLYMERS

Some of the physical characteristics of polymers cause special problems when one attempts to apply the usual analysis of linear elastic fracture mechanics or elastic-plastic fracture mechanics as used in metals and ceramics. One of the most important is visco-elasticity in that it introduces time dependent phenomena and renders energy based methods difficult to apply. For example, it is not easy to define an energy release rate in a generally dissipative system and the usual Griffith type of energy balance presents problems. In many cases, however, polymers are only slightly visco-elastic (tan $\Delta \sim .01$) and are essentially linear and conventional LEFM can be used to define toughness in terms of G_c and K_c without serious problems being incurred and with the enormous advantage that all the available linear solutions may be employed. There are important consequences of vicso-elasticity, however, and in particular, the fact that K_c or G_c will be <u>time dependent</u> or are often expressed as functions of the crack speed å. This dependence is the basis of the methods used to predict long term performace.

Polymers do undergo shear yielding and this can be described in terms of the familiar yield criteria though perhaps with some modifications such as time dependence. The plasticity theory based analyses of fracture mechanics can thus be applied sensibly to polymers but some caution is necessary. This is because thermoplastic resins also undergo an alternative form of deformation mode in which energy is dissipated via planar arrays of voids called crazes (see article by Kausch in this volume). These have somewhat different governing criteria and they are formed in the high stress regions around crack tips along with other microvoiding processes in some cases. This means that it is by no means certain that methods developed for metals will apply to polymers, or at least not necessarily under the same conditions. Examples will be discussed in the LEFM size criteria and the J Integral method.

It is interesting to note that rubbers, which of course are polymers, are often highly elastic and very non-linear but they present no fundamental problem providing G is evaluated for the non-linear system. There is a well developed body of knowledge in this area which is used in the design of rubber components such as springs and tyres.

4. SPECIAL ANALYTICAL RESULTS

The usual LEFM solutions for G and K can be utilized for polymers and some geometries are of special importance. In particular, the various forms of the cantilever

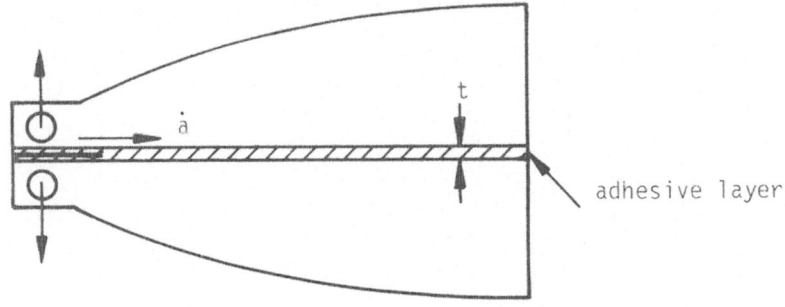

a) Contoured D.C.B. specimen - shown as an adhesive test.

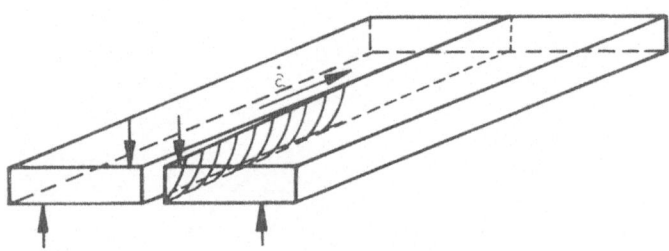

b) Double Torsion test geometry.

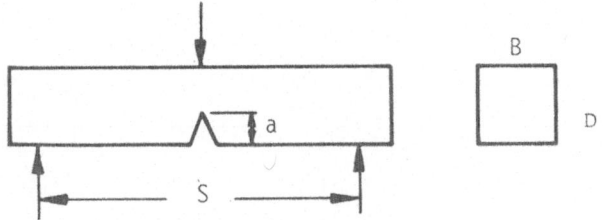

c) Charpy Test geometry

Fig. 1.   Testing Geometries

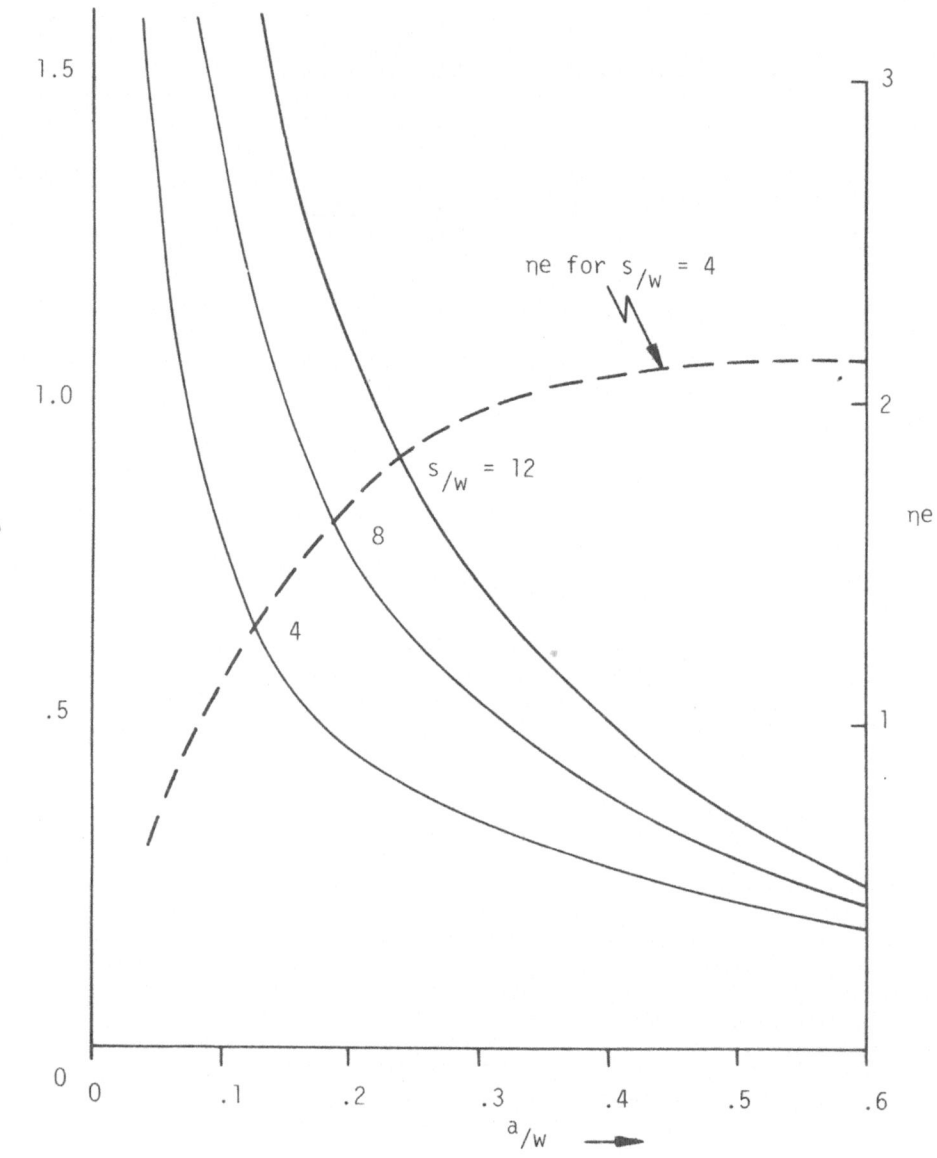

Fig. 2.   Energy Calibration Factors - Charpy Test shown
in Fig. 1.c.

beam geometry are widely used, particularly in testing adhesives. The forms using contoured depth and width are frequently employed in order to achieve stable, steady crack growth. In addition, the double torsion test, which also gives K (and G) independent of crack length is widely used. These two geometries are shown in Figs. 1 (a) and (b) and in both cases constant crack speeds result in constant loads and displacement rates and vice versa. Some special results relevant to adhesives should also be noted and, in particular, the fact that in the K-G relationship, $K^2 = EG$, the modulus E is the <u>local</u> value at the crack tip and thus that of the adhesive and <u>not</u> of the substrate.

In polymers quite wide use has been made of fracture mechanics to analyse the conventional energy measuring impact test. Here it is relevant to point out that the usual expression for G;

$$G = \frac{P^2}{2B} \cdot \frac{dC}{da}$$

where P is the load, B the thickness and C the compliance can also be written as;

$$G = \frac{U}{B} \frac{1}{C}\left(\frac{dC}{da}\right)$$

where U is the stored energy. This can be rewritten as;

$$U = G.BW\phi, \quad \phi\left(\frac{a}{W}\right) = C/dC/d(a/W) \tag{1}$$

(W is the specimen depth)

so the G_c may be found by measuring U at fracture for specimens with various crack lengths for which ϕ can be calculated. A graph of U versus $BD\phi$ will thus give G_c. Fig. 2 shows ϕ versus a/W for three (S/W) values for the Charpy test geometry shown in Fig. 1(c). It is also particularly important for the J tests to be discussed later that the energy form may be written in terms of the ligament area $B(w-a)$ and in this case we use the elastic η factor η_e since;

$$G = \frac{U}{B(W-a)}\eta_e; \quad \text{and} \quad \eta_e = \frac{(1-a/W)}{\phi} \tag{2}$$

Fig. 2 shows η_e for $S/W = 4$ and it is important to note that $\eta_e \sim 2$ for $.3 < a/W < .6$. The line zone or Dugdale model shown in Fig. 3 is frequently used for polymers since

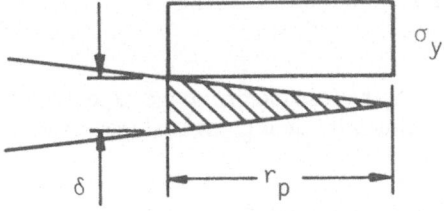

Fig. 3.   The Dugdale model.

crazes at crack tips are often in the form of linear zones. Their length and displacement at the tip, i. e;

$$r_p = \frac{\pi}{8}\left(\frac{K}{\sigma_y}\right)^2 \quad \text{and} \quad \delta = \left(\frac{K}{E\sigma_y}\right)^2 = \frac{G}{\sigma_y} \tag{3}$$

where σ_y is the zone stress, have a wide utility since they describe the crazes quite accurately.

The fracture criterion $\delta = \delta_c$, the critical crack opening displacement, COD, has also been found to be very useful since it is often insensitive to time and temperature and provides a basis for predicting time dependent behaviour. For example, a crack subjected to a constant K value will fail at $\delta = \delta_c$ when E and σ_y decrease with time sufficiently to allow δ to increase to δ_c and t_c is obtained from,

$$\delta_c . E(t_c) . \sigma_y(t_c) = K^2 \tag{4}$$

In many cases in polymers the yield strain $e_y = \sigma_y/E$ is approximately constant with time and temperature so various useful results may be determined from Eq. 4.

For example, we have;

$$G = \delta_c . \sigma_y(t) = (\delta_c e_y) . E(t) \tag{5}$$

which will be used in adhesives testing later and if we use a power law dependence of the form;

$$E = E_0(t/t_0)^{-n}, \quad E_0, \ t_0 \text{ and } n \text{ constants}$$

we have;

$$K^2 = (\delta_c e_y E_0^2).(t/t_0)^{-2n} \tag{6}$$

Thus for initiation time t_i we would expect relationships of the form $K \propto t_i^{-n}$. For running cracks we must use a time scale appropriate to the local processes at the crack tip and this is given by;

$$t = \lambda r_p/\text{å}$$

where λ is a constant dependent somewhat on n, but for small values of n, is $\sim 1/5$. Using this we have a K - å relationship;

$$K \propto \text{å}^n \tag{7}$$

In general n is small and $\sim \tan \Delta$, the loss factor, so we expect log K - log å curves to be approximately linear and rising with their greatest slopes around loss peaks.

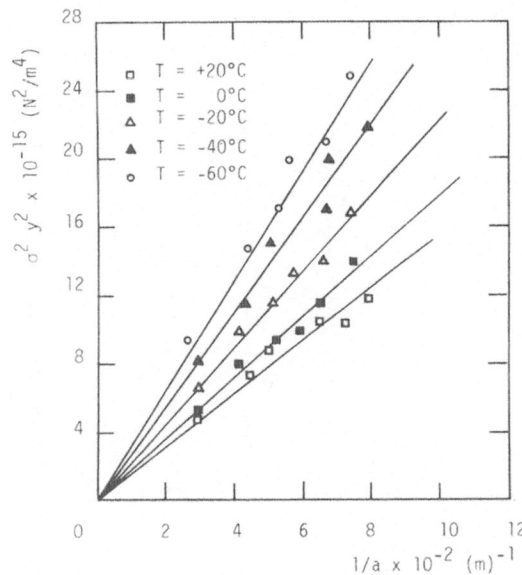

Fig. 4. Single-edge notch bend data for polyacetal
at temperatures of +20°C to -60°C.

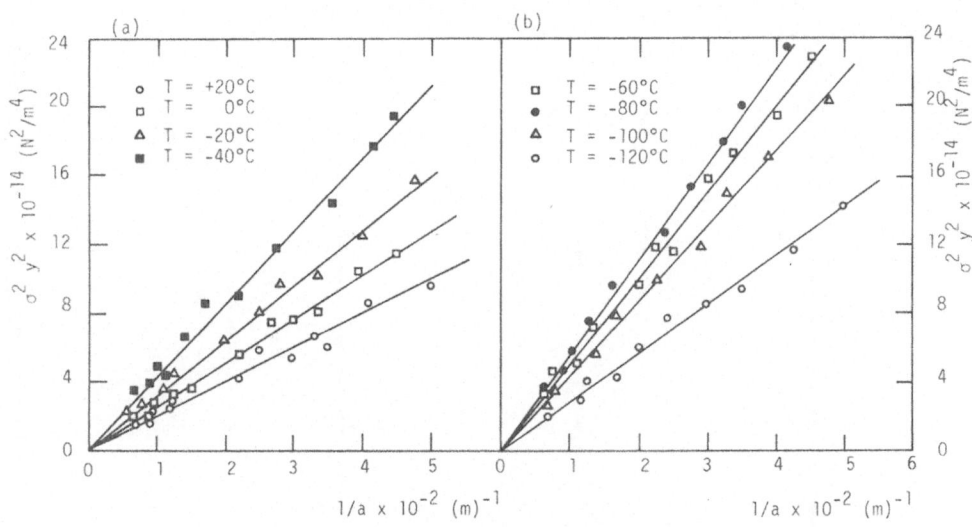

Fig. 5.' Single-edge notch tension data for PMMA over a temperature range of
(a) +20°C to -40°C, and (b) -60°C to -120°C, at a constant crosshead
speed of 0.5 cm/min.

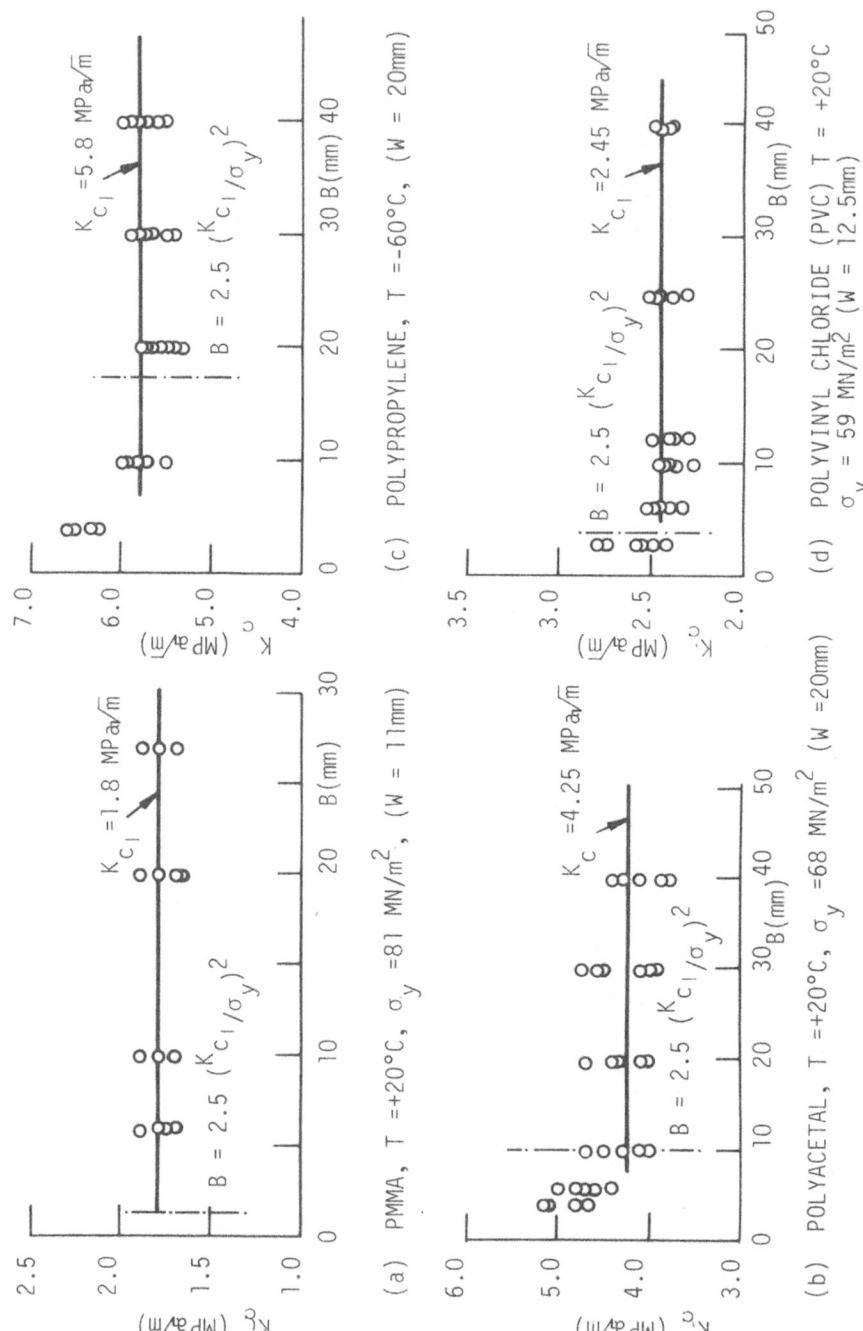

Fig. 6. Effect of Specimen Thickness on the Fracture Toughness in Three-Point Bend.

5. INITIATION OF CRACK GROWTH

The characterization of materials in terms of their toughness at the initiation of crack growth is generally the simplest way to utilize fracture mechanics. Much use has been made of LEFM for the more brittle fractures and is expressed as a K_c or G_c value at same loading rate. In the LEFM conditions, the small plastic zone at the crack tip is under plane strain conditions and this can only be maintained if the other specimen dimensions (thickness, width, crack length) are much larger than the zone size. The ASTM criteria for metals in the bend test is:

$$B > 2.5 \cdot \left(\frac{K_c}{\sigma_y}\right)^2 \tag{8}$$

and it has been shown that this is sensible for polymers and that smaller thicknesses give higher values because of a loss of constraint. A further requirement is that;

$$W > 5\left(\frac{K_c}{\sigma_y}\right)^2 \tag{9}$$

since too small a width leads to plasticity curtailing the load level achieved and thus giving a low K_c value. A convenient way of determining K_c in slow tests is from the gross failure stress using the usual relationship;

$$K_c^2 = Y^2 \sigma^2 a \tag{10}$$

where Y^2 (a/W) is the calibration factor and a the crack length. Thus one may plot $\sigma^2 Y^2$ versus a^{-1} and K_c may be computed from the resulting line. Figs. 4 and 5 show typical data for brittle failures in polymers; in Fig. 4 for polyacetal in 3 point bending from -60°C to +20°C and in Fig. 5 for PMMA in single edge notch tension from -120°C to +20°C. It can be seen that the method works well in these cases and the dimensions are well within the limits in Eqs. (8) and (9). The effects of thickness in bend tests is illustrated in Fig. 6 for four polymers at various temperatures where the K_c value is plotted versus B together with the ASTM suggested minimum value from Eq. (8). The condition is, perhaps, somewhat conservative but is a useful working limit. Below this value K_c rises because of loss of constraint in the stress field. The effect of W in bending is shown in Fig. 7 where values which are below the limit are now low because of the lower load carrying capacity of the yielded ligament. The ASTM limit is again seen to be a useful criterion.

Figs. 8 and 9 show the variation of K_c with temperature for several polymers. In Fig. 8 we have low toughness materials; PMMA, a generally glassy polymer and two grades of low density polyethylene. Note that for the latter materials there are no values above -20°C because of the onset of ductile tearing and the consequent inability to use valid LEFM specimens. In Fig. 9 we have tougher materials with K_c values up to 5MPa√m and it can be seen that polyacetal (PA) and PVC give rising curves as the temperature decreases. The polypropylene is rubber modified as is the second type of

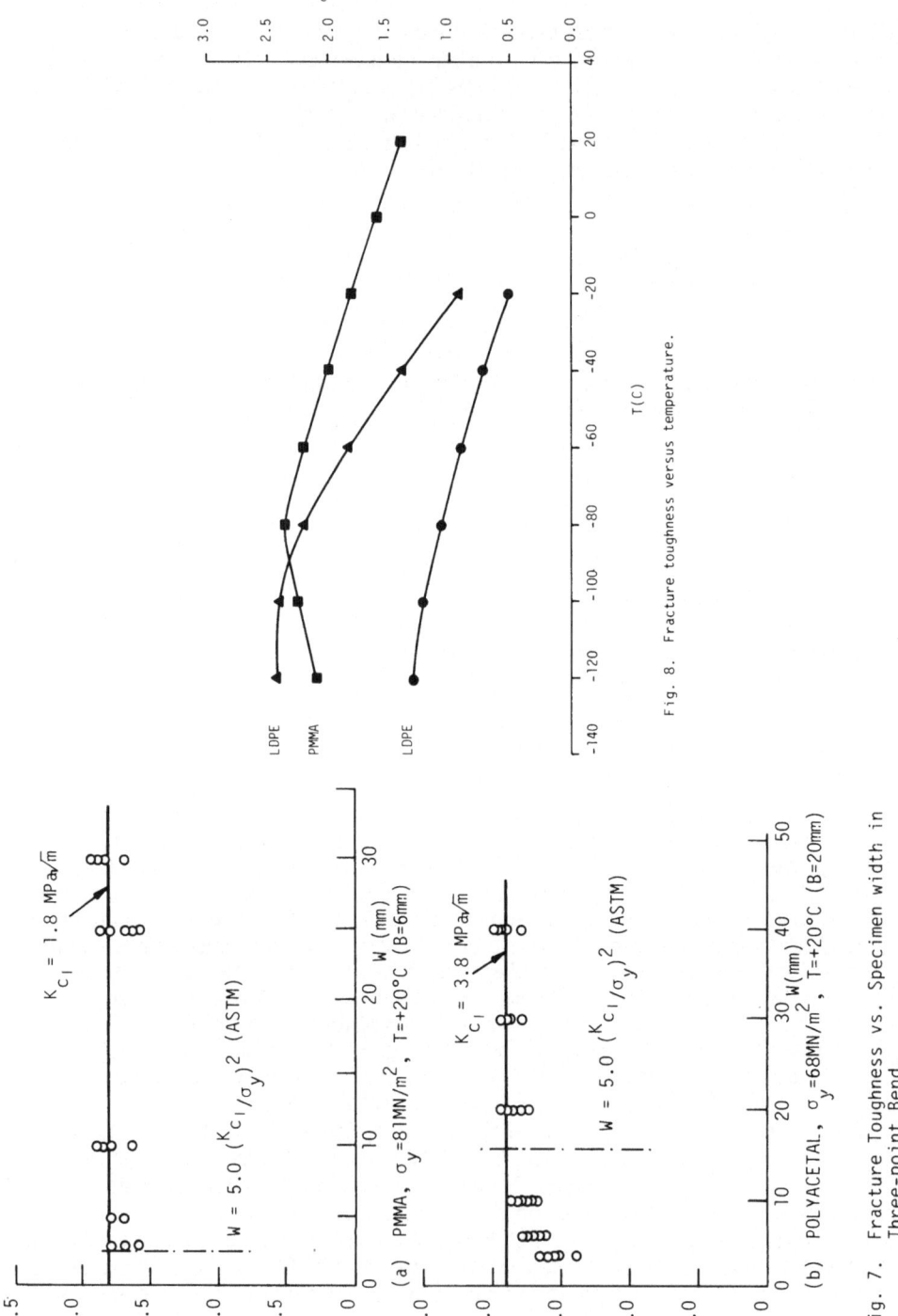

Fig. 8. Fracture toughness versus temperature.

Fig. 7. Fracture Toughness vs. Specimen width in Three-point Bend.

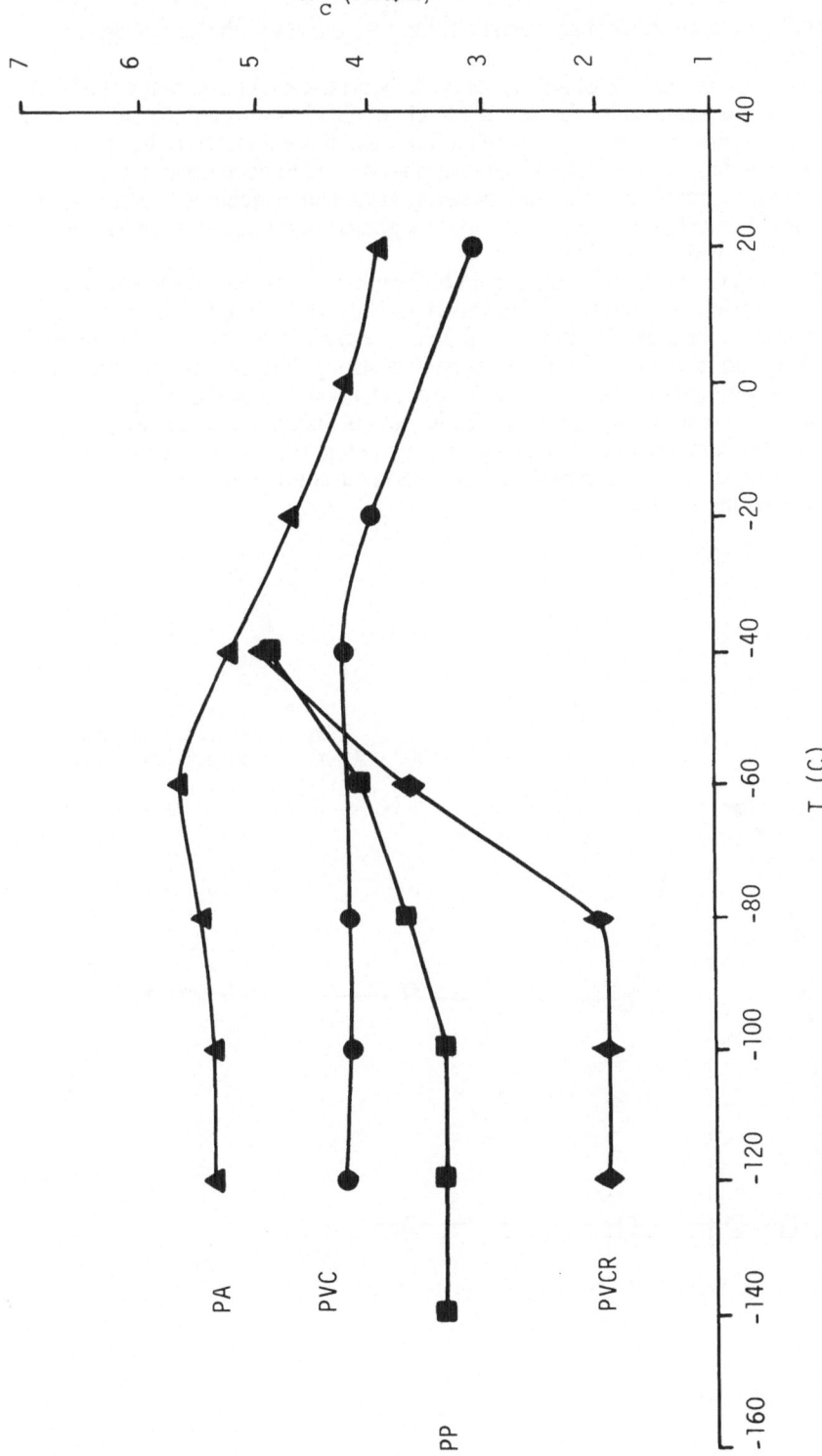

Fig. 9. Fracture toughness versus temperature.

PVC, PVCR, and both show characteristic falling K_c curves with decreasing temperature.

An extra factor in the case of adhesives is the bond thickness and changes of this with respect to the plastic zone size and the thickness lead to substantial changes in the constraint on the material. For very tough adhesives, it is necessary to have thick bond layers in order for the energy absorbing plastic zone to develop and in some cases toughnesses greater than the bulk material value can be achieved. Such factors are important in determining the interlaminar toughness of composites where the interfibre spacing is the critical factor.

Fig. 10 shows G_c for a joint made from both rubber modified epoxy resin and from the base resin using the specimen shown in Fig. 1(a). Here the bond thickness, t, is varied and for the unmodified material gives the same value as the bulk material. For the rubber modified resin, however, there is a strong thickness dependence and a value considerably greater than the bulk material at a bond thickness of t_m as shown. The essential difference compared to the basic resin is that the increased toughness results in a much larger energy absorbing zone r_p so that there is considerable interference with the stress state at the crack tip and curtailing of the zone size by the material outside the zone.

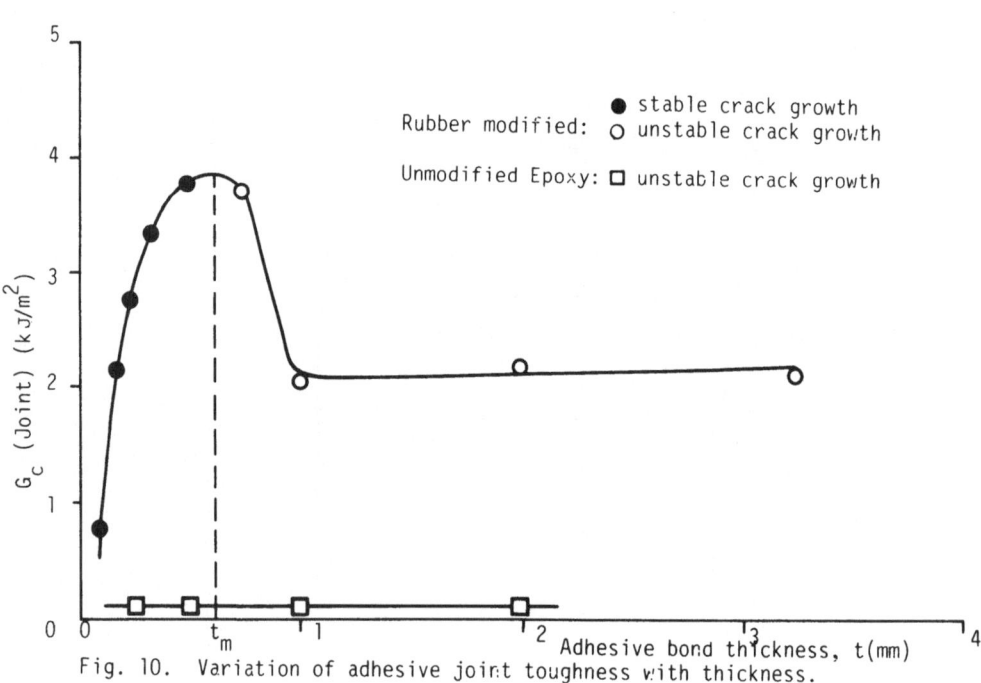

Fig. 10. Variation of adhesive joint toughness with thickness.

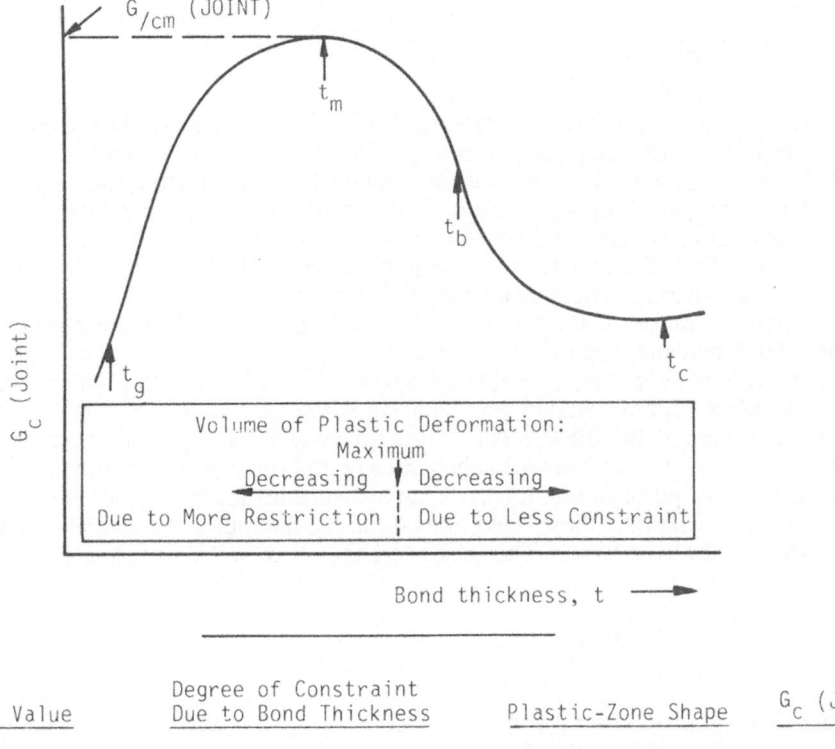

t - Value	Degree of Constraint Due to Bond Thickness	Plastic-Zone Shape	G_c (Joint)
t_a $(< 2r_p)$	High		Comparatively Low.
t_m $(= 2r_p)$	Moderate		At Maximum.
t_b $(> 2r_p)$	Low		Below G_{cm} (Joint) Value.
t_c $(\gg 2r_p)$	Almost Nil		Approx. Equal to G_{Ic} (Bulk).

Fig. 11. A Summary of Bond Thickness-Toughness Interactions.

Fig. 11 summarises these interactions and shows that the maximum in G_c is obtained when t is equal to $2r_p$, i.e;

$$t_m = \frac{EG_c}{\pi \sigma_y 2}$$

This sort of effect emphasises the importance of defining the stress state under which G_c or K_c is determined. In this layers, as in joints or in the matrices of composites the constraint is high and the value achieved is that of a true plane strain value. When comparing composite or joint performance with that of the bulk resin it is essential that both values are obtained with the same stress state; i.e. plane strain for the worst case. Much disappointment in the performance of composites made from apparently tough matrices would be avoided if this were done.

It is difficult to make satisfactory polymer specimens of B>12mm because of the thermal effects and the size criteria limits $(K_c/\sigma_y)^2$ to values of 5mm. For the somewhat tougher polymers, $K_c \sim 3MPa\sqrt{m}$ so the lower limit on σ_y which can be tested is 43 MPa. Many tougher polymers have values below this and it is thus difficult to test them in the LEFM mode. The J test provides a way of overcoming this problem in that a fully plastic, deep notched three point bend test has a constraint factor very close to that of plane strain. It is possible to determine the energy per unit area to initiate crack growth under these conditions (J) using the multiple specimen method devised for metals by ASTM. This has been shown to work well and the J size criterion,

$$B, \frac{W}{2} \quad a > 25 \, (J_c/\sigma_y) \tag{11}$$

is about a factor of three lower than that for LEFM thus giving a much extended range of properties which can be covered.

In the J method the high constraint of the deep notched bend specimen allows crack initiation under plane strain conditions to be obtained with extensive plastic deformation. This is very useful but would be of limited value if it was necessary to separate the elastic and plastic parts of the energy measured because of the inherent difficulty of such a process in polymers. J for plastic deformation is given by a total energy form;

$$J = \frac{\eta_p U_p}{B(W-a)} \tag{12}$$

and for three point bending $\eta_p = 2$. Thus, if we use $S/W = 4$ we know, from section 4, that $\eta_e \sim 2$ so J may be found from;

$$J = \frac{2U}{B(W-a)} \tag{13}$$

where U is the total energy absorbed. Thus if U at crack initiation can be found then J_c may be deducted and, since it is in plane strain; this should be the value one would obtain in an LEFM test, if it could be performed.

Defining initiation in this type of test is not easy since it often occurs near the centre

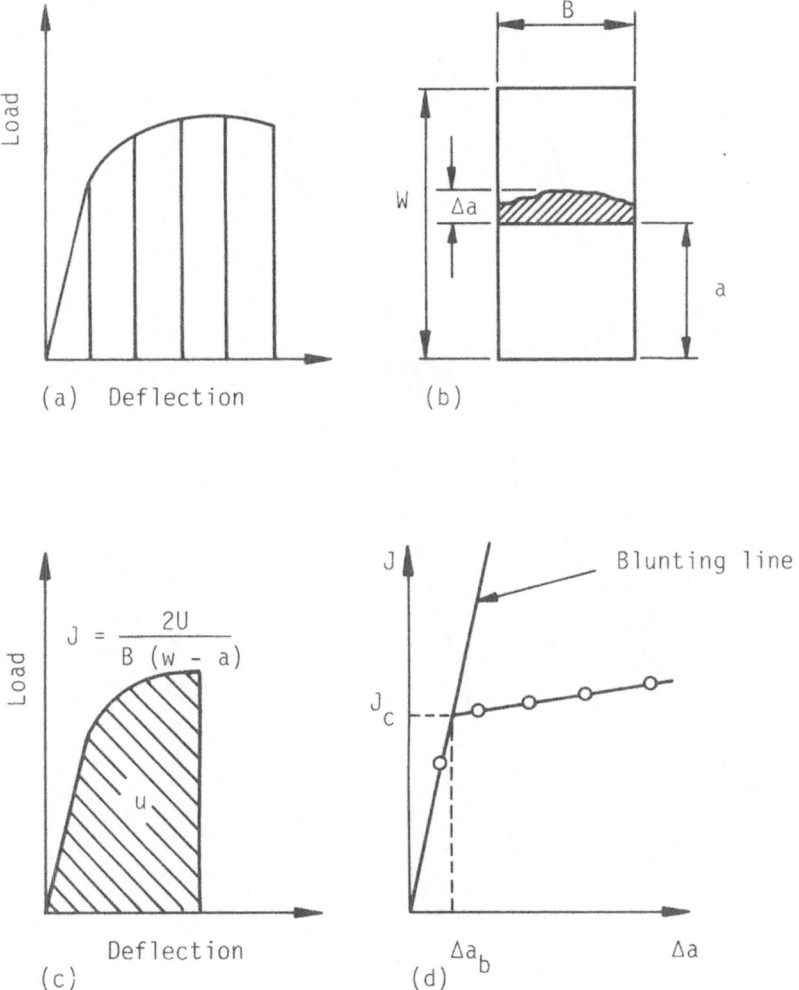

(a) Deflection

(b)

$$J = \frac{2U}{B \, (w - a)}$$

(c) Deflection

(d)

Blunting line

Fig. 12. Procedure for measuring J_c.

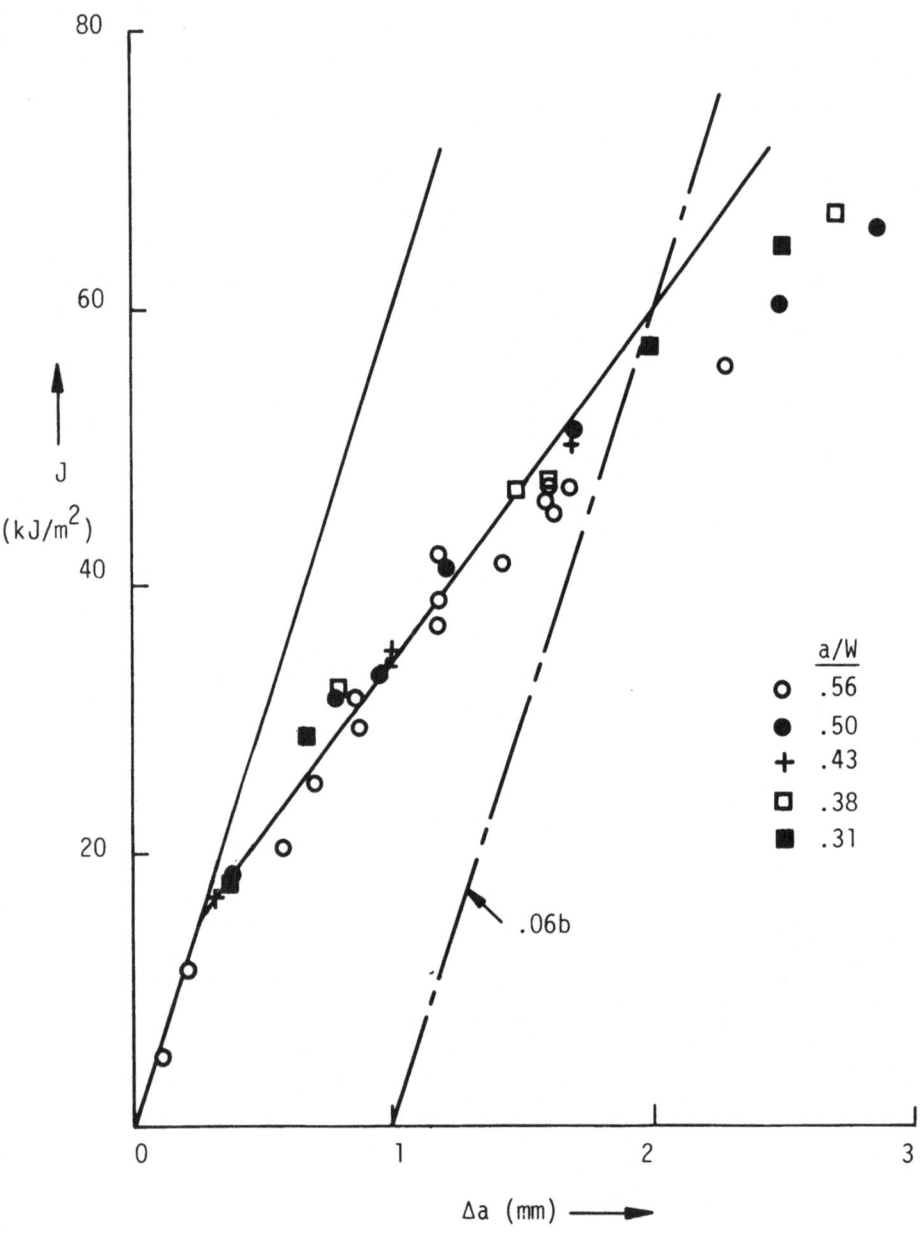

Fig. 13. J versus Δa at 20°C with various a/W values, σ_y = 30 MPa, W = 40 mm (PP).

of the specimen so a multiple specimen method has been suggested by ASTM. The procedure is outlined in Fig. 12. A set of identical specimens with $S/W = 4$ and $a/W \sim 0.5$ are made and loaded to various deflections of less than complete failure, Fig. 12 (a). The specimens are then cooled to a very low temperature and broken open in impact to reveal the slow growth on the surface, Δa, which can be measured, Fig. 12 (b). From the load-deflection curve may be found and from Eq. (13), J, see Fig. 12 (c). J is then plotted versus Δa as shown in Fig. 12 (d) but allowance must be made for the blunting of the crack prior to initiation. This may be modelled as an apparent crack growth of $\delta/2$ so that a blunting line is given by;

$$J = (2\sigma_y)\Delta ab \qquad (14)$$

and it is the intersection of this and the line through the measured Δa values which gives the true J_c value at initiation; again as shown in Fig. 12 (d). There are some limitations on Δa and it should not be greater than .06b, where b = W-a.

Figs. 13, 14 and 15 show typical data for three tough polymers; rubber modified Nylon, polypropolyne and a PVC. Generally quite good lines are obtained and providing there is sufficient difference between the slopes of the lines J_c can be found accurately. As the system becomes more elastic and LEFM is approached the growth line become flatter and eventually becomes horizontal when this method is not necessary. When the slope increases to equal that of the blunting line then there is, of course, no growth.

Comparisions with LEFM data would indicate that the J_c values are indeed equivalent to G_c and this is demonstrated in Fig. 16 where K_c is plotted as a function of temperature for several tough polymers. The low temperature values are from LEFM tests while those at higher temperatures are found using J_c and computed via $K_c^2 = EJ_c$. The agreement and continuity of trends is generally good. Table I lists K_{c1} (plane strain) values, G_c (from K_{c1}) and J_c values for a range of materials.

TABLE I K_{c1} (plane strain) values for several polymers

MATERIAL	T (°C)	K_{c1} (MPa√m)	G_c kJ/m^2	J_c kJ/m^2
PMMA	+20	1.80	1.1	
PVC	+20	2.40		1.70
PVCR	+20	3.35		3.00
ABS	+20	3.00		3.0
PA	+20	4.00	5.3	
NYLON (modified)	+20	5.00		10.5
PP (copolymer)	+20	5.00		15.5
HDPE	+20	1.20		0.78
LDPE	-20	0.93		1.14
MDPE	-40	6.46		19.4
LLDPE	-60	8.12		33.0
PC	+20	2.20	1.6	
EXPOXIES	+20	0.60	0.12	

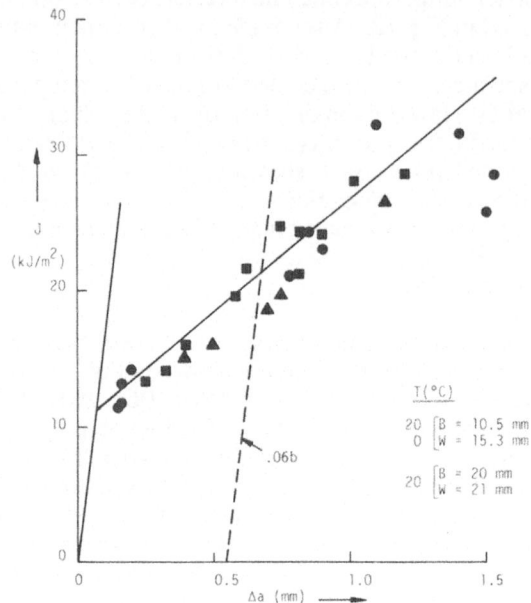

Fig. 14. J - Δa data for Toughened Nylon

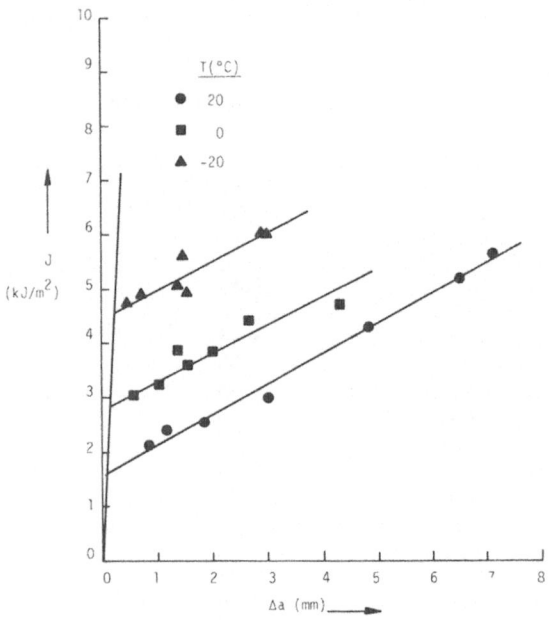

Fig. 15.   J - Δa for modified PVC.

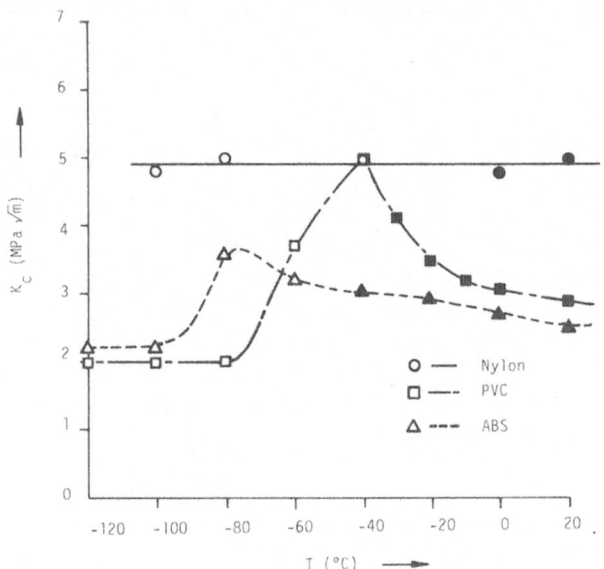

Fig. 16. K_c as a function of temperature for three toughened polymers, - closed points converted from J_c values $(K^2 = E\ J_c)$ - open points from LEFM tests.

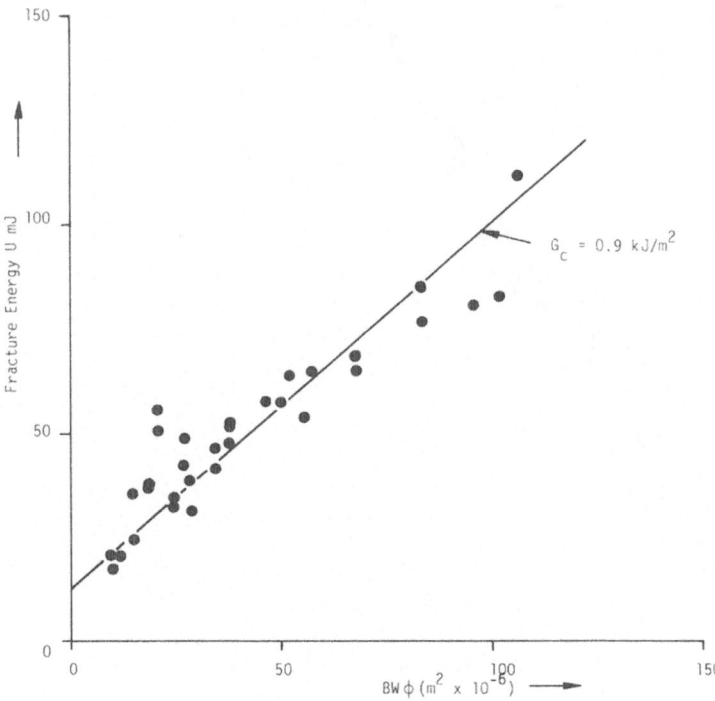

Fig. 17. Energy versus BWφ for Charpy Specimens - PMMA at 20°C.

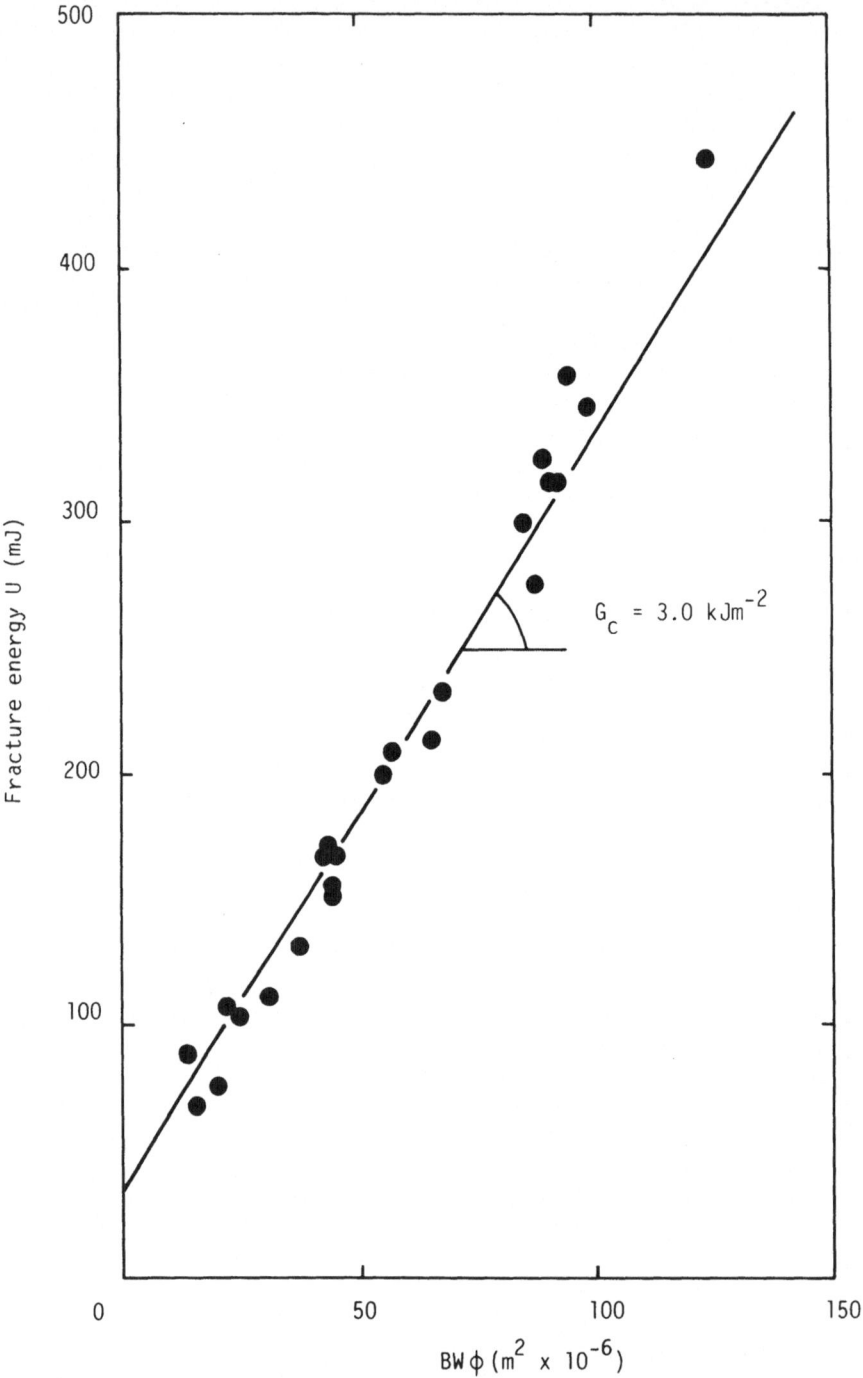

Fig. 18.   Energy as a function of BWϕ for a polypropylene-
rubber blend at 23°C.

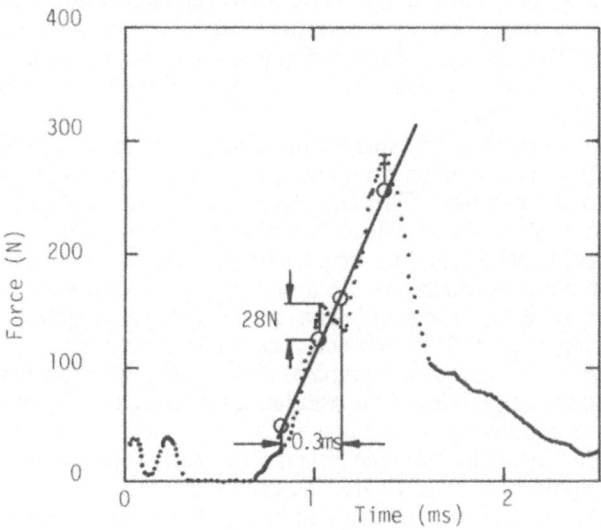

Fig. 19. Load-time Trace for a Charpy Test - HDPE

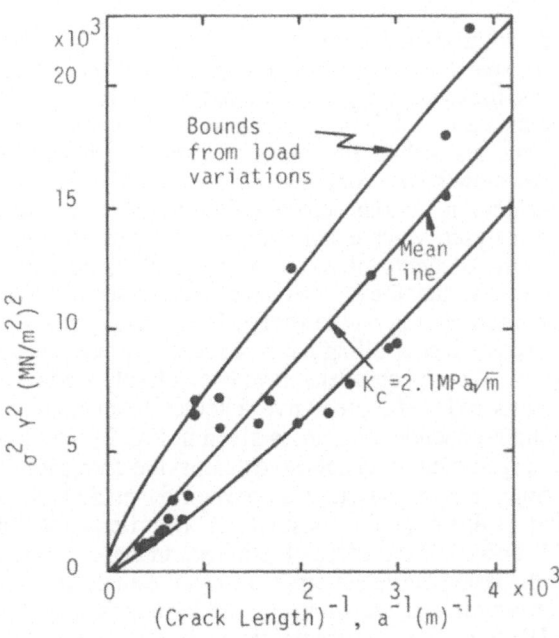

Fig. 20. The determination of K_c showing Calculated Bounds.

Initiation at high loading rates is also important and this can be measured in various forms of impact test. The traditional energy measuring system is the most simple to perform and the use of the BWϕ analysis discussed in Section 4 can be applied to measure G_c. The same size constraints as in the lower rate tests apply, of course, but it has proved to be a useful technique.

Figs. 17 and 18 show U versus BWϕ plots for the Charpy impact test on two polymers. Both have a clear positive intercept on the energy axis which is due to the kinetic energy imparted to the specimen. This arises from the dynamics of the system and as loading times become shorter such effects become larger. Fig. 19 shows a load-time diagram from an impact test at 2m/s on a Charpy specimen. There is clear evidence of oscillations in the load which arise because the system of striker, contact stiffness and specimen responds to the impact load as a dynamic system giving oscillating loads. Quite large variations in stress can occur which give rise to considerable scatter in the $\sigma^2 Y^2$ versus a^{-1} graph as show in Fig .20. The bounds shown are calculated from the amplitudes of the oscillations observed on the trace in Fig. 19. If the loading time becomes too short then neither G_c nor K_c can be found because the readings are dominated by dynamic effects. Typically such a time limit would be about 100µs for polymers. Fig. 19 also illustrates another important point in that the BWϕ method relies on a sudden, unstable failure giving a triangular diagram. Clearly in this case this is not so and the total energy, which is what a conventional test would measure, would be much too large. It is only that energy up to the initiation point which should be used to find G_c at initiation and this can only be determined in this stable type of fracture from some form of load record.

6. CRACK PROPAGATION

(See article by Döll in this volume for further details.) In a perfectly brittle fracture, G_c is independent of crack speed and it is possible to grow cracks at a wide range of speeds by controlling G as, for example, in the double torsion test. As discussed in Section 4, the visco-elastic nature of polymers results in a relationship between G_c (or K_c) and å in which G_c increases with å. In most circumstances, this promotes stable growth so that cracks grow slowly over long periods of time. In addition, they may remain stationary for considerable periods before initiation. Both of these phenomena may be incorporated into analyses which enable time to failure or life predicions to be made. In addition, many environments interact with the crack tip zone and promote growth by local plasticisation so that the K_c - å curve is depressed to lower K_c values and crack growth rates increase giving rise to shorter lives. In addition, the flow of these environments can become a controlling factor since they must penetrate into the very small crack tip region. These phenomena have been closely studied for both polymers and adhesive joints and useful predictive schemes are available.

A good example of a time-dependent failure is given in Fig. 21 which plots the G value applied to a cracked specimen of an adhesive against the corresponding modulus value E(t) at the failure time. Eq. (5), based on a constant δ_c predicts the linear relationship observed and, in this case, the constant C.O.D. criteria is valid over eight decades of time. Fig. 22 shows K-å data for polyethylene in water at three temperatures and shows the rising curve anticipated from the analysis. Such data can be used to predict the stress-time data shown in Fig. 23 for a pressurised pipe made from the same material.The lines shown were predicted by intergrating the growth law of the form given in Eq. (7) form a small initial flaw to a much larger size and the slopes are the negative of those in Fig. 22 and are fitted to an a_0 value of 100µm.

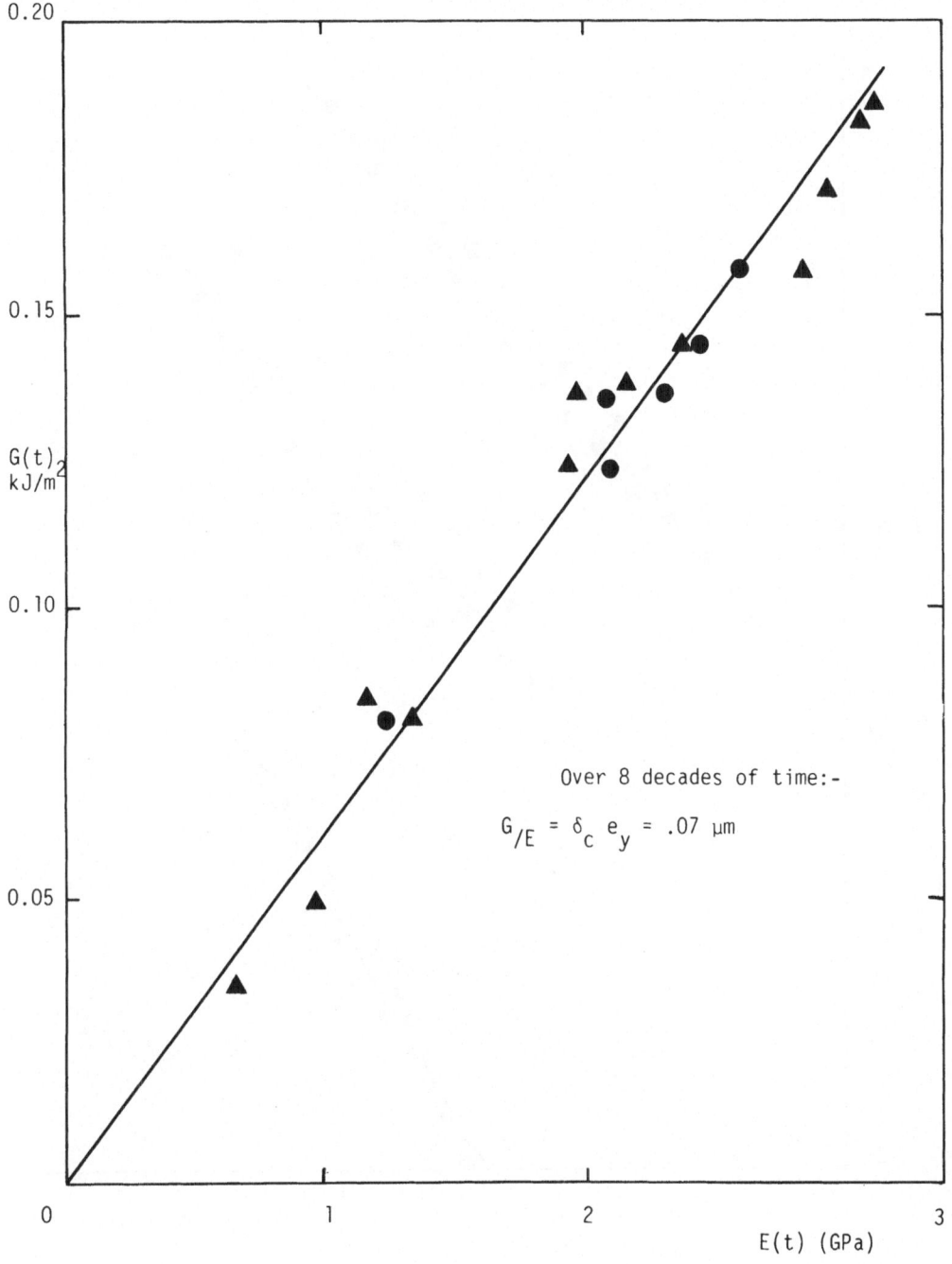

Fig. 21. Time dependent failure of an adhesive G_c versus modulus.

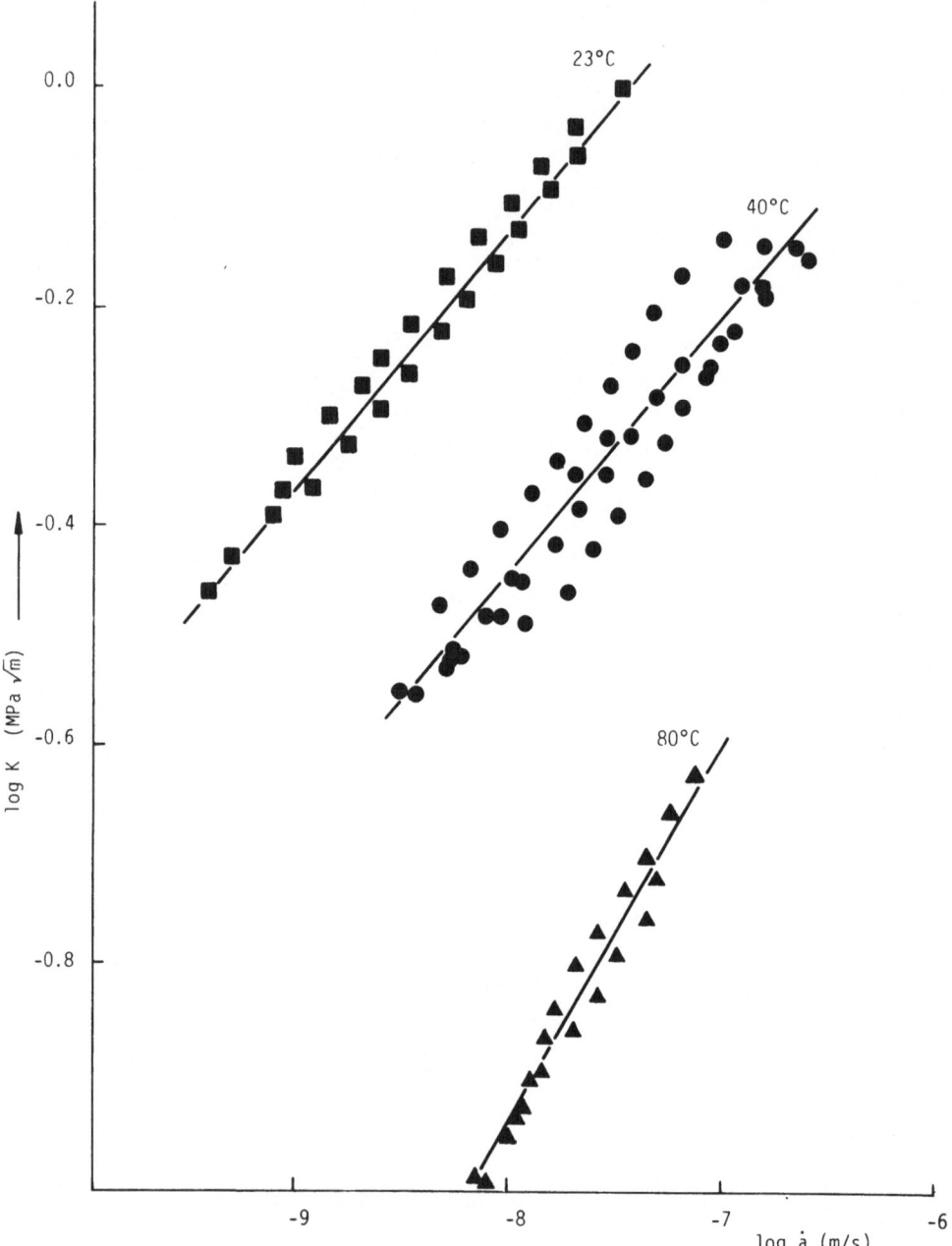

Fig. 22. Crack growth data for HDPE in water. (Data kindly supplied by BP Chemicals).

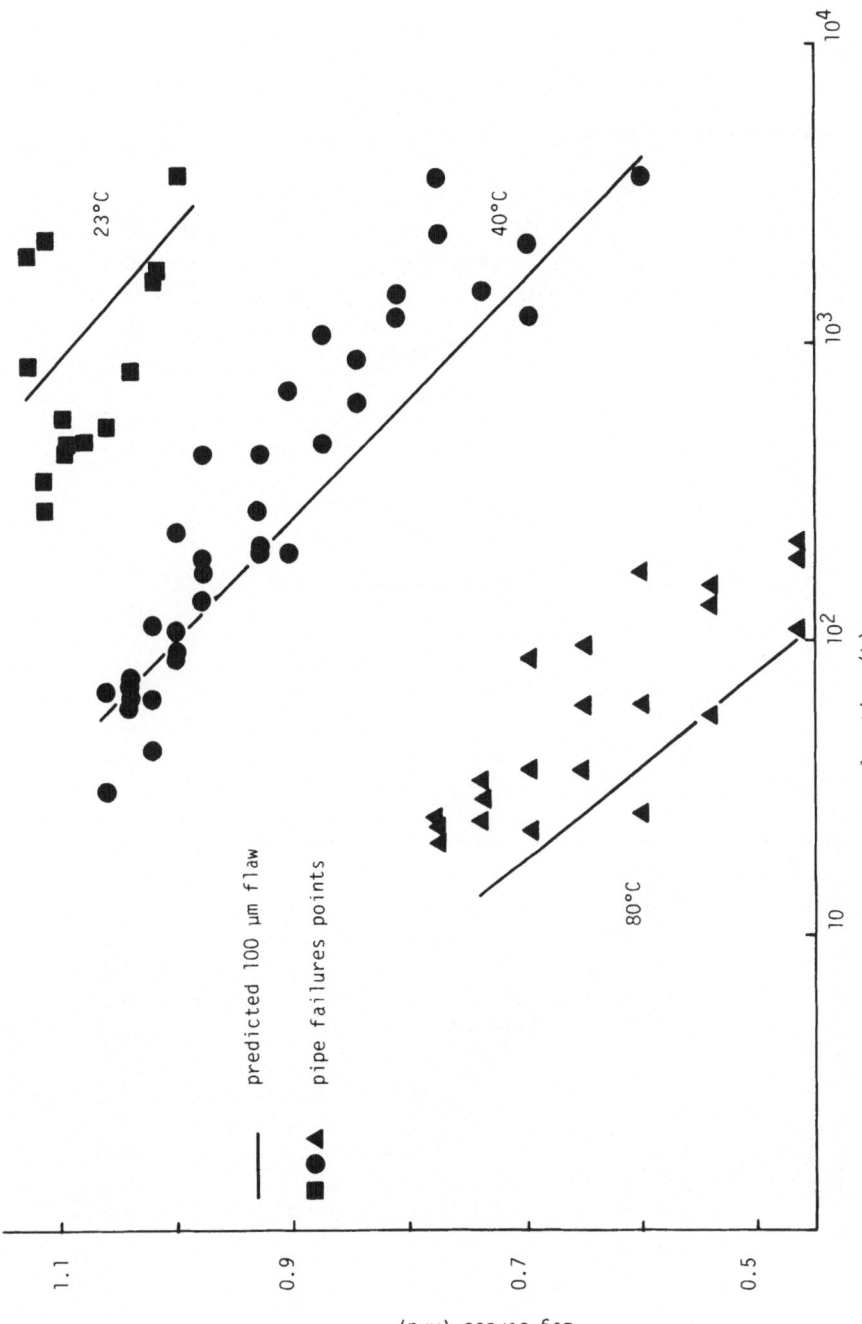

Fig. 23. Stress-lifetime curves for HDPE pipes in water.

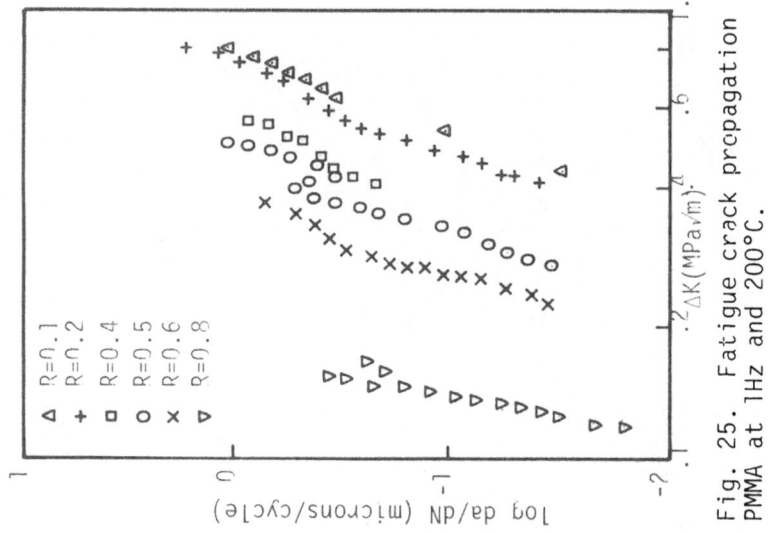

Fig. 25. Fatigue crack propagation
PMMA at 1Hz and 200°C.

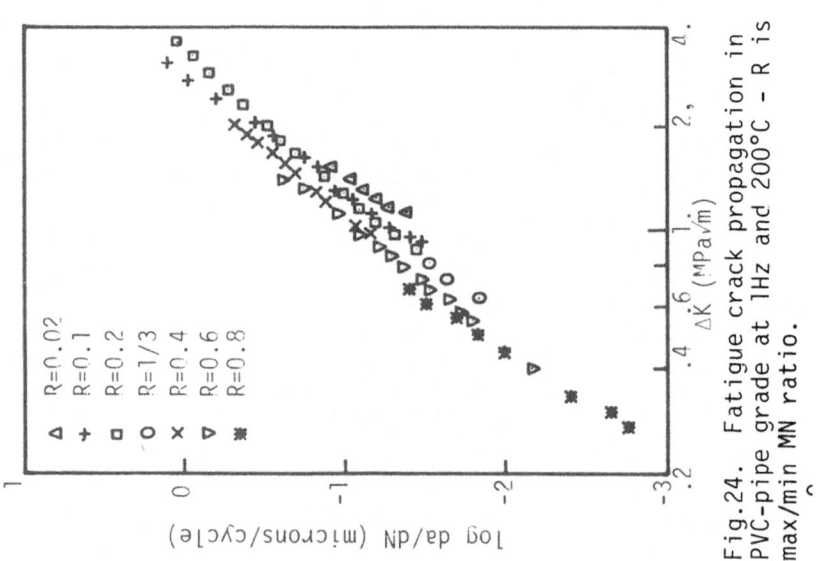

Fig.24. Fatigue crack propagation in
PVC-pipe grade at 1Hz and 200°C - R is
max/min MN ratio.

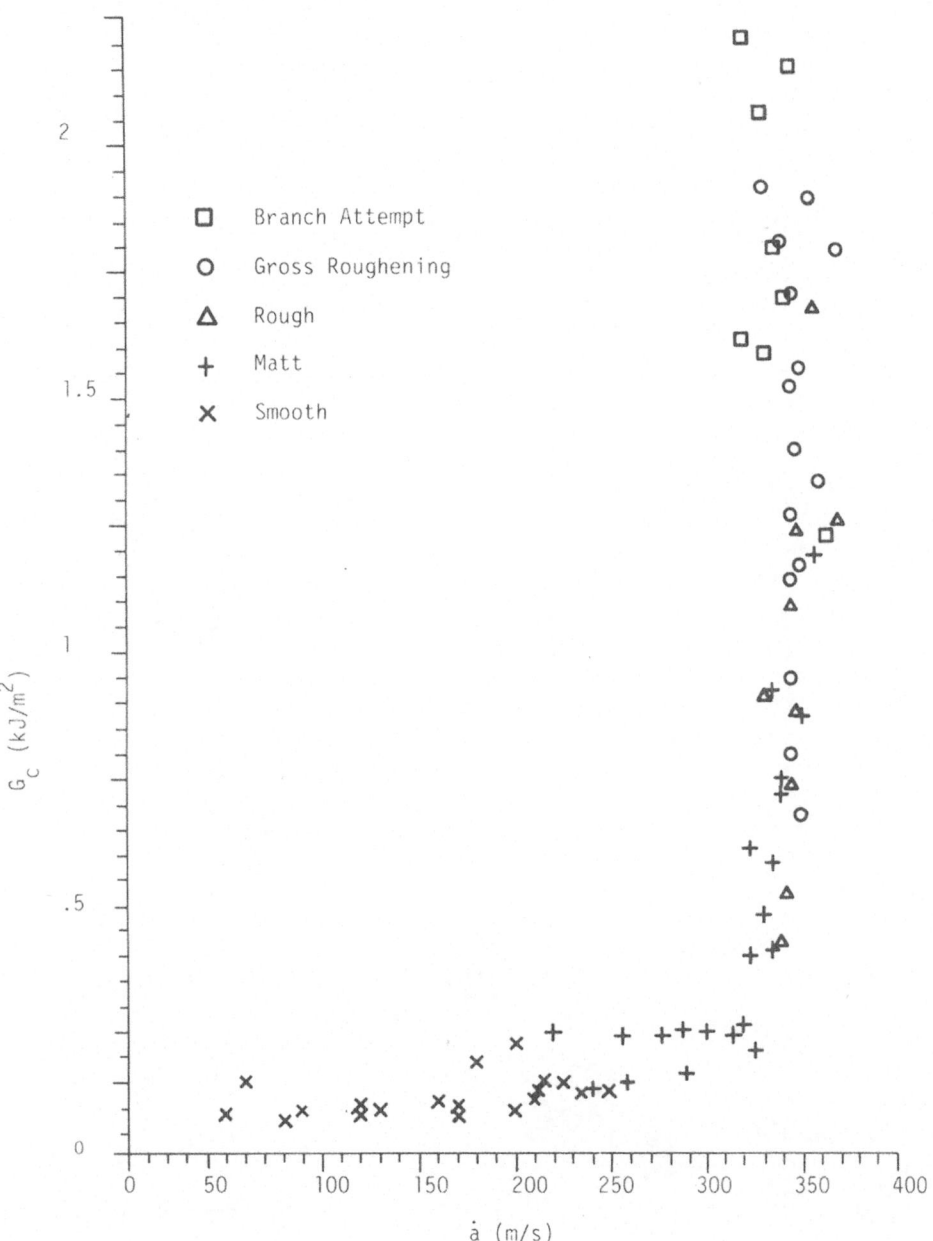

Fig. 26. High speed crack data for epoxy at 20°C

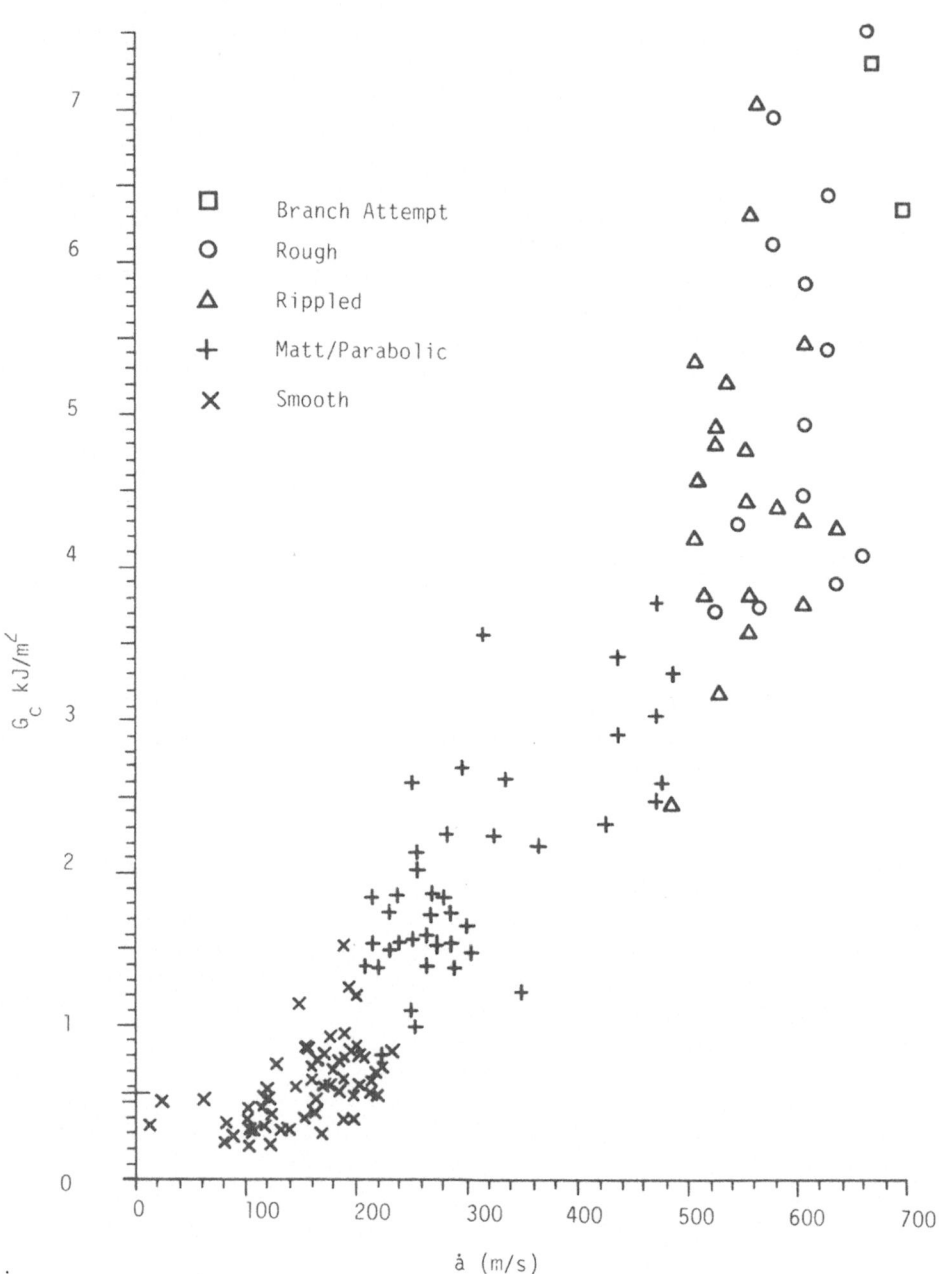

Fig. 27. High speed crack data for PMMA at 20°C.

Fatigue crack propagation arising from load cycling also occurs in polymers (see article by Döll in this volume) and the conventional Paris law representation is widely used. Similar methods to the time dependent case may be used for life predictions.

Fig. 24 shows a log ΔK - log da/dN curve for PVC at 20°C where ΔK is the range applied and is typical of data obtained on tough polymers. For more brittle materials there is no unique relationship on a ΔK basis and the maximum K seems to be the controlling factor as shown in Fig. 25 for PMMA.

At the other end of the scale is the occurrence of high speed propagation usually arising from an unstable situation. Here, crack speeds can reach 700 m/s and again it is possible to describe the process as a K_c (or G_c) versus å relationship. Dynamic effects are of major importance in these circumstances since these high speed growths can give rise to the majority of the energy being kinetic. This behaviour is important in areas such as piping because self-propelling, high speed cracks do occur and material resistance under such conditions is important.

Figs. 26 and 27 show high speed data for PMMA and an epoxy resin respectively. Both show a marked rise of toughness with å with increasing roughening of the fracture surface at the higher speeds. For the epoxy the rise is very rapid at about 350 m/s but is more gradual in PMMA over the range 100-700 m/s. The toughness in both cases increases enormously at these higher speeds; a factor of about 20 in both cases. The origins of such rises are not clear as yet and may be related to inadequacies in the dynamic analysis or may be mechanisms such as thermal effects or microcracking.

7. CONCLUSIONS

This is a very brief review of how fracture mechanics is being used to describe failure in polymers and covers a wide range of topics. Much of the detail is lost, of course, but it is hoped that there is sufficient here to encourage the reader to pursue the subject in the bibliography and hence the extensive literature in the field.

BIBLIOGRAPHY

1. Fracture Mechanics of Polymers, J. G. Williams, Ellis Horwood, (1984).

2. Fracture Behaviour of Polymers, A. J. Kinloch and R. J. Young, App. Sci. London (1983).

3. 'A Fracture Mechanics Approach to the Failure of Structural Joints'
   A. J. Kinloch and S. J. Shaw. Chap.3 in Developments in Adhesives -2,
   Ed. A.J. Kinloch, App. Sci. (1981).

4. Fatigue of Engineering Plastics, R. W. Herzberg and J. A. Manson,
   Academic Press (1980).

FRACTURE TESTING OF BRITTLE POLYMERS

W. Doell
Fraunhofer-Institut für Werkstoffmechanik
Woehlerstrasse 11
D-7800 Freiburg

ABSTRACT. Three optical methods will be presented which, together with
adequate measuring devices of short time physics, are powerful tools in
the investigation of relevant fracture parameters during crack propagation. By reflection optics, crack speeds can be measured in a very large
range. By microscopic optical interferometry, the size and shape of the
craze zone at the crack tip and the crack opening can be determined thus
leading to an insight into the micromechanics in the crack tip region.
By the shadow optical method of caustics, information on the stress
field around the crack tip is used in the determination of stress-intensity factors. The methods are applied to fracture induced by different
types of loading such as quasistatic, fatigue and impact. For PMMA, life
times measured in creep rupture will be compared with the fracture mechanics approach. Critical craze dimensions of PMMA measured during
crack propagation will be given and results of the micromechanics of fatigue crack propagation in PMMA and PVC will be presented. For dynamic
loading some recent results will be reviewed.

1. INTRODUCTION

Polymers include a large number of differently behaving materials due to
the great variation in their composition and structure. By their time-
and temperature-dependence the field of application and investigations
is further enlarged. Dealing with deformation and fracture, nonlinear
behaviour has to be considered in addition and also the type of loading
such as quasistatic, fatigue or impact loading which will in general
cause different material responses. In this context also the parameter
crack speed has to be mentioned due to the close relation between time-
and crack speed-dependent material behaviour. Crack propagation, may
occur in a large range of crack speeds, depending on material and testing conditions, from about 10^{-7} mm/s up to the maximum crack speed
(some hundred meters per second). Thus crack speed reflects a very wide
range of loading time which spans about twelve decades. This must be
seen in close connection with the high stress field acting in the crack
tip region, especially in and around the plastic zone.

K. P. Herrmann and L. H. Larsson (eds.), Fracture of Non-Metallic Materials, 257–290.
© *1987 by ECSC, EEC, EAEC, Brussels and Luxembourg.*

In polymers two basically different plastic deformation processes may occur: shear yielding and normal yielding (crazing).

Clearly, the area of problems has not even been mentioned in total and in this contribution even further restrictions have to be made. Hence, the interested reader is referred to some reviews [1-5]. The interest here will be focussed on brittle fracture and crazing of polymers [6], especially of transparent thermoplastics. Although a relatively narrow field is considered this, nevertheless, offers the opportunity to present some optical methods which are unique in fracture testing.

By the method of total reflection, crack speed measurements are performed to establish $\dot{a}$- K_I curves which may be the basis for lifetime predictions.

Using the interference optical method insight may be given into the crazing process at the crack tip. Characteristic craze dimensions and critical displacements at the crack tip may be derived before and during fracture which has been induced by different loading types. Of special interest is the length of stretched molecular bundles (fibrils) since it is one of the essential parameters governing the resistance against crack propagation in glassy thermoplastics. The other essential parameter, the characteristic stress which the fibrils can sustain, may also be derived from interference optical measurements by application of the Dugdale model to them.

The third optical method to be presented is the shadow optical method of caustics which is based on the deflection of light by the elastic stress field around the crack tip. It gives direct information on the stress intensity factor or the strain energy release rate. This method has proved to be very powerful in investigating the difficult problems of impact loading. Since it also works in reflected light the restriction of material transparency does not apply.

2. SOME OPTICAL TEST METHODS

The three optical methods applied here in brittle fracture testing of polymers are schematically indicated in Fig. 1 by their optical paths and the resulting registration pattern.

Light beam L1 shows the well known interference optical method in reflection for measuring craze zone sizes at the crack tip. The method can easily be applied to static or very slowly moving cracks. With increasing crack speed even at a relatively slow speed of 1 mm/s problems of short time physics arise. They are induced by the application of microscopy to a moving object.

Working at a visual magnification of 400x with a field of 0,3 mm, which is needed for adequate measurements of a craze zone length of about 35 μm, it must be established for the registration of the interference fringe pattern of craze zone and crack opening that the movement blur caused by the running crack is in the order of only 1 μm. The solution with its technical device is indicated in the block diagram of Fig. 2.

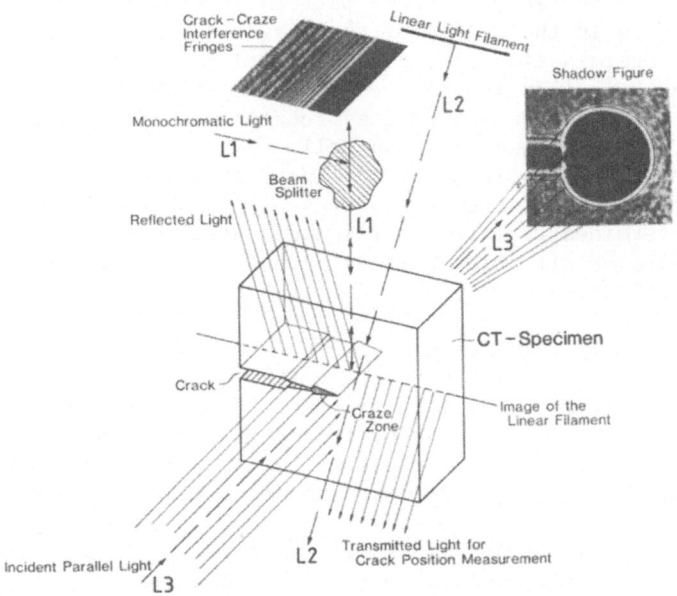

Fig. 1: Schematic diagram of three optical methods and their principal
light paths L to measure characteristic crack propagation parameters

    L1: Interference optics for determination of craze zone sizes
    L2: Total reflection optics for measurement of crack speed
    L3: Shadow optics for determination of stress intensity factors

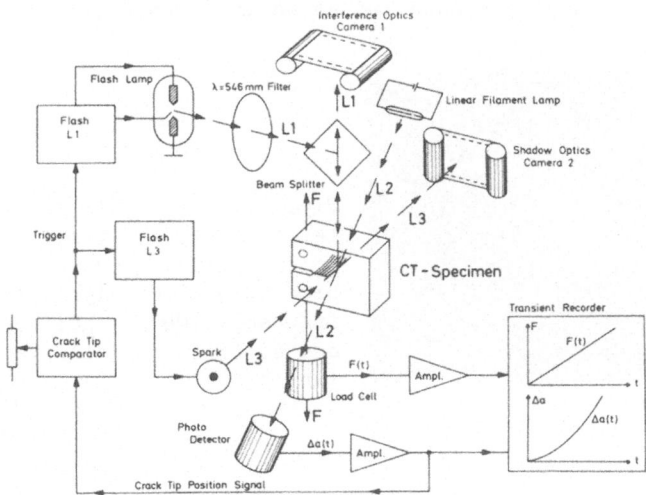

Fig. 2: Measuring and recording devices associated with optical methods
of Fig. 1

The problem of triggering the flash for the photographic registration of the interference fringe pattern in that instant when the running crack is in the middle of the field of view is solved together with the determination of the crack speed by light beam L2. From the registered signal as function of time the crack speed is derived. The applied load F can be measured as function of time using a fast-response quartz transducer load cell.

Light beam L3 offers the virtue of shadow optics by which stress intensity factors at fracture especially in dynamic crack propagation can be determined from the stress field near to the crack tip.

Details of all three optical methods will be given in the following sections.

2.1. Crack speed measurement

In characterizing the brittle fracture of materials the relationship of crack speed to stress intensity factor gives valuable information. As will be shown in Sect. 3.1 this may be used to predict life times of components. Hence, accurate measurement devices of crack speed are needed.

In the slow speed range up to about 10^{-2} mm/s a travelling microscope is usually used to observe the increase in crack length under fatigue or quasistatic loading either on the side of the specimen or − if the specimens are transparent − at the mid thickness. At higher crack speeds especially in the maximum crack speed region special methods have to be used which are adequate to the special problem or material, such as ultrasonic ripple marking [7], high speed cameras or grid method [8]. In the medium crack speed range the following technique [9] schematically indicated in Fig. 1 has proved to be very useful in transparent materials. The crack length is measured photooptically by focusing onto the crack plane a line source of light which is parallel to the crack path and at an incident angle of 45° to it.

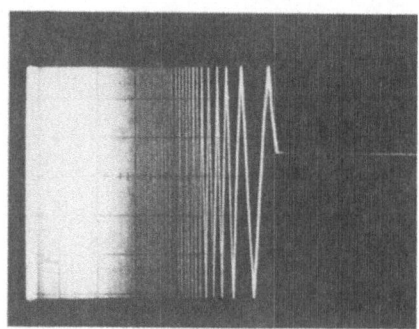

Fig. 3: Oscillogram of a time-crack length curve at transition from slow to fast crack propagation in PMMA of high molecular weight; triangular frequency 100 Hz; horizontal scale one division per millimeter crack length [24]

The amount of light transmitted through the uncracked portion of the specimen (i.e. not reflected by the crack surface) is detected with a photo-diode, so that the signal from the diode is inversely proportional to crack length. By feeding this signal to the X-plates of an oscilloscope, while at the same time modulating the Y-plates with a triangular wave form of known frequency a trace is obtained from which the crack length can be determined as a function of time. A typical example for PMMA is shown in Fig. 3 where also the abrupt transition from slow to fast crack propagation can be clearly seen. The associated change in fracture surface appearance due to transition from slow to fast crack propagation is shown in Fig. 4.

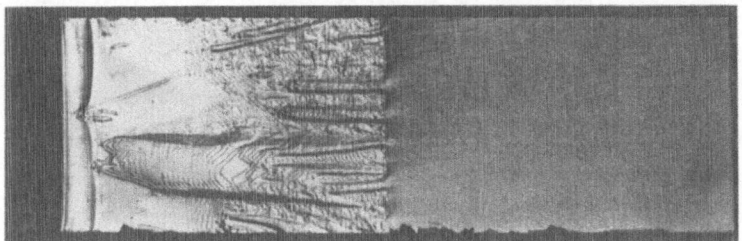

Fig. 4: Fracture surface of high molecular weight PMMA showing transition from slow to fast crack propagation. Specimen thickness shown corresponds to 4.16 mm (crack propagation direction from left to right) [24]

For PMMA Fig. 5 shows the thus measured crack speed å versus K_I curve in double logarithmic scale.

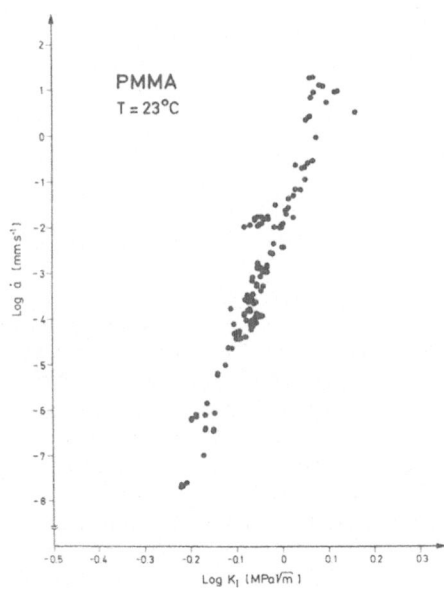

Fig. 5: Crack speed å as function of stress intensity factor K_I [25]

It is perhaps worth mentioning that this technique allows practically continous determination of crack speed to be made up to and
even beyond the transition point, if a higher time resolution is used.

2.2. Interference Optics

In transparent materials the classical methods of light optical interference are well known in the determination of small dimensions which
are of the order of the wave length of light. Their application to investigations of the craze behaviour at crack tips in transparent glassy thermoplastics has led to information on the micromechanics in the
crack tip region [10-13]. A review on this topic has been published
recently [14].

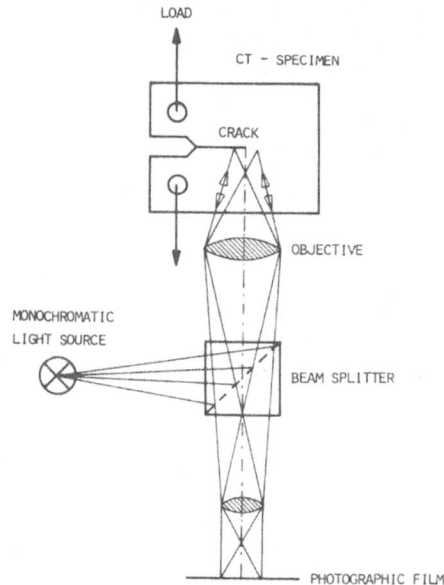

Fig. 6: Sketch of optical interference arrangement for displacement
measurements at the crack tip

A typical arrangement for interference investigations is schematically shown in Fig. 6. The fracture mechanics specimen is illuminated
in reflection with monochromatic light under normal incidence. Typical
fringe systems are shown in Fig. 7 for an unloaded and a loaded crack
in a PMMA-specimen. The crack propagation direction is from left to
right; two fringe systems are apparent, both with decreasing spacing
towards the crack tip: those to the left arise from interference between reflections from the two crack faces, while those to the right
arise from reflections at the two boundaries between crazed and un-
crazed material. In order to calculate the crack openings and craze
thicknesses the positions of the individual fringes are determined

accurately by scanning in a microdensitometer (Fig.7).

The thikness $2v(x)$ of craze and crack opening is given by basic interference theory

at a bright fringe by

$$2v(x) = \frac{\lambda}{2\mu} (n - 1/2) \tag{1a}$$

and

$$n = 1,2,3...$$

at a dark fringe by

$$2v(x) = \frac{\lambda}{2\mu} n \tag{1b}$$

where n is the order of fringe at position x, λ is the wavelength of monochromatic light and μ is the refractive index.

The crack opening can be determined ($\mu = 1$), if the fringe order is known. This method has been used in investigations of stationary cracks to determine static [15] and dynamic [16] stress intensity factors, latter being induced by a shock wave.

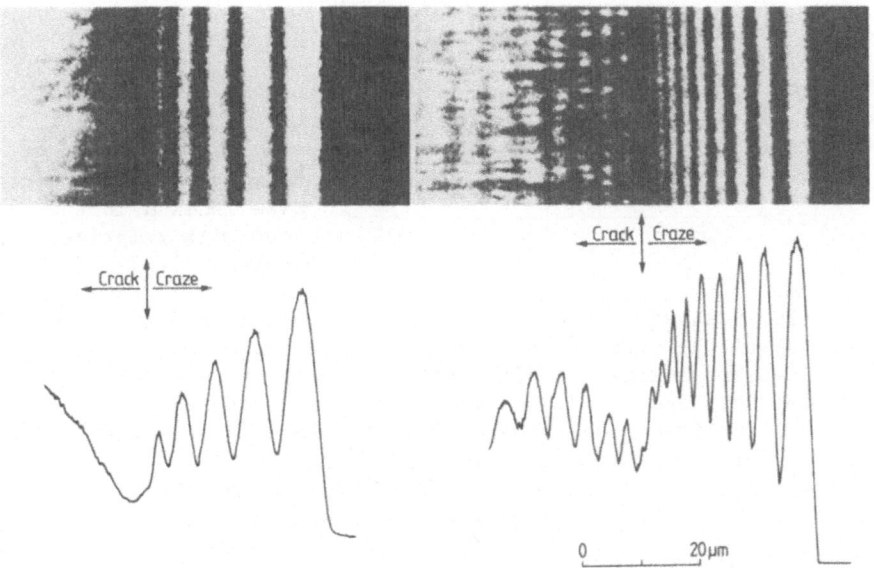

Fig. 7: Interference fringe patterns (λ = 546 nm) and microdensitometer traces of an unloaded (left side) and loaded (K_I = 0,66 MPa $\cdot$ $\sqrt{m}$) crack and craze in PMMA

For the calculation of the craze thickness knowledge of the refractive index of the craze and its variation with strain is required. Refractive indices of unloaded crazes have been determined in different glassy polymers by measuring the critical angle of total reflection of light at the craze/bulk polymer interface [17-19], thus e.g. in PMMA the measured craze refractive index was μ_o = 1.32. Using the Lorentz-Lorenz equation, which relates polarizability to the refractive

index, the density of the craze material and the craze refractive index
μ have been determined as a function of strain. For the evaluation of
interference fringe patterns it is more convenient to use the numbers
of fringes n_1 and n_o of the loaded and unloaded craze respectively
instead of strain or extension ratio. Fig. 8 shows the decrease of craze
refractive index μ and hence of the craze density with n_1/n_o starting
with different values of craze index μ_o in the unloaded state. Experimental results show typical values of n_1/n_o at break in PMMA,
in the range of 2-3 and in PC of 1.4-1.5 leading to craze refractive
indices μ of 1.15-1.09 and 1.19-1.12, respectively. A similar restriction on the range of values of n_1/n_o is found in the experimental
results for quasistatic and cyclically loaded cracks. Hence, with
reference to Fig. 8, the variation in craze refractive index is relatively restricted for the polymer under investigation.

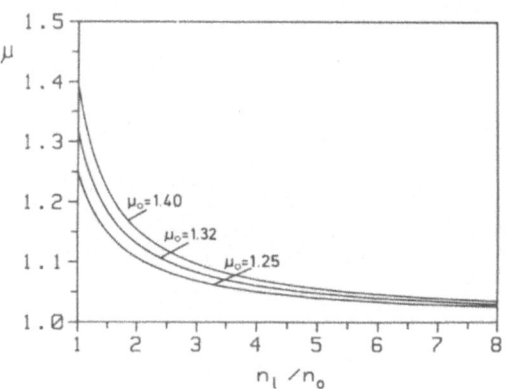

Fig. 8: Variation of refractive index μ in the craze
zone with relative fringe
number n [14]

Fig. 9 shows an example of the experimentally determined shapes
(points) of the crack opening and the craze zone in PMMA of high molecular weight. Taking the different scales of the vertical and horizontal axes into account it becomes evident that the craze is a long thin
wedge. In addition the arrangements of molecules and of stretched and
broken fibrils are schematically indicated. In Fig. 9 the measured
points indicate that the crack tip is blunted and that in this position
the craze width 2v is larger than the crack opening.

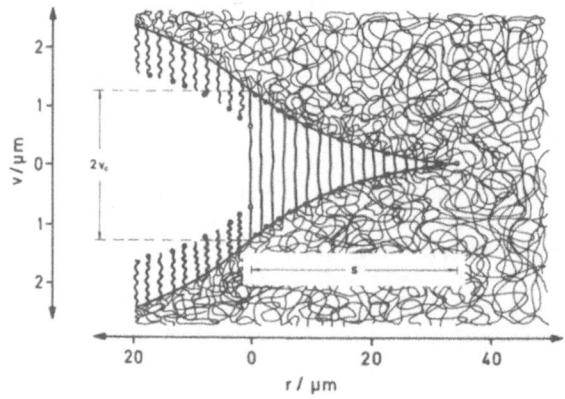

Fig. 9: Measured crack and
craze opening (points) in
PMMA (arrangements of
molecules and deformation
behaviour of fibrils are
schematically indicated)

The similarity between the measured craze zone and the calculated Dugdale plastic zone [20] is apparent, but, some points of detail should be noted. While the position of the crack tip is known fairly precisely, the location of the craze tip and hence the craze length s cannot be measured directly, but can - in a similar way to the maximum craze width 2v and the crack opening COS - only be obtained by extrapolation [14], which is shown in Fig. 9 as the open circle. Also shown are the lines corresponding to the displacements of the Dugdale model. The analytical expression for the displacement 2v of the elastic-plastic boundaries in front of and behind the crack tip was derived by Goodier and Field [21]:

$$2v(x) = \frac{2c\sigma_c}{\pi E^*} \cdot \left\{ \cos\theta \cdot \ln \frac{\sin^2 \left(\frac{\theta_2}{2} - \theta\right)}{\sin^2 \left(\frac{\theta_2}{2} + \theta\right)} \right. \tag{1}$$

$$\left. + \cos\theta_2 \cdot \ln \frac{\left(\sin\frac{\theta_2}{2} + \sin\theta\right)^2}{\left(\sin\frac{\theta_2}{2} - \sin\theta\right)^2} \right\} \tag{2}$$

x = coordinate parameter ($x < |c|$), θ = arc $\cos\frac{x}{c}$, θ_2 = arc $\cos\frac{a}{c}$ = $\frac{\pi}{2}\frac{\sigma}{\sigma_c}$; the position of the crack tip is

a = crack length, c = a + s
σ_c = craze stress, $E^* = E/(1-\nu^2)$ in plane strain, ν = Poisson's ratio)

The width of the plastic zone at the crack tip will be denoted by 2v and in fracture mechanics terms it can be expressed as

$$2v = K_I^2/(\sigma_c E^*) \tag{3}$$

Here the displacement 2v at the crack tip is governed not only by material properties (E^*, σ_c) but also, and more severely, by the stress intensity factor K_I.

Using the relationship between K_I and G_I Eq. (3) may be rearranged in the form:

$$G_I = 2v \cdot \sigma_c \tag{4}$$

which is of special importance for the interpretation of fracture in polymers on molecular basis.

The length s of the Dugdale zone is given by

$$s = a \cdot \left[\frac{1}{\cos \frac{\pi}{2}\frac{\sigma}{\sigma_c}} - 1 \right] \tag{5a}$$

For small scale yielding (i.e. $\sigma \ll \sigma_c$) this can be expressed in fracture mechanics terms as:

$$s = \frac{\pi}{8} \frac{K_I^2}{\sigma_c^2} \qquad (5b)$$

Fitting Eq. (2) to the measured craze profile, values of E* and σ_c may be derived. From Fig. 9 the so calculated displacement values $2\bar{v}(x)$ may be compared with the measured ones and it can be seen that there is a good agreement in the craze zone (as it might have been expected) but the calculated curve for the crack opening is displaced from the experimental points. This apparent discrepancy between the Dugdale model and the experimentally determined crack opening is removed, however, if it is taken into account that the model provides the locus of the displacements of the elastic plastic boundary not only ahead of the crack tip but also behind it. This means that, in thermoplastics, to the measured crack opening there must be added the thickness of the layers of craze material which remains on the fracture surfaces. This thickness is of the same order as the apparent discrepancy; e.g. in PMMA the thickness of the surface layer was determined to be 0.58 μm (although this does vary with molecular weight).

In a thermoplastic material it is, therefore, important to distinguish between the crack opening stretch (= COS) and the maximum craze width 2v. To characterize plastic deformation and fracture behaviour of a thermoplastic material the maximum length of streched fibrils and hence the maximum craze width 2v is a more fundamental parameter than the crack opening stretch. The latter, in addition, depends on the relaxation behaviour of the broken remnants on the fracture surface.

While it has been reported that in PMMA the Dugdale model provides a good quantitative description of the craze zone at the crack tip, the description by the Dugdale model seems not to be so good in some other materials. In Fig. 10 some shapes of crack tip crazes are compiled [14]. Together with the measured points the individual fits of the Dugdale model are given as lines. Clearly the fits in Fig. 10 are not as good as for PMMA in Fig. 9. PES and plasticized PMMA show a still reasonable agreement, whilst PVC indicates and PC exhibits deviations. It seems to be typical that in these materials the application of the Dugdale model with a constant craze stress leads to an overestimation of the maximum craze width at the crack tip which amounts in PC to about 15 %.

Using a modified Dugdale model with a variable craze stress along the craze zone this effect has qualitatively been interpreted. At positions where the constant stress Dugdale model gives displacements higher than the measured ones the actual craze stress must be higher.

The behaviour of the material under high stresses and strains in the micro region at the crack tip should also reflect specific features determined in macroscopic experiments. In thermoplastic materials the dependence on strain and time is of prime importance.

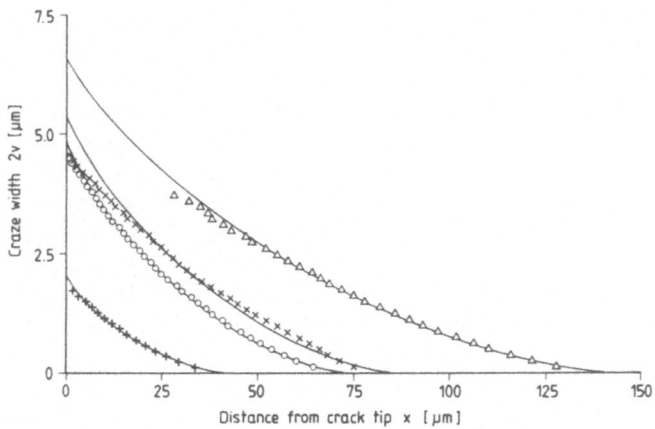

Fig. 10: Measured craze zones at the crack tip and fits by the Dugdale
model (−) for different polymers: + PES; o plasticized PMMA, x PC;
plasticized PVC; [14]

2.3. Shadow Optics

The shadow optical method of caustics is an optical device for measuring stress concentrations and especially stress intensity factors.
Thus strain energy release rates may also be derived.
    The method has been developed by Manogg [22] and later on extended
and refined. Here only a short introduction will be given and for more
detailed information reference should be made to the recent review article of Kalthoff [23].
    The physical principle of the method is illustrated in Fig. 11.
A precracked specimen under tensile load is illuminated by parallel
light with the direction of the incident light beam also parallel to
the crack plane. In Figs. 11 b and c, cross-sections through the specimen
are shown for the two basic different situations of transmission and
reflection referring to a transparent and non-transparent material,
respectively. Due to the stress concentration the physical conditions
at the crack tip are changed. For transparent specimens both the thickness of the specimen and the refractive index of the material are reduced.
Thus, the area surrounding the crack tip acts as a divergent lens
and the light rays are deflected outwards. As consequence, on a screen
(image plane) at a distance z_0 behind the specimen a shadow area is
observed which is surrounded by a region of light concentration, the
caustic (see Fig. 12). Fig. 11c shows the situation for a non-transparent specimen with a mirrored front surface. Due to the surface de-

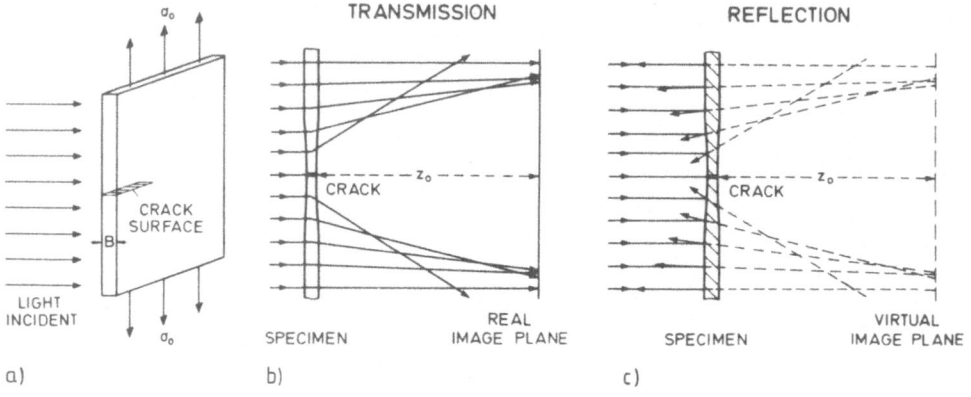

Fig. 11: Physical principle of the shadow optical method of caustics
e.g. [26]

formation light rays near the crack tip are reflected towards the centre line. An extension of the reflected light rays onto a virtual image
plane, at the distance z_0 behind the specimen, results in a light configuration which is similar to the one obtained in transmission
(compare Fig. 12).

For mode-I-loading conditions the following quantitative correlation between the diameter D of the caustic and the stress intensity factor K_I has been derived:

$$K_I = M \cdot D^{5/2} \qquad (6)$$

The factor M is depending on the geometrical arrangement of the experimental set-up and also either on the elasto-optical constants (in case
of transmission) or on the elastic constants (in the case of reflection)
of the material:

$$M = \frac{2\sqrt{2\pi}}{3 \cdot f_{o,i} \cdot c \cdot d_{eff}^{5/2} \cdot z_0} \qquad (7)$$

where: d   = geometrical thickness of the plate,
  d_{eff} = effective thickness of the plate,
       = d for transparent specimens,
       = d/2 for reflecting, non-transparent specimens,
  z_0 = distance between specimen and image plane,
  c   = photoelastic constant,
  $f_{o,i}$ = numerical factor for outer/inner caustic.

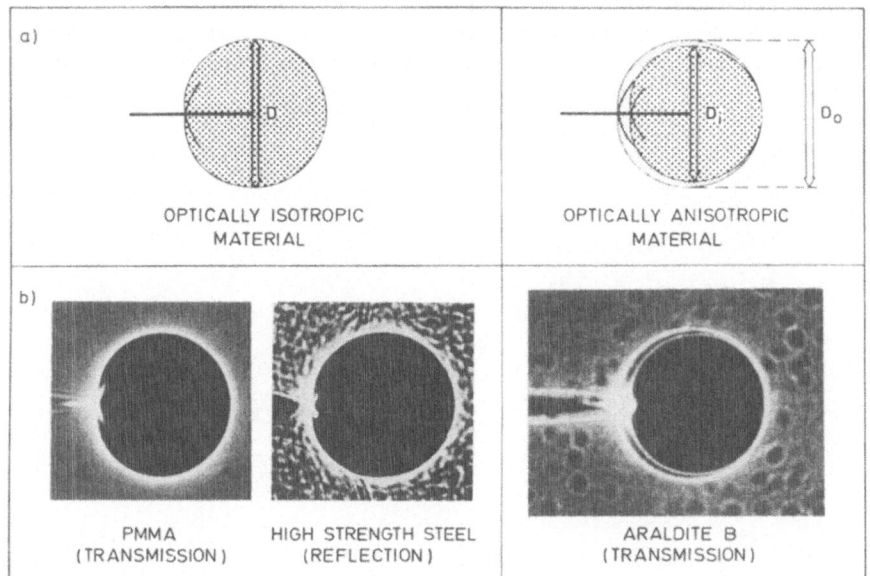

Fig. 12: **Caustics around a crack tip due** to different optical conditions under Mode-I-loading
a) calculated, b) measured (after Kalthoff [26])

The single caustic curve (D) obtained for isotropic materials (like PMMA) splits up into a double caustic ($D_{o,i}$) for optically anisotropic materials (like Araldite). Fig. 12 compares the theoretical results with experimentally observed caustics in transmission and reflection.
     It should be noted that during transient processes (e.g. crack propagation, impact loading) in polymers the time dependence of the elastic and elasto-optical constants has to be taken into account.

3. APPLICATION TO FRACTURE

In this section the optical methods described above will be applied to fracture which has been induced by different types of loading: quasi-static, fatigue, impact.

3.1. Quasi-static loading

Here some effects of static and slowly increasing tensile loading will be discussed.

3.1.1. Creep-rupture and fracture mechanics. In order to determine the long-time behaviour of polymeric components under static tensile load creep-rupture-tests are performed. Fig. 13 shows fracture stresses against life-time t_L for PMMA measured (points) at elevated temperatures. Since such measurements are very time consuming the question arises of long-time prediction and thus the extrapolation of these curves.

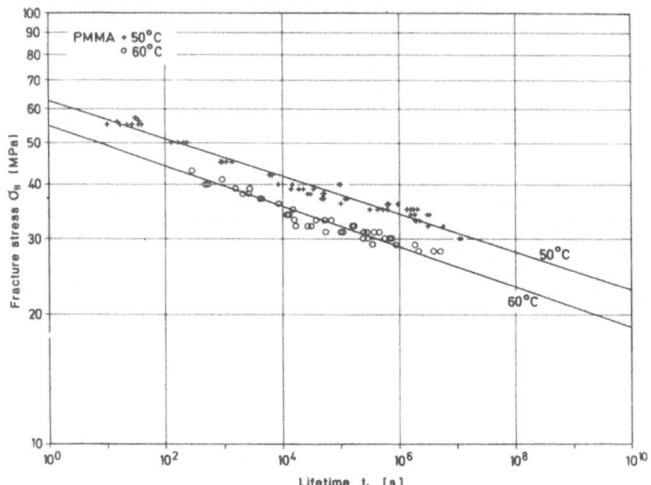

Fig. 13: Creep rupture curves for PMMA comparing experimental (+,o) and calculated (-) results [30]

For brittle fracturing materials like ceramics a new concept based on a fracture mechanics approach has been developed to predict lifetimes [27,28] and it has been extended [29] and after further refinement successfully applied [30] to brittle fracture of PMMA.

It may be briefly characterized as follows: preexisting flaws will grow under the influence of load up to that critical size at which the components fail catastrophically. Thus the basic assumption in this concept is that the total life-time is determined by the time of crack propagation. Knowing the material specific relation (see Fig. 5)

$$\dot{a} = AK_I^n \qquad (8)$$

its integration with respect to time and crack length leads to the time for crack propagation t_c

$$t_c = \frac{2 \cdot a_o^{(1 - n/2)}}{f^n \cdot A \cdot (n - 2)} \; \sigma_B^{-n} \qquad (9)$$

with σ_B applied stress, f correction factor for crack- and specimen geometry; a_o initial flaw size.

The determination of a_o is, however, a major problem. This has been handled by fitting Eq. (9) to measured data (σ_B, t_L) at short times, using following rearrangement of Eq. (9)

$$\sigma_B = B \cdot t_c^{\,m}$$

with

$$B = \left[\frac{2a_o \dfrac{(1 - n/2)}{}}{f^n \cdot A \cdot (n - 2)} \right]^{-m} \quad ; \quad m = -\frac{1}{n} \tag{10}$$

The thus determined life-time curves are shown in Fig. 13 as solid lines which indicate a good agreement with the measured points.

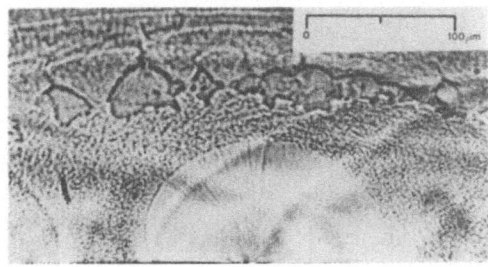

Fig. 14: Details of a fracture surface in PMMA revealing the origin of fracture induced at 50°C [30]

It may be argued, that this agreement might be fortuitous. Therefore it is worth-while to look into the different time contributions of the damage mechanisms involved: first initiation and growth of crazes, followed by initiation and growth of cracks. It has been demonstrated [30] that at least for some thermoplastics at long lifetimes the times needed for craze initiation and growth are negligibly small compared to the total time. Secondly, a range of initial crack length a_o may be derived from the data (B in Eq. (10)) leading to (100 – 140) μm at 50 °C in PMMA. An inspection of fracture surfaces indeed reveals structures of this size as is shown in Fig. 14.

3.1.2. Craze zone at moving crack tips. In order to get information on the craze size in that crack speed range (Fig. 5) which is relevant for life time predictions microscopic interference optics have been applied [25]. It should be emphasized that in these experiments the following relevant parameters were measured simultaneously: interference fringe pattern, load, crack length and crack speed. Thus a complete characterization of any instant during the fracture process could be obtained.

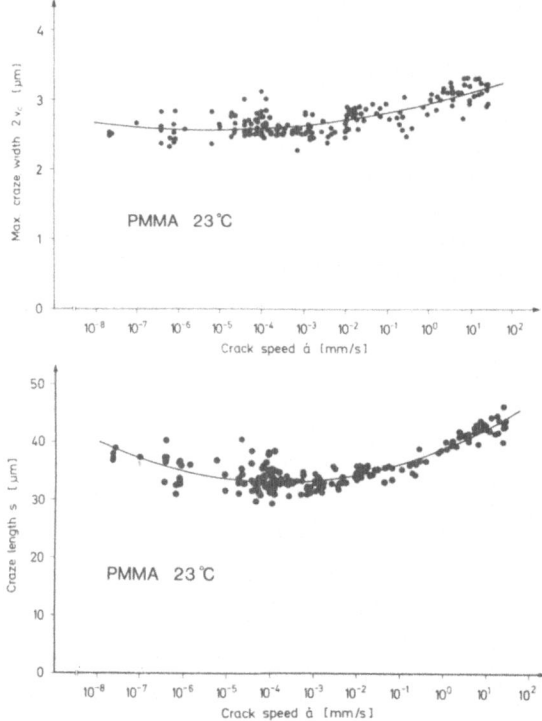

Fig. 15: Characteristic dimensions of the craze zone at the crack tip and their variation with crack speed [25]

a) max. craze width $2v_c$
b) craze length s

Figure 15a shows the maximum craze width $2v_c$ as a function of crack speed $\dot{a}$ and Fig. 15b the corresponding craze length s. The mean values obtained are $s = (34.9 \pm 3.1)$ µm and $2v_c = (2.7 \pm 0.2)$ µm. A closer inspection of these results reveals that both of these craze dimensions go through a minimum at speeds of about $10^{-4} - 10^{-3}$ mm/s. The effect is more pronounced for the craze length (Fig. 15b) than for the craze width (Fig. 15a). The increase in craze dimensions during crack propagation is to be expected at lower speeds due to a long-time-effect (visco-elastic- behaviour) and at higher speeds due to a temperature effect.

In PMMA the Dugdale model gives a good description of the craze shape at the crack tip. For the calculation of the Dugdale plastic zone size, in addition to an accurate K_I-value, Young's modulus (E) and craze stress (σ_c) must be known. These mechanical parameters are time-dependent in a visco-elastic material like PMMA and hence for propagating cracks a time-dependence, or equivalently, a crack speed-dependence of these "moduli" will be involved. By fitting the Dugdale model to the measured craze-zones, as shown in Fig. 15a,b, σ_c and E are found to be dependent on crack speed $\dot{a}$ as shown in Fig. 16. In the investigated crack speed range σ_c and E increase from about 60 to 120 MPa and from 2000 to 3400 MPa, respectively. This increase in σ_c is the reason, that at nearly constant maximum craze width the strain energy release rate G_I increases with crack speed.

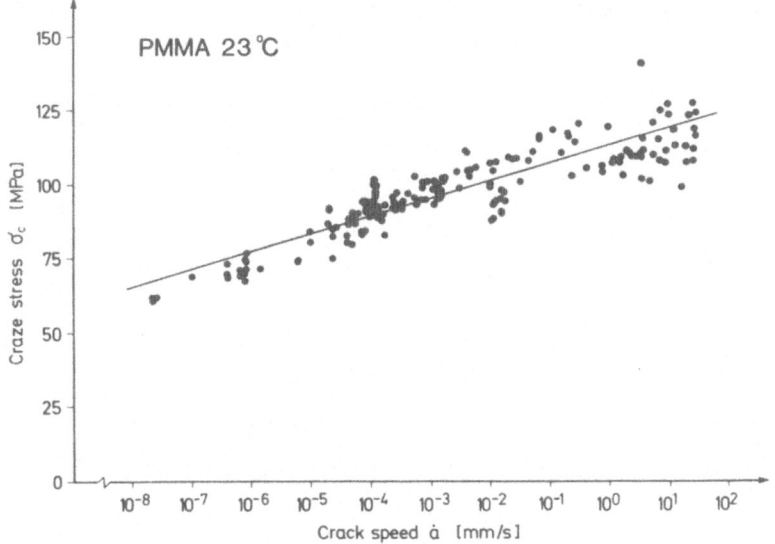

Fig. 16: Craze stress σ_c derived by the application of the Dugdale model to measured craze sizes [14]

Attempts have been made to correlate crack speed $\dot{a}$ with time t. An analysis of the fracture behaviour of thermoplastics shows that it is essentially determined by craze formation and stretching of the fibrils up to fracture. Therefore, the time involved in this process is considered to be the relevant time t which may be calculated by:

$$t = s/\dot{a} \qquad (11)$$

Thus the crack speed axis can be converted into a time scale ranging from about 10^4 to 10^{-4} seconds. On this basis a comparison has been made with data determined in macroscopic experiments. Fig. 17 represents the time dependent creep moduli E as compiled in [14]. It can be seen that the creep modulus curve obtained from the material behaviour in the micro region near the crack tip corresponds to a curve of about 1 % strain. Using the fracture mechanics approach, the strain ε_y near the crack tip can be calculated from

$$\varepsilon_y = \frac{1}{E(t)} \quad \frac{K_I}{\sqrt{2\pi \ r}} (1 - \nu - \nu^2) \qquad (12)$$

in plane strain. Thus, for a moving crack the strain at the craze tip (r = s) may be estimated to be 0.8 − 0.9 %. Hence, in the field of macroscopically determined creep moduli in Fig. 17 the position of that curve which characterizes the strain field near and around the craze zone seems to be reasonable.

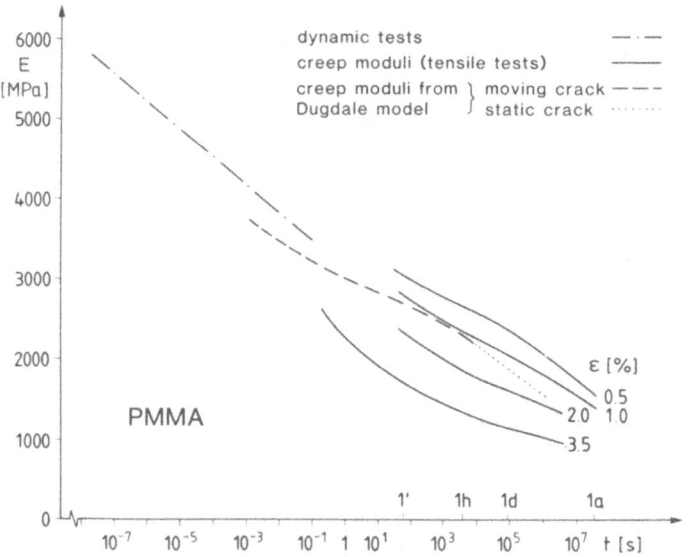

Fig. 17: Creep moduli E at different strains ε as function of loading time t determined in macroscopic tensile tests and in the micro region at the crack tip [14]

3.2. Fatigue

3.2.1. Some introductory remarks. The crack growth behaviour of polymers under cyclic loading has been intensively studied [31–33]. Here attention will be concentrated on the relationship between crack and craze growth in fatigue, and there will be only some brief introductory comments on macroscopic aspects of fatigue crack propagation in polymers. The main points will be the microscopic aspects of the so-called "continuous" and "discontinuous" crack propagation in fatigue. These two types of fatigue crack growth differ in their formation and growth mechanisms. In "continuous" crack propagation each loading cycle causes an increment in crack length, whilst in "discontinuous" crack propagation the crack tip remains stationary for many loading cycles and then at some particular cycle it advances by an increment.

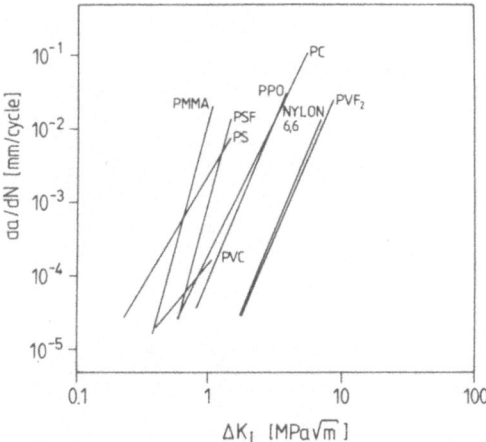 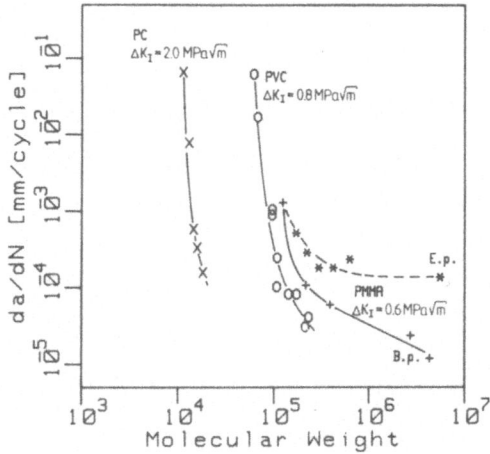

Fig. 18: Fatigue crack propagation rate da/dN as a function of relative stress intensity factor ΔK_I in different polymers [31]

Fig. 19: Effect of molecular weight on crack propagation rate da/dN in different thermoplastics
E.p. = Emulsion polymerized,
B.p. = Bulk polymerized (e.g. [14])

It is now well established that fracture mechanics is an excellent tool to characterize the growth of a fatigue crack from an existing crack under mode-I-loading conditions. It allows the determination of the increase in crack length da per number of cycles dN as a function of relative stress intensity factor ΔK_I (see Fig. 20). Examples of the relationship da/dN versus ΔK_I are given in Fig. 18 for different polymers. In addition to the chemical composition of the polymer the configuration of the macromolecule plays an important role as can be seen from the effect of molecular weight on the fatigue crack resistance in Fig. 19 [14]. Although the constant ΔK_I-values applied had different amplitudes, the fatigue crack propagation behaviour is quite similar for the three polymers. They all show a sharp decrease in the crack propagation rate with increasing molecular weight and - if an appropriate linear scale is taken - a levelling off at higher molecular weights. Copolymerisation and plasticizing can also effect fatigue crack propagation as can be seen from the data on PMMA and PVC. Additives, e.g. stabilisers, may also have an influence. It is often due to differences in these intrinsic parameters that test results in the literature on commercially available polymers from different manufacturers are not fully comparable.

The influence of mechanical parameters such as mean stress or ratio between minimum and maximum load on fatigue crack propagation has been investigated for various polymers. These effects are usually incorporated into the well known Paris law using empirical parameters. It is also interesting that an influence of test frequency has been observed for different polymers in which the fatigue crack propagation rate decreases with increasing frequency [31].

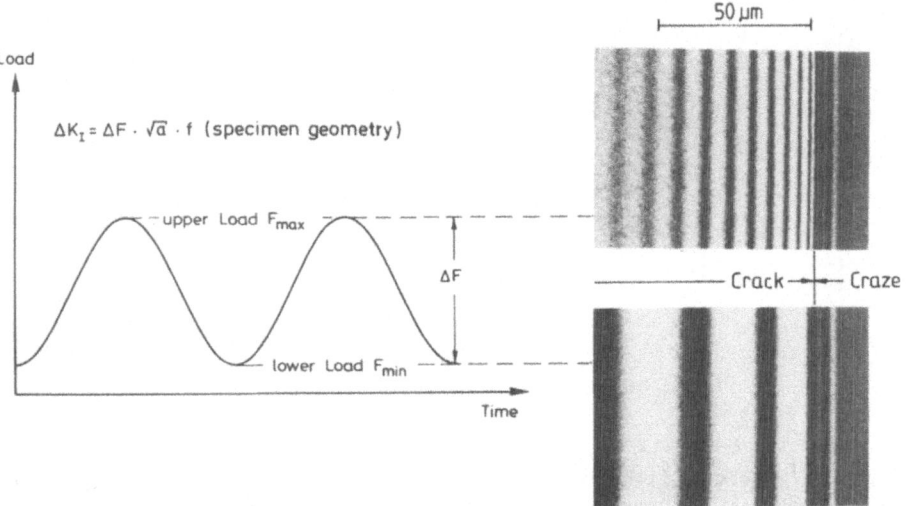

Fig. 20: Example of interference fringe patterns measured during fatigue crack propagation (ω = 50 Hz; da/dN = 1.2 10^{-5} mm/cycle) in PMMA at upper and lower loads [36]

It is interesting to note that a new model has been developed which combines concepts of fracture mechanics and rate process theory [34]. Effects of molecular weight, molecular weight distribution, frequency, R-ratio are incorporated and have been sucessfully applied to PS [35].

In the optical interference set-up used in fatigue by the author's group the following relevant parameters can be simultaneously recorded during fatigue crack propagation: the interference fringe pattern, the load acting ΔF, the crack length a, and the number of load cycles N. From this the craze zone dimensions (length, width and contour), the stress intensity factor ΔK_I, the crack propagation rate da/dN and crack speed $\dot{a}$ can be derived. Thus a full characterization of crack and craze growth under fatigue loading can be obtained. Figure 20 shows an example of the interference fringe systems generated during fatigue crack propagation at a loading frequency of 50 Hz in PMMA photographed at the upper and the lower load. The evaluation of the interference fringe systems is performed as described above.

3.2.2. Continuous Crack Propagation. Fig. 21 shows the material response in the micro region at the crack tip by representing the maximum craze width $2v_c$ and the craze length s as functions of crack propagation rate da/dN. It can be seen that during fatigue crack propagation in high molecular weight PMMA both, the maximum craze width $2v_c$ (and hence also the COS) and the craze length s are not constant but increase by a factor of nearly 5 in the predominantly investigated crack propagation range between 10^{-5} and 10^{-2} mm/cycle. It should be mentioned that the plotted points were obtained at four frequencies in a range from 0.4 to 50 Hz.

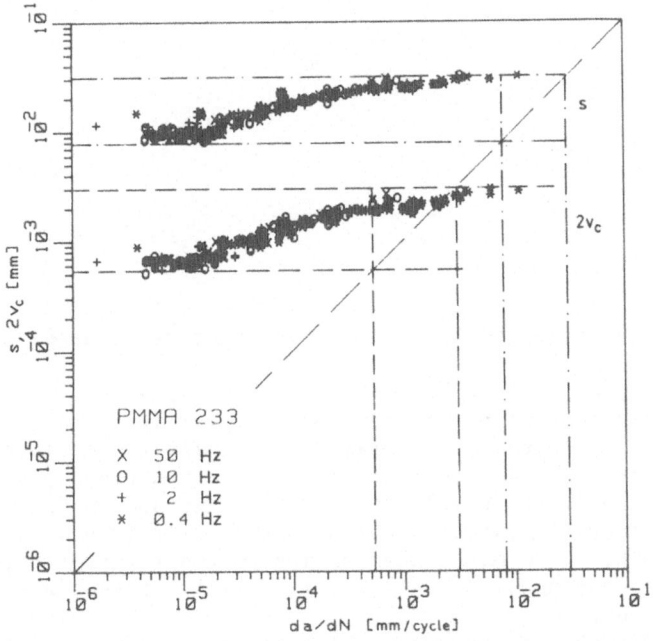

Fig. 21: Characteristic craze dimensions (length s, maximum width $2v_c$) versus crack propagation rate da/dN [36]

Figure 21 suggests lower limits of $2v_c$ and s at crack propagation rates smaller than 10^{-5} mm/ cycle. As has been shown [36], an upper limit can be derived by comparing the curve with the results for continuously moving cracks under constant load using the crack speed à as a common measure. Since à is equal to da/dN · ω (ω = frequency) the data will be separated into different curves according to their different frequencies (Fig. 22).

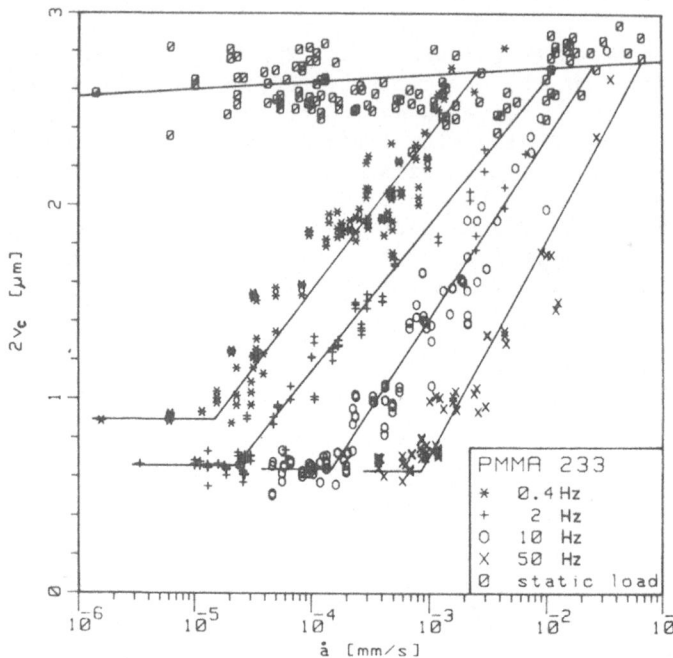

Fig. 22: Maximum craze width $2v_c$ as function of crack speed á [36]

Information on the craze growth and on the fibrillation process are obtained by direct comparison of the two characteristic craze dimensions, maximum width and length. They remain practically proportional to one another up to the values determined for continuous crack propagation under constant load. It is evident that the increase in craze length must occur by fibrillation of the amorphous bulk polymer. However, the thickening of the craze may either be due to additional fibrillation or to enhanced stretching of already formed fibrils or even a combination of both processes.

Further information on craze thickening in fatigue can be obtained by comparing the width $2v_c$ of the strained craze with the craze width $2v_o$ at the lower fatigue load. The ratio of these is assumed to be a measure of the amount of fibrillation. From such results it has been deduced that both processes (i.e. new craze material being formed and also an increased stretching of the fibrils) occur with increasing crack propagation rate. If the strain in the craze zone is defined by $\varepsilon = (2v_c - 2v_o)/(2v_o)$ then the strain on the fibrils is nearly doubled from about 70 % at low crack propagation rates of 10^{-5} mm/cycle up to 120 % at 10^{-2} mm/cycle, thus approaching the magnitude measured for static cracks. This clearly demonstrates, that the assumption of a constant ratio between the thicknesses of relaxed and extended crazes is not generally valid in fatigue.

Applying the Dugdale model to the measured craze zone sizes in high molecular weight PMMA under fatigue the thus derived craze stresses σ_c increase with crack propagation rate or crack speed respectively from about 70 to 100 MPa, exhibiting a general tendency which is quite similar to that measured under static loading (see Fig. 16).

The magnitude of Young's modulus as determined from the craze dimensions lies in the same range as that measured for crack propagation under constant load; the fatigue results do not show any significant change of E with crack speed. There is, however, a slight increase in E with frequency which has also been reported [43] for moduli measured under small strains. Since the values of the latter lie about 50 % higher than of those determined here, one has to conclude that the moduli here must be characteristic for large strains. This conclusion becomes even more evident if it is recognized that the material around the craze zone is highly strained due to the high stresses around the crack tip and that the Dugdale model strictly applies to the contour of the elastic/plastic boundary in that region. Thus, a similar result is derived as that already obtained for continuously moving cracks.

In the literature on fatigue crack propagation in metals and, later, in polymers it has been well documented that fracture surfaces can exhibit ripple markings which are due to the cyclic loading process.

In polymers two sets of markings on fatigue fracture surfaces have been observed which differ distinctly in their loading genesis. One set is called striations, where each striation is found to correspond to the incremental advance of the crack front caused by just one loading cycle (this has also, somewhat confusingly, been called continuous crack growth).

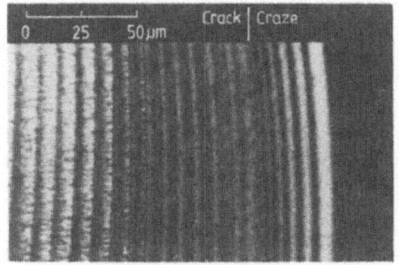

Fig. 23: Optical interference photograph showing fatigue striations on the fracture surface (left side) and craze zone in a partially broken PMMA specimen [41]

The spacing between fatigue striations have often been correlated with the plastic zone size and especially its length. A direct comparison between striation spacing and length of the craze zone in PMMA can be made through the interference optical photograph shown in Fig. 23 [37]. It can be seen that in this case the craze length is significantly larger than the striation spacing. An indirect comparison between craze dimensions and line spacing is obtained from Fig. 21 since in high molecular weight PMMA the striation spacing reflects the crack growth during one cycle. It can be seen from Fig. 21 that up to about $2 \cdot 10^{-3}$ mm even the craze width is always larger than the increase in crack length during one cycle. Only at high crack propagation rates ($> 3 \cdot 10^{-2}$ mm/ /cycle) the striation spacing will be larger than the craze length.

It can, thus, be concluded that during the so-called continuous mode of fatigue crack propagation there is no simple correlation between fatigue striation spacing and craze dimensions.

3.2.3. Discontinuous crack propagation. The so-called discontinuous fatigue crack growth was first observed in PVC [38] and subsequently in many other polymers. It has been pointed out [31] that the formation of discontinuous growth bands is dependent on the molecular weight and the molecular weight distribution of the polymer and on mechanical loading conditions and that low molecular weights and low stress intensity factors ΔK_I seem to favour their formation. For example, low molecular weight PMMA may exhibit discontinuous band growth at low ΔK_I-levels, while high molecular weight PMMA shows striations at all ΔK_I-levels. In PVC, on the other hand, discontinuous growth bands are found at all ΔK_I-levels.

It has been noted [39] that in discontinuous crack growth the number of cycles between two successive bands depends on the polymer and strongly decreases with increasing relative stress intensity factor ΔK_I. An extrapolation of these data to the value of one cycle/band – that is to continuous crack growth – gives ΔK_I-levels which are too large, thus indicating that the mechanisms operating in continuous and discontinuous crack growth are fundamentally different. The interpretation of the band spacing as the plastic zone length has been favoured by different authors [31,38].

A deeper insight into the growth behaviour of the craze zone at the crack tip during retarded crack growth in PVC has been obtained by recent optical interference measurements [14,40].

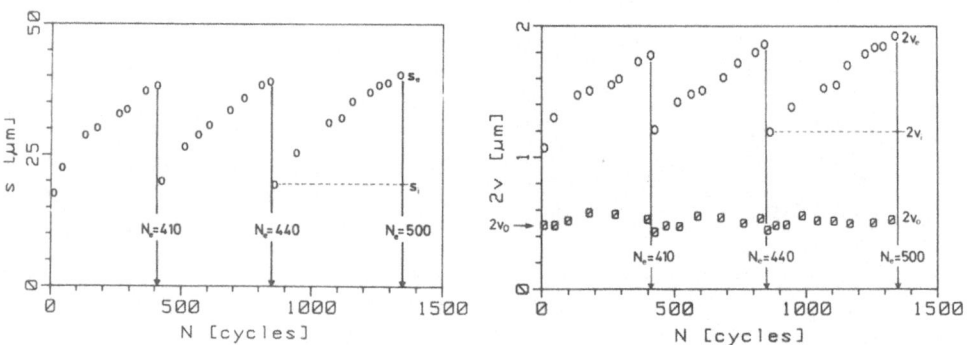

Fig. 24: Craze growth as function of the number of loading cycles N between successive crack jumps in PVC at ΔK_I = 0.48 MPa $\sqrt{m}$
(a) craze length s, (b) craze width 2v directly at the crack tip under upper ($2v_i$ to $2v_e$) and lower ($2v_o$) load [40]

The quantitative development of craze length s and maximum craze width 2v at the crack tip are given in Fig. 24 as functions of the cycle number N for several successive crack jumps. In this experiment ($\Delta K_I = 0.48$ MPa $\sqrt{m}$, R < 0.1) the crack remained stationary up to a critical cycle number N_e which is connected with a critical end length s_e or end width $2v_e$, respectively. Together with the result of a directly observed crack jump [14] it is now obvious that the crack does not go through the craze all the way up to the craze tip as was usually assumed.

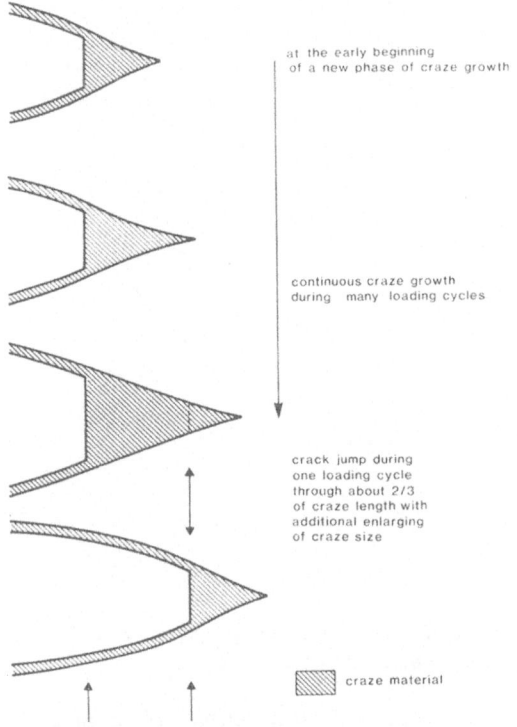

at the early beginning
of a new phase of craze growth

continuous craze growth
during many loading cycles

crack jump during
one loading cycle
through about 2/3
of craze length with
additional enlarging
of craze size

craze material

crack tip before and after jump

Fig. 25: Schematic representation of optical interference results of continuous craze and discontinuous crack growth in PVC [14]

As a consequence of these results the discontinuous crack and continuous craze growth may be schematically represented as shown in Fig. 25. In order to obtain information on the sudden breakdown of the craze after a long sustaining period of several hundred loading cycles, it is helpful to analyse the craze growth in detail between two successive crack jumps. Obviously the craze lengthening takes place by fibrillation of amorphous bulk material. The thickening of the craze may be due to an additional fibrillation of bulk material or due to enhanced stretching of already formed fibrils or possibly due to both processes.

To investigate this question, optical interference measurements were performed on PVC and the craze width $2v_o$ at the crack tip was determined at the lower load. $2v_o$ can be regarded as a direct measure of the amount of fibrillated material. The data suggest a constant value $2v_o$ (Fig. 24). This result is very instructive since it implies that the craze thickening at the crack tip does not occur by additional fibrillation of new bulk material and hence it must take place by an enhanced stretching of the already formed fibrils. An estimate on the increase in strain in the craze between two successive crack jumps gives a factor of 2 at breakdown.

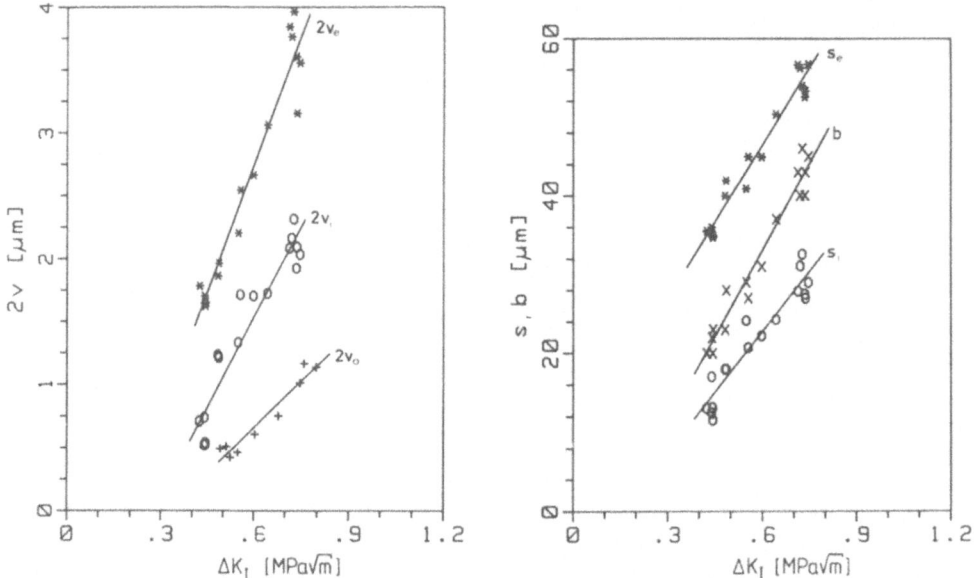

Fig. 26: Influence of ΔK_I on craze width 2v, craze lengths s and band width b in PVC during discontinuous crack growth [40]

In the interval between two successive crack jumps the crack length remains unchanged and hence the stress intensity factor is constant. In order to investigate the influence of the stress intensity factor on the craze dimensions, crazes under different loads have been compared. Fig. 26 shows the craze width 2v measured directly at the crack tip as a function of ΔK_I. The maximum craze widths $2v_e$ and $2v_i$ were measured under upper load just before and after a crack jump, and $2v_o$ is the craze width under lower load between two successive crack jumps. For all three characteristic craze widths a linear increase with ΔK_I can be seen. This behaviour is quite similar to that already found for continuous crack growth in PMMA. However, the slopes of the curves differ to such a degree, that a quite different straining behaviour of the craze fibrils must be assumed. This is brought out more clearly when the data are expressed in terms of the previously defined strain. In the situation where the individual fibrils are most severely stretched, just before a crack jump occurs, the strain decreases from about 400 % at ΔK_I = 0.44 MPa $\sqrt{m}$ to about 250 % at ΔK_I = 0.76 MPa $\sqrt{m}$.

The influence of ΔK_I on craze length s in PVC is shown in Fig. 26. Both characteristic craze lengths, s_e just before and s_i just after a crack jump, show a linear increase with ΔK_I, thus leading to a similar result as that observed during continuous fatigue crack growth (see Fig. 21).

In Fig. 26 the distances b of the crack jumps measured as growth bands on the fracture surface are also given. For a given ΔK_I the sum of band width b and initial craze length s_i should, in principle, be equal to the craze end length s_e. At low ΔK_I-levels this is almost the case. With increasing ΔK_I, however, the sum of b and s_i increasingly exceeds s_e, so that at ΔK_I = 0.76 MPa $\sqrt{m}$ the excess amounts to nearly 20 µm. This effect might be explained as follows. As already noticed in Fig. 24 and 25 an enlargement in craze length occurs in the very moment of the crack jump. The increase in craze length with ΔK_I has been quantitatively explained by application of the Dugdale model [14,40]. From Eq.(5b) it follows that the stress on the craze boundary and hence on the fibrils is proportional to ΔK_I and inversely proportional to the square root of craze length. From Fig. 26 it would be expected that the initial craze length as the difference between s_e and b would be nearly constant. With increasing ΔK_I a constant initial craze length requires that the fibrils would have to bear an increasing stress. The material, therefore, responds by a redistribution of stress which results in an enlarged craze length under lower stress. It is interesting to note that the measured initial craze lengths s_i lead to a nearly constant stress on the fibrils [40].

In this context it should be mentioned that the optical interferences results on PVC indicate that the "simple" Dugdale model with a constant stress is not appropriate to describe the measured craze contour sufficiently accurate. In contrast to PMMA a variable stress distribution along the craze as determined for PS seems to give a better fit to the measured craze contour in PVC. This "second order correction", however, does not change the principal conclusion of the preceding discussion.

3.3. Dynamic Loading

The impact fracture toughness K_{Id} is usually determined with prenotched bend specimens in instrumented impact tests. The specimens are loaded by a drop weight or by a pendulum type impact tester. Strain gages at the tup of the striking hammer measure the load during impact. From the critical load at the moment of crack initiation the impact fracture toughness is derived.

Using the advantages of the shadow optical method of caustics the mechanical behaviour of cracks under impact loading was investigated in Araldite B and a high strength steel by measuring the dynamic stress intensity factors directly at the crack tip by Kalthoff and coworkers [23,26,42,44]. Here, mainly their results will be reviewed.

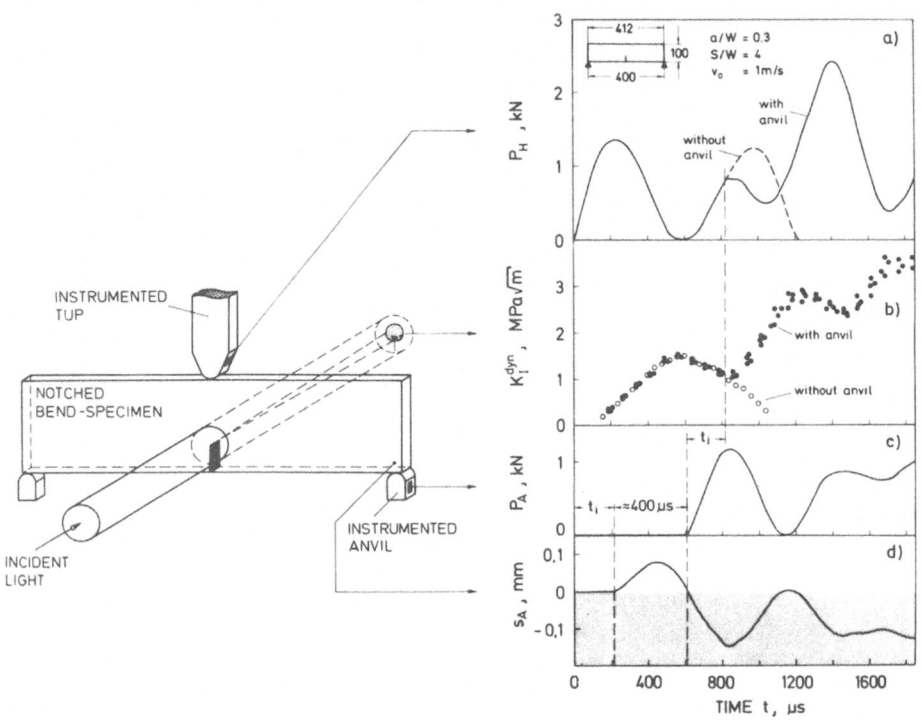

Fig. 27: Mechanical response of a prenotched bend specimen to impact loading [42]

Fig. 27 gives a schematic view of the experimental set-up with the shadow optical arrangement. The influences of dynamic effects were evaluated by comparing the dynamic stress intensity factors, K_I^{dyn}, with equivalent static stress intensity factors, K_I^{stat}. The static values were determined from the measured hammer load P_H utilizing conventional static stress intensity factor formulas from ASTM E 399. Quantitative data are given in Fig. 28. The specimens were impacted at a velocity of 0.5 m/s by a hammer with a mass of 4.9 kg. The dynamic stress intensity factors K_I^{dyn} (experimental points) and the corresponding static stress intensity factors K_I^{stat} are plotted as functions of time. The times are given in absolute units and also in relative units by normalization with the period τ of the eigen-oscillation of the impacted specimen.

The K_I^{stat}-values show a strong oscillation behaviour, whereas the actual dynamic stress intensity factors K_I^{dyn} show a more steadily increasing tendency. In the small time range of $t < \tau$ these differences are very pronounced, while they become smaller with increasing time. But even for times larger than 3 τ the influences of dynamic effects are obviously acting and there are still marked differences between K_I^{stat} and K_I^{dyn}.

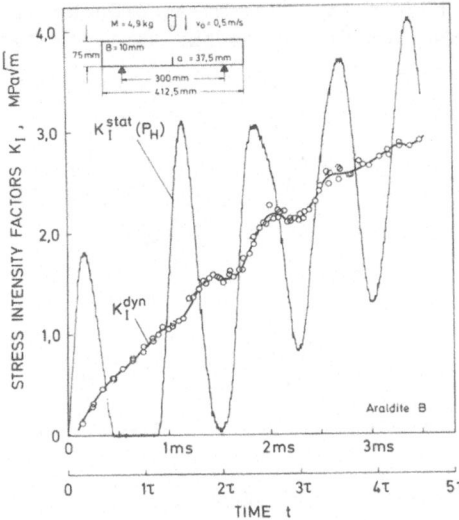

Fig. 28: Stress intensity factor for crack under drop weight loading Araldite B specimen [26]

The dynamic responses not only of the specimen but also of the striking hammer and the anvils were investigated simultaneously as shown in Fig. 27 [42]. It compares the load measured at the striking hammer (a), the stress intensity factor measured at the crack tip (b), the load measured at the anvils (c), and the position of the specimen ends with regard to the anvils (d). The data were obtained with Araldite B specimens impacted at 1 m/s. The τ-value of the specimens utilized for these investigations is about 700 μs. A comparison of the four signals indicates that non-zero loads at the anvils were registered only after a rather long time of about 600 μs. This time is about three times larger then the time it would take the slowest wave, i.e. a transverse wave, to travel from the point of impact to the anvils. This unexpected behaviour is explained by diagram (d) in Fig. 27. A loss of contact is observed between the specimen ends and the anvils. The loss of contact starts at about 200 μs. This time is in agreement with the above consideration of wave propagation times. For about 400 μs the specimen is completely free and only after this time, i.e. at a time of about 600 μs total, the specimen ends come into contact with the anvils.

In accordance with this observation, load values are then recorded at the anvils. With different test conditions this loss of contact can occur later for a second time and loss of contact can also take place between the hammer and the specimen. These processes are illus-

trated in the schematic representation of Fig. 29. Since in the reported experiments the anvils obviously were of no influence during the
early phase of the impact process, additional experiments have been
performed with unsupported specimens. The results are represented by
the dashed curve and open data points in Fig. 27. In accordance with
speculation, the early specimen reaction ($t < \tau$) is the same for both,
the supported and the unsupported specimen.

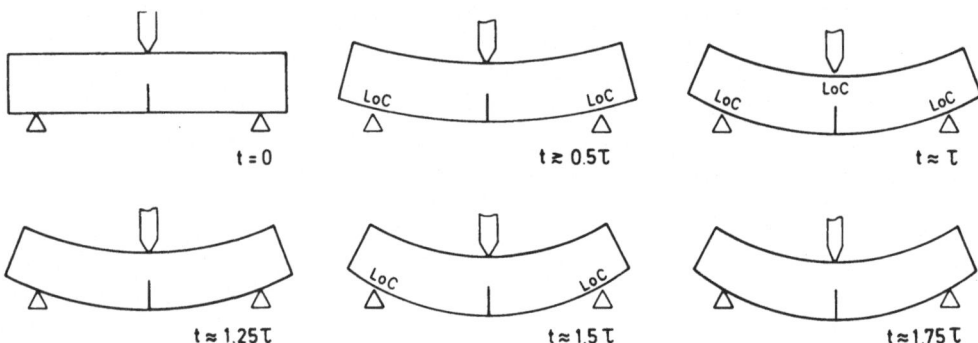

Fig. 29: Loss of contact effects observed with prenotched specimens
under drop weight loading [26]

The measured data demonstrate the strong influence of dynamic
effects on the mechanical behaviour of cracks under impact loading. A
method proposed to ASTM for measuring the dynamic fracture toughness
value K_{Id} in instrumented impact tests assumes that for times larger
than 3τ K_I^{stat}-values would represent a good approximation of the
actual dynamic stress intensity factor K_I^{dyn}. However, data from
these experiments indicate that a static analysis is not adequate to
describe the loading condition in the specimen under the proposed conditions except at much later times during the event. Very large times
to fracture, however, cannot always be achieved. While the time to
fracture is increased if the impact velocity is reduced, the τ-value of
a specimen, on the other hand, is reduced when the specimen dimensions
are reduced. Often both goals cannot simultaneously be reached due to
size requirements which in general demand large specimen dimensions.
The conditions of course become very unfavorable for testing of brittle
materials at high impact speeds. Consequently the determination of
reliable impact fracture toughness data with freedom of choice of test
conditions requires a fully dynamic evaluation procedure.

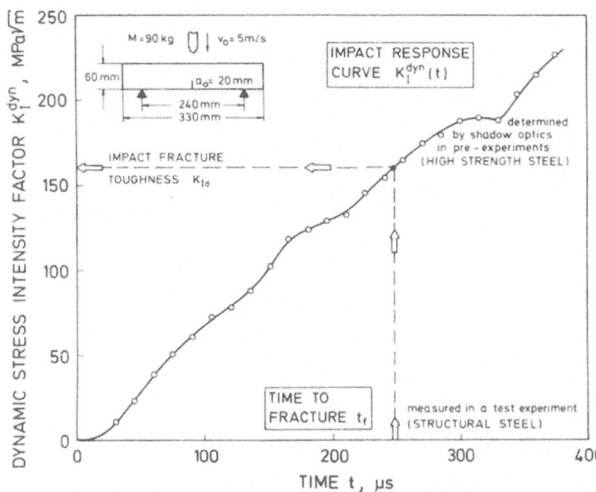

Fig. 30: The concept of impact response curves for measuring the impact fracture toughness K_{Id} [26]

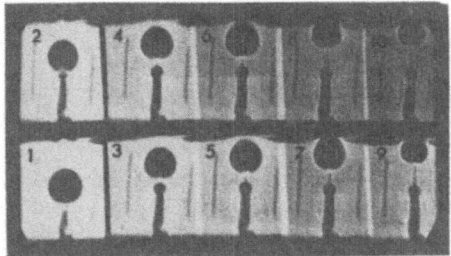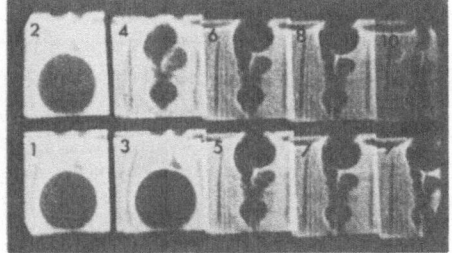

Fig. 31: Shadow optical observation of crack propagation induced by impact loading in high molecular weight PMMA (left side) and in rubber toughened PMMA. Frame rate 10 µs.

Therefore, by Kalthoff et al. [44] the dynamic concept of impact response curves has been developed. The principle of the measuring procedure is illustrated in Fig. 30. For fixed test conditions (i.e. specimen geometry, hammer mass, impact velocity, etc.) the dynamic stress intensity factor versus time relationship is determined in a pre-experiment by means of the shadow optical method of caustics with an edge notched specimen. The obtained curve, $K_I^{dyn}(t)$, describes the response of the specimen to the impact process. This curve, called impact response curve, is controlled by the elastic properties of the system. The dynamic fracture toughness of the material is then determined by performing an impact experiment and measuring the resulting time to fracture. Examples referring to this procedure are shown in Fig. 31 and 32 for rubber toughened PMMA.

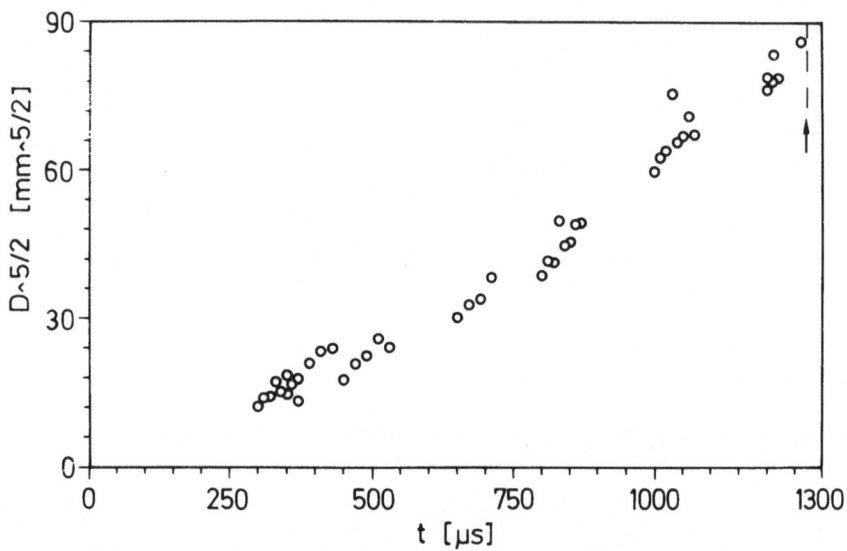

Fig. 32: Diameter of caustics in impact loaded PMMA (rubber toughened) as function of loading time up to fracture initiation (↑)

Acknowledgement:
The financial support by the Deutsche Forschungsgemeinschaft (DFG) which funded much of the work is gratefully appreciated.

References

[1] Williams, J.G., Fracture Mechanics of Polymers, Ellis Horwood Ltd, Chichester (1984)

[2] Kinloch, A.J., Young, R.J., Fracture Behaviour of Polymers, Applied Science Publishers Ltd, London/New York (1983)

[3] Andrews, E.H. (Ed.), Developments in Polymer Fracture -1, Applied Science Publishers Ltd., London (1979)

[4] Kausch, H.H., Polymer Fracture, Springer, Berlin/Heidelberg/ New York (1978)

[5] Bucknall, C.B., Multiphase Polymers and Polymer Composites, in Advances in Polymer Sci., 27, Springer-Verlag, Berlin/Heidelberg/ New York, p. 121 (1978)

[6] Kausch, H.H. (Ed.), Crazing in Polymers, in Advances in Polymer Sci., 52/53, Springer-Verlag, Berlin/Heidelberg/New York/Tokyo (1983)

[7] Kerkhof, F., Naturwiss., 40, 478 (1953)

[8] Doell, W., Int. J. Fract., 12, 595 (1976)

[9] Richter, H., IFKM 9/74, Institut fuer Werkstoffmechanik, Freiburg (1974)

[10] Bessenov, M.I., Kuvshinskii, E.V., Soviet Phys. (Solid State) 3, 1957 (1961)

[11] Kambour, R.P., J. Polym. Sci., A-24, 349 (1966)

[12] Brown, H.R., Ward, I.M., Polymer, 14, 469 (1973)

[13] Weidmann, G.W., Doell, W., Colloid and Polym. Sci., 254, 205 (1976)

[14] Doell, W., in [6] p. 105 (1983)

[15] Sommer, E., Eng. Fract. Mech., 1, 705 (1970)

[16] Sommer, E., Soltész, U., Eng. Fract. Mech., 2, 235 (1971)

[17] Kambour, R.P., Nature 195, 1299 (1962)

[18] Kambour, R.P., Polymer 5, 107 (1964)

[19] Kambour, R.P., J. Polym. Sci. A 2, 4159 (1964)

[20] Dugdale, D.S., J. Mech. Phys. Solids, 8, 100 (1960)

[21] Goodier, J.N., Field, F.A., Fracture of Solids, (ed.) Drucker, D.C., Gilman, J.J., p.103, New York/London, John Wiley (1963)

[22] Manogg, P., Dissertation, Universitaet Freiburg (1964)

[23] Kalthoff, J.F., The Shadow Optical Method of Caustics, in Handbook on Experimental Mechanics, Kobayashi, A.S., (Ed.), Chap. 9, Prentice Hall, Englewood Cliffs, New Jersey (in press)

[24] Doell, W., Weidmann, G.W., J. Mater. Sci., 11, 2348 (1976)

[25] Schinker, M.G., Doell, W., Inst. Phys. Conf. Ser., 47, 224, (1979)

[26] Kalthoff, J.F., in Application of Fracture Mechanics to Materials and Structures, Sih, G.C., Sommer, E., Dahl, G., (Eds.) Martinus Nijhoff Publishers, The Hague/Boston/Lancaster, p.107 (1984)

[27] Davidge, R.W., Mc Laren, J.R., Tappin, G.J., J. Mater. Sci., 8, 1699 (1973)

[28] Evans, A.G., Wiederhorn, S.M., Int. J. of Fract., 10, 379 (1974)

[29] Beaumont, P.W.R., Young, R.J., J. Mater. Sci., 10, 1334 (1975)

[30] Doell, W., Koenczoel, L., Kunststoffe, 70, 563 (1980)

[31] Hertzberg, R.W., Manson, J.A., Fatigue of Engineering Plastics, New York/London/Toronto/Sydney/San Francisco, Academic Press (1980)

[32] Sauer, J.A., Richardson, G.C., Int. J. Fracture 16, 499 (1980)

[33] Radon, J.C., Int. J. Fracture 16, 533 (1980)

[34] Michel, J.C., Manson, J.A., Hertzberg, R.W., in Deformation, Yield and Fracture of Polymers, p. 21.1, The Plastics and Rubber Institute, London (1985)

[35] Michel, J.C., Manson, J.A., Hertzberg, R.W., Polymer Preprints, 26.2, 141 (1985)

[36] Doell, W., Koenczoel, L., Schinker, M.G., Polymer, 24, 1213 (1983)

[37] Schinker, M.G., Koenczoel, L., Doell, W., Colloid a. Polym. Sci, 262, 230 (1984)

[38] Elinck, J.P., Bauwens, J.C., Homès, G., Int. J. Fracture Mech., 7, 277 (1971)

[39] Skibo, M.D., et al., J. Mater. Sci., 12, 531 (1977)

[40] Koenczoel, L., Schinker, M.G., Doell, W., J. Mater. Sci., 19, 1605 (1984)

[41] Schirrer, R., et al., Colloid a. Polymer Sci., 259, 812 (1981)

[42] Boehme, W., Kalthoff, J.F., Int. J. Fract., 20, R 139 (1982)

[43] Hertzberg, R., Skibo, M.D., Manson, J.A., J. Mater. Sci., 14, 1745 (1979)

[44] Kalthoff, J.F., et al., in Advances in Fracture Research, Francois, D., et al. (Ed.), Pergamon Press, Oxford/New York, p.363 (1980)

STRUCTURAL EFFECTS IN POLYMER FRACTURE

H.H. Kausch, B. Stalder
Swiss Federal Institute of Technology
Polymer Laboratory
32, chemin de Bellerive
CH-1007 Lausanne

ABSTRACT. On a molecular and microscopic level, all polymer materials are strongly anisotropic and non-homogeneous.
These structural peculiarities, such as molecular weight, crystallinity orchain orientation, give rise to and strongly influence quite different fracture phenomena which are described in this paper. Important variations of polymer strength are also caused by defects, inclusions, chemical degradation and other production related defects. Typical examples concerning thermoplastic (pipe) materials are presented.

1. INTRODUCTION

In many fracture mechanics applications the given material is treated as an isotropic and homogeneous elastic-plastic solid. Taking the proper precautions as discussed by J.G. Williams such an approach may also be valid in the case of polymers. The principal polymer parameters already introduced by Williams are the nature and extent of plastic deformation at the crack tip and the viscoelastic behavior of the material. In this talk structural effects will be treated.

On a molecular level (1-10 nm) all polymers are strongly anisotropic and two basically different failure phenomena can occur : chain slippage and chain scission. The considerably different stress levels (50 MPa vs. 10-20 GPa) at which local failure occurs and the parameters of influence (degree of chain orientation, presence of entanglements and/or cross-links) will have to be considered [1].

The **non-homogeneity** of polymer materials must also be taken into account. Crystal lamellae (5-50 µm) influence stress distribution, crack initiation and propagation and the ultimate behavior. The tendency of these composite materials to change their fracture pattern as a function of stress level and loading time is especially pronounced; consequently, careful analysis of short time test data is required if they are to be used in the prediction of long time behavior.

K. P. Herrmann and L. H. Larsson (eds.), Fracture of Non-Metallic Materials, 291–307.
© *1987 by ECSC, EEC, EAEC, Brussels and Luxembourg.*

2. STRUCTURAL ELEMENTS

In view of the fact that macromolecules are long, flexible and highly anisotropic elements very complex structures can arise [2]. The principal organizational elements and their typical sizes are indicated in Table 1:

Table 1

Elements of polymer structure*	Scale
- Spherulites	10-400 μm
- Defects, voids	1- 20 μm
- Microfibrils	20-100 nm
- Crystal lamellae	5- 20 nm
- Extension of chain segments	1- 20 nm
- Entanglement network	3- 10 nm
- Chain orientation	-

* In this table no mention is made of the constituants of composites or polymer blends.

The structural elements influence sample strength in several ways since they determine :

- Deformability
- Rigidity (increases with crystallinity)
- Level of stored elastic energy
- Stress or strain concentration
- Efficiency of stress transfer
- Inhomogeneity of fracture resistance

The strength of a sample will be the higher the more homogeneously **"in space and time"** stresses can be transferred onto the **molecular chains** .

3. FRACTURE PHENOMENA

The major part of the results discussed in the following have been obtained under quasistatic sollicitation from stress-strain curves or from the analysis of data in terms of fracture mechanics. For the definition and correct application of the terms stress intensity factor and frac ture toughness in relation to polymers the reader is referred to the presentation by J.G. Williams.

Among the types of sollicitation currently encountered in practice, fatigue and abrasion are also very important for polymers. Although the mechanisms of failure might be peculiar, the dominant structural parameters and their qualitative influence are the same. Those cases will not be treated here. It is sufficient to notice that under fatigue loading a specimen may rupture through fatigue crack development at a load

which is smaller than the quasi-static strength. This phenomenon has
been treated in the preceding paper by W. Döll.

Specimens in sliding or rolling contact with each other mutually
exert forces on their surface regions which may easily go beyond the
elastic limits and lead to damage formation and local rupture. Small
particles can be torn off the surface which is thus abraded and worn. As
mainly brittle fracture will be treated in the following, these
mechanisms are beyond our present concern. Before discussing in more
detail the mechanical effects of structural elements, it seems to be
advisable to define the different forms of fracture.

3.1. Brittle fracture at low temperatures, at high loading rates or at high strain energy content

Brittle fractures are the most conspicuous fracture events. They occur
in highly loaded samples, often in an unstable manner (rapid crack
propagation) and at very small additional strains. Figures 1 and 2
underline some characteristics of brittle fracture behavior of different
polymers : thermosetting resins and thermoplastic polymers at low
temperature show an almost linear increase of stress with strain and
fracture occurring at strains of a few percent, without noticeable
plastic deformation, resulting in macroscopically smooth fracture
surfaces. Elastomeric networks on the other hand need to be strained by
several hundred percent before the strain energy becomes sufficient

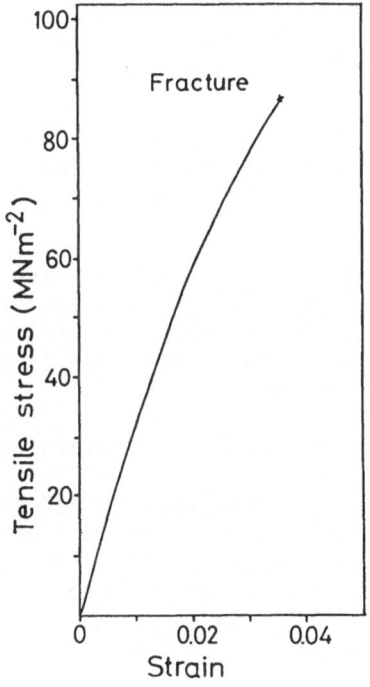

Fig. 1 : Stress strain curve
         of an unfilled,
         highly cross-linked
         epoxy resin

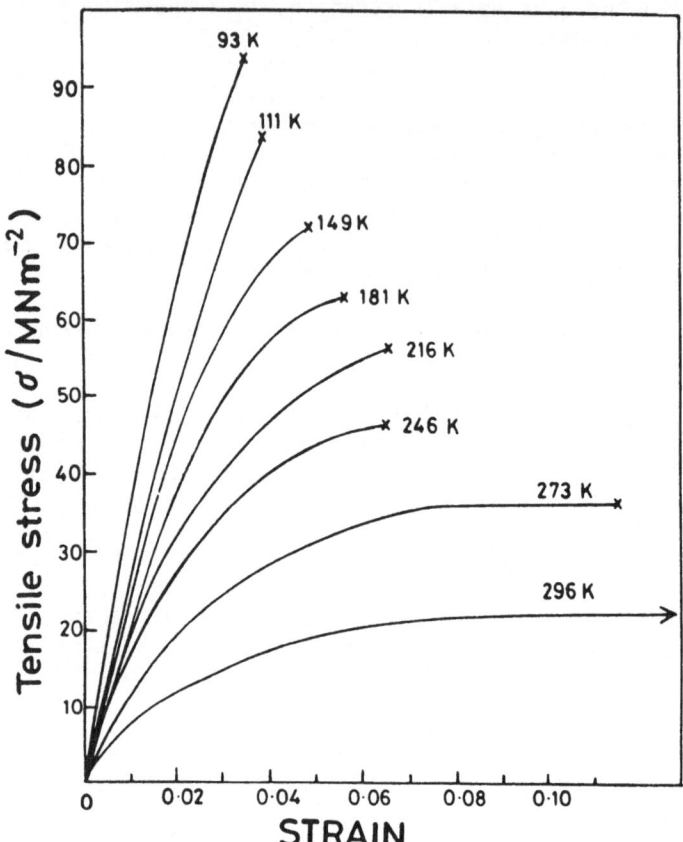

Fig. 2: Stress-strain curves of polyethylene at different temperatures
(from N. Brown / I.M. Ward, J. Materials Sci. 18, 1405 (1983)).

for unstable rapid crack propagation to occur. Such a rupture event is
also classified as **brittle fracture** because the sample is not subjected
to any noticeable additional deformation between the moment of fracture
initiation - generally at some flaw or defect - and termination (see [1]
for an extended discussion of fracture phenomena).

3.2. Dry crazing, stress whitening and environmental stress cracking

Many glassy polymers (stressed well below their breaking stress) develop
large numbers of straight silvery zones which are traversed by fibrillar
matter and are called crazes [3]. Taking polycarbonate (PC) as an
example, the characteristic features of crazing are presented as
follows. During a tensile test at temperatures below the glass
transition conventional crazes are observed well below the yield point

(Figure 3). These crazes are largely separated from each other (Figure 4). Surface defects may frequently be identified as craze initiators. Crazes of this type also appear under comparable circumstances in other glassy polymers such as PS or PMMA and in semi-crystalline polymers such as PE or PP.

At higher strains many polymers (e.g. PVC, PE, PP) have a tendency to deform by opening up small voids throughout the whole volume. The deformed regions appear to be white, the reason why this phenomenon is called stress-whitening (intrinsic crazing).

Crazing of polymers will be greatly enhanced by certain active environments (liquids, greases, even gases). It is generally agreed that the "crazing agent" must be able to diffuse into a polymer in

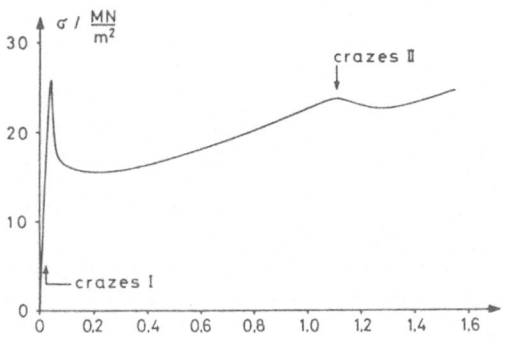

Fig. 3: Engineering stress-strain curve for polycarbonate at T = 129°C, ε = 3.5 %/min (from 3a).

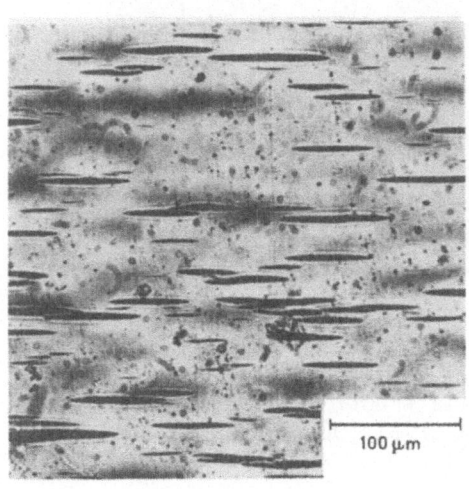

Fig. 4: Extrinsic craze in polycarbonate.

order to influence craze initiation. The crazing agent thus acts through its presence within the polymer matrix. In increasing the chain mobility it facilitates craze initiation but also craze breakdown.

3.3. Creep crazing (fissure-like brittle rupture)

Many thermoplastics exposed to low stresses for extended periods of time fail through the development of a mirror-like fracture zone generally initiating from a defect at a boundary. As becomes evident by comparison of the resulting fracture surface brittle fracture at low loading rates and creep crazing (at constant load) frequently lead to similar fracture surface patterns.

3.4. Yielding, necking, creeping, flow

As opposed to brittle fracture, which is a **localized event,** rupture through ductile deformation generally concerns a more extended region, occasionally the whole specimen volume (Figure 5). Ductile deformation phenomena leading to polymer fracture are yielding with and without neck formation, creep and flow. The latter two phenomena certainly do not complywith the general definition of fracture used in this course, since neither creep nor flow would give rise to the formation of new surfaces within the body. There are two important reasons, however, why all of these phenomena must be discussed in the context of polymer fracture : in some cases the homogeneous ductile deformation only precedes and modifies a subsequent brittle fracture event; in other cases slight modi-

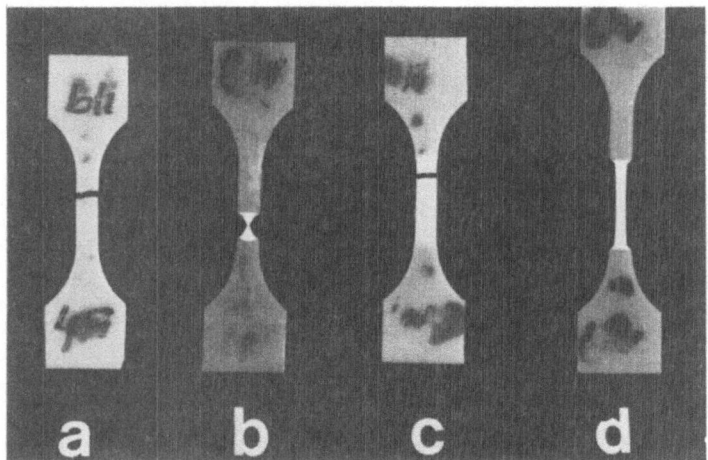

Fig. 5: Tensile rupture of PE.
        a) Brittle fracture at low temperatures (T < 95 K).
        b) Necking/rupture (100 < T < 240 K).
        c) Non-necking ductile fract. of slow cooled low mo-
           lecular weight material (90 < T_0 < 300K).
        d) Necking/drawing (T > 0 C). (Courtesy Prof. I.M. Ward)

fications of internal (molecular or microscopic structure) or external
parameters (temperature, rate of loading) cause the change from brittle
to ductile internal behavior [4].

4. EFFECTS OF MOLECULAR AND SUPRAMOLECULAR STRUCTURE ON FRACTURE

4.1. Molecular weight and molecular weight distribution

Without any doubt the length of their chains is the dominant characteristic of macromolecules. It greatly influences the mechanical properties
and among them strength and toughness.
    At very low molecular weights, glassy polymers exhibit the brittle
behavior of frozen liquids. Increasing the molecular weight to some critical value M_C (which is 30 000 for PMMA) permits the build-up of entanglements between different molecules. Toughness increases more or
less steeply. As it is shown in Figure 6 a plateau is reached at about 4
M_C. This transition is also clearly revealed by fractographic investigations

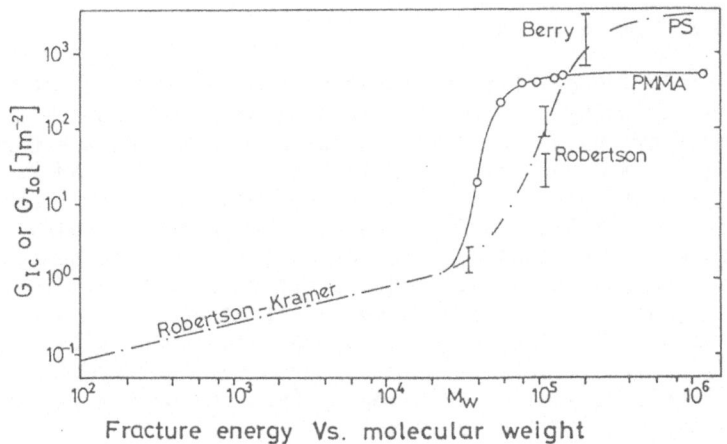

Fracture energy Vs. molecular weight

	Chain separation (= free surface energy)	Energy of elastic straining and scission of chains	Plastic deformation (fibrillation and fibril deformation)
Material resistance	$R = 2\gamma +$	$\dfrac{\partial U_{es}}{B\partial a} + \dfrac{\partial U_{cs}}{B\partial a} +$	$\dfrac{\partial U_{pl}}{B\partial a}$
PMMA $[Jm^{-2}]$:	0.080	~ 2	300

Fig. 6: Fracture energy of PMMA and PS.

which show noticeable plastic deformation at this stage in contrast to the much smoother surfaces of low M_w polymers. This is linked to the apparition of crazes initiated via the entanglements as soon as the molecular weight is higher than 2 M_c. Longer chains participate in more entanglements which increases the orientation hardening capability of the microfibrils. Preceded by a craze the crack tip is then able to withstand higher stresses. Fracture stress and fatigue life consequently become greater for high molecular weight polymers.

The behavior of semi-crystalline polymers is more complicated because of the superimposed effects of the degree of heterogeneous morphology. The sole effect that can be mentioned here is that the higher the M_w the higher the number of interlamellar tie molecules which link the microfibrils formed during straining. The draw ratio hence decreases with M_w and the stresses can increase.

As an overall consequence, the Izod impact fracture toughness increases rapidly in the range of molecular weight of 10^5 to $5 \cdot 10^5$. It then decreases slightly above 10^6. A broad molecular weight distribution is generally found to ameliorate impact strength. Such an effect, however, is occulted by the more drastic influence of other parameters such as the molecular preorientation (see below).

4.2. Molecular mobility

The influence of molecular mobility on fracture strength σ_b is directly shown by its temperature dependence. Obviously slip and disentanglement of chains and sample ductile deformation are enhanced by chain mobility and generally lead to a decrease of σ_b and K_c. At a given temperature chemically different molecular chains can have widely different mobilities. If, however, the respective glass transition temperatures T_g are chosen as reference temperatures, some common characteristics can be established : brittle behavior at $T \ll T_g$ once the elastic modulus $E > 4.5$ GPa, semi-ductile at $T < T_g$ with $2 < E < 4$ GPa, tough at $T < T_g$ with $E < 1.5$ GPa, and rubber elastic behavior at $T > T_g$. In this context it should be noted that the transition from brittle to semi-ductile **impact** fracture generally seems to be correlated with the β-relaxation peak.

4.3. Chain branching and tacticity

The molecular architecture of a chain influences its mobility and the nature and extent of the crystalline superstructure. As a rule the degree of nrystallinity and the rigidity of a semi-crystalline polymer decrease with chain branching and increase with chain regularity.

4.4. Degree of crystallinity

The internal variables already discussed (chain length, architecture, and mobility) but also external variables such as the thermal history (e.g. the molding conditions) influence the degree of crystallinity (Figure 7).

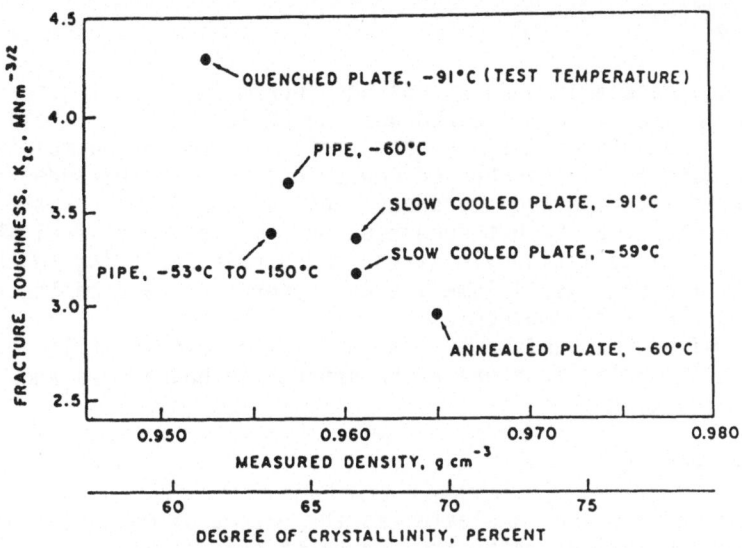

Fig. 7: Influence of thermal treatment on crystallinity and stress intensity factor of polyethylene pipe material (from Mandell, Roberts, McGarry [5]).

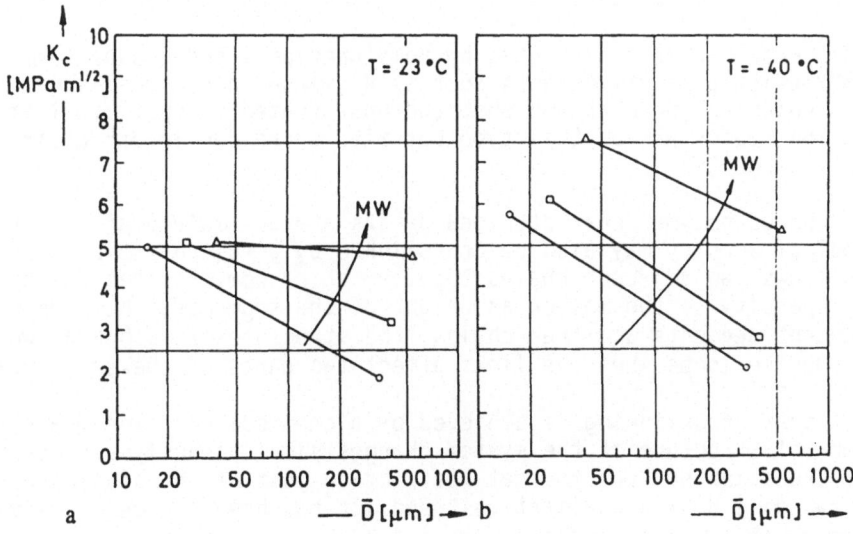

Fig. 8: Fracture toughness K_C as a function of both, molecular weight M_W and morphology (sperulite diameter D) of PP 1120; a) at room temperature and b) at -40°C (from Friedrich [3b]).

4.5. Morphology

By morphology one generally understands the supermolecular structure of crystallizable polymers. Such a structure can be formed by crystalline lamellae containing more or less defects, folds and emanating molecules, spherulites (Figure 8), shishkebab or row structures, and extended chain crystallites with interspersed amorphous regions [2]. A semi-crystalline polymer is, therefore, quite heterogeneous on a microscopic level **which leads to large local variations in stress and strain** depending on the size and perfection of crystal lamellae and spherulites and on the number and orientation of tie molecules.

Possible failure mechanisms are the ductile or semi-brittle failure of the sample as a whole, fracture along spherulite boundaries and intercrystallite cracking.

4.6. Crosslink density

In thermosetting resins and in elastomers the extent of molecular deformation is seriously influenced by the degree of crosslinking. The latter is defined by the average length or molecular weight between two crosslinks or simply by the number of crosslinks per unit volume. This quantity is not easily defined nor measured in view of the possible configurations of the network :

- The simplest network would be formed by segments of equal length, linked at points having the same functionality. No real system has such a regular structure.

- Monodisperse chains crosslinked by polyfunctional end-groups form a network which presents defects such as dispersed non-linked segments, chains which are free at one end, and unsaturated groups. Usual thermosetting resins are of this type but with a broader length distribution of segments.

- An amorphous polymer characterised by its average molecular weight and its polydispersity may also be crosslinked by a vulcanization process which links two adjacent chains together at random. Besides the strong polydispersity in the molecular weight of the segments, the presence of pendent segments and free chains, the structure of such a network may comprise loops when the links affect two parts of the same chain.

The fact that crosslinking is achieved by a chemical reaction means that the chemical structure of the system changes simultaneously with the degree of crosslinking. Due to that restriction most of the studies performed can only draw a qualitative trend. It has been noticed, however, that the mobility of segments is decreased by crosslinking and that the elongation at break tends to decrease with the degree of crosslinking. The featureless and mirror-smooth fracture surfaces of highly crosslinked resins comfort this opinion, viscous flow, large scale plastic deformation and crazing at the tip of a crack are inhibited.

Except for very low degrees of crosslinking, the tensile strength also progressively decreases as crosslinking increases, due to local stress concentration on peculiar segments. The stress intensity factor K_{IC} consequently follows the same trend.

More conclusively, Fischer, Lohse and Schmid [6] prepared networks of variable crosslink density and constant chemical structure. They found that the Izod impact strength, SB, is strongly influenced by the average number of atomic distances between crosslinks n_c (see Figure 9). The highest mechanical stability was obtained when n_c is low. However, only long molecular segments can improve fracture toughness by increasing the extensibility.

The effect of crosslink density on the strength of elastomers will not be discussed at this point since fracture will generally only occur after considerable orientation and strain-hardening of the sample [1].

4.7. Molecular orientation

The possibility to obtain very high values of modulus and failure strength is the basic aim for producing drawn fibers from crystalline polymers. As an example, PE and aromatic polyamide (Kevlar) fibers can reach a Young's modulus of respectively 200 GPa and 144 GPa and a fracture strength of up to 5 and 3 GPa. Although the orientation effect for preoriented glassy polymers is less impressive it is also of concern here because of the unintended or intended anisotropy of extruded or injection molded plastic articles.

Figure 10 shows the dependency for PMMA of the fracture stress σ_f, the elastic modulus E and the Poisson's ratio ν on the hot stretch ratio λ. Just a two-fold increase in σ_f is observed resulting from the molecular stress redistribution. At higher preorientations some polymers show a marked tendency for fibrillar fracture.

Crack propagation in hot-stretched polymers has as yet received little attention. Published results permit to give a qualitative analysis. Evidently, the modulus in the hot stretch direction is drastically increased. For low draw ratios K_{IC} also increases rapidly for a crack perpendicular to the stretch direction and decreases only slightly when it is parallel. However, the values of K_{IC}, reported in some studies, are questionable in view of the very strongly anisotropic moduli. Moreover, the crack tends to propagate in an oblique direction.

Conversely, the energy dissipated by a moving crack is lowered as a result of the inhibited slippage which reduces local yielding or crazing at the crack tip. Fracture toughness decreases with molecular orientation leading to brittle unstable fractures when craze initiation is suppressed. Parallel to the orientation direction crazing occurs more easily while G_{IC} decreases sharply to a value some 6 to 10 times less than the original one already for a low degree of orientation. It then stays almost stable.

Values of impact strength of PS and PMMA parallel and perpendicular to the preorientation direction have been measured by Curtis [8]. The trends are the same as in tensile tests, although more marked. Such data only have a qualitative significance : crack bifurcation already occurs at low degrees of orientation in response to the fall of transverse

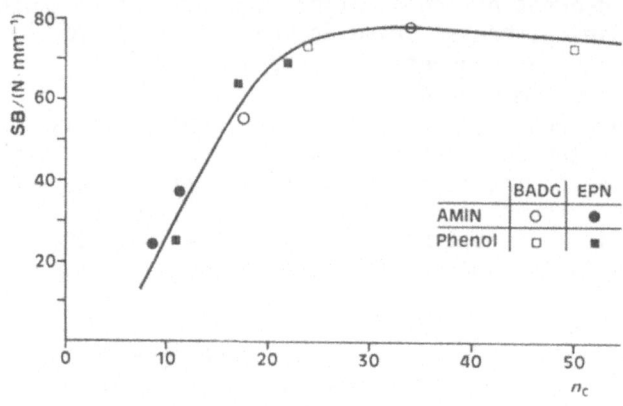

Fig. 9: Impact·bending energy SB as a function of number of main chain
bonds n_c between crosslinks BADG: Bisphenol-A-diglycidylether;
EPN Epoxide-novolaque (from Fischer et al.[6]).

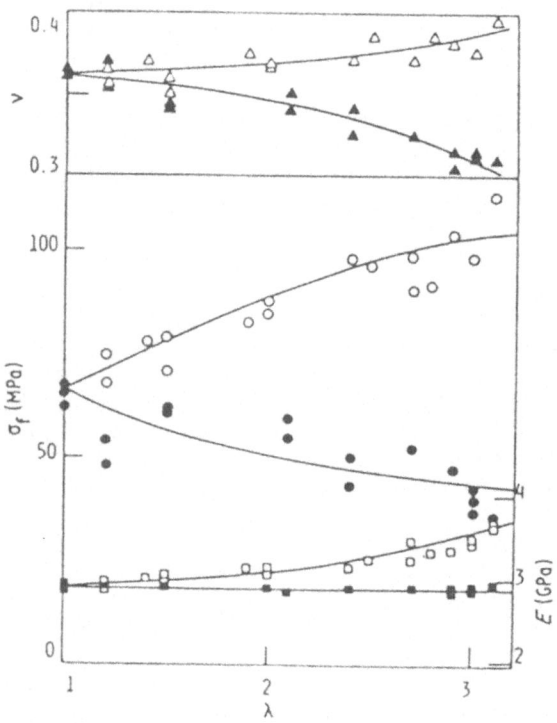

Fig. 10: Dependence of fracture stress σ_f, elastic modulus E and
Poisson's ratio v on hot stretch (from Kitagawa et al. [8]).

fracture strength. The angle formed by the two propagating cracks varies from some 90 degrees to almost 180 for highly oriented specimens.

5. DEFECTS

5.1. Defects, flaws and inclusions

The influence of defects on the brittle strength and time-to-failure of load bearing materials and also of polymers has "always" been known and is generally considered as the main source of variation of such experimental data. If it is assumed that the stress concentration caused by a flaw is directly responsible for the premature brittle fracture of a specimen then the size a_m of the largest intrinsic flaw can be determined from the breaking stress σ_b in short time loading according to

$$a_m = K_{Ic}^2 / \sigma_b^2 \pi .$$

Thus, intrinsic flaw sizes between 62 and 95 μm for PMMA and of 260 μm for PS have been observed. These values seem to be very high since no direct optical observation of such flaws was possible.

It must be concluded, therefore, that the flaw sizes inferred from a tensile experiment or some other mechanism are **by no means inherent** but the result of stable crack growth during the experiment.

Starting from this consideration Döll and Könczöl [9] carefully studied the fracture surfaces of a large number of long-time tested PMMA dumbbell specimens. The fracture surfaces showed the usual pattern: a semi-circular mirror zone (slow crack growth) and a subsequent region covered with parabolae (rapid crack propagation). In the center of the mirror regions they frequently found semi-circular markings which could very well correspond to former craze zones. The development of an existing defect into an unstable crack in PMMA would then occur in the following manner:

- a preexisting defect (a few μm or even less) develops within Δt_1 into a craze zone (fairly rapid development up to a temperature dependent size of 70 to 160 μm) followed by period Δt_2, the growth of the mirror zone (slow growth up to several 100 μm, depending on stress);
- Sandilands et al. [10] also confirm the fracture mechanics concept that decreasing the flaw size should increase stress rupture lifetime. In a controlled experiment they ascertained the influence of flaw size by either purposely adding flaws or removing them by fine melt filtration. For the polyethylene pipe resin investigated, it has been clearly shown that the pipe stress vs. rupture lifetime diagram depends on flaw size, and that fine filtration improves pipe performance [10].

5.2. Inclusions and chemical degradation

In order to identify systematically the nature and position of defects in (LDPE) pipes Stockmayer and Wintergerst [11] developed a cutting technique which permitted to transform the entire pipe wall into a continuous thin peeling of 0.06 to 0.15 mm thickness. By inspecting this film they were able to correlated nature, frequency and position of defects with time to failure t_b of the pipes. The irregularities they found were clear zones not mixed with carbon black in the form of points, linear marks, hooks and parallel striations. They claim that with a certain preference the final creep crazes originated in those areas of the pipe wall where the irregularities were more frequent. It was particularly noted by them that the lifetimes t_b of those pipes where the creep crazes had started close to the inner or outer surface of the pipe wall were an order of magnitude smaller ($\bar{t}_b = 740h$)- than those where the crack had originated in the center $\bar{t}_b = 7$ 400 h). Although the differences in stress distribution within the pipe wall, due to frozen-in tensions (see next section), may influence crack initiation it seems to be more probable that network defects are mostly responsible for this observation. As discussed in the preceeding section the inner walls of extruded pipes down to a depth of 0.5 mm are the first to experience an eventual network degradation. It was precisely in this region that Stockmayer and Wintergerst found the origins of the most rapid failures.

To test the influence of the observed irregularities further the authors performed tensile drawing and static loading experiments on samples cut from the peelings. In these experiments, there was practically no correlation between yield stress and time to failure on the one hand and kind and concentration of flaws and inclusions on the other. However, the effective elongation of films containing defects was much smaller since the necking samples generally broke whenever the neck had reached the defect. This behavior confirms that creep craze nuclei must be considered as network defects which are not detectable in short time loading tests unless the defective zone is locally subjected to larg deformations.

Evidently the question on the influence of inclusions added in the form of blending components or fillers is of vital importance to the fabrication and use of polymer composites. This problem, however, cannot be discussed at this point.

5.3. Other production related defects

Geometry. A simple variation in sample cross-section can turn out to be a decisive defect. Thus, the **ductile failure** of uniaxially or biaxially loaded samples generally takes place at the smallest cross-section. It should be noted that the failing section may be perfectly safe at higher stresses (brittle fracture) or longer times (creep crazing), because the breakdown in those cases could be initiated by network defects or flaws at other sites.

Internal tensions. Any viscoelastic body cooled from outside from above its solidification temperature (T_g or T_m) to a lower temperature T_c deforms non-homogeneously and either bends or contains frozen-in tensions. The distribution and absolute value of such tensions can be determined from a layer removal experiment [12]. In pipes they generally show a parabolic distribution, compressive at the outside, smaller and tensile at the wall inside. Williams [12] derived the thermal, circumferential stresses $\sigma_R(s)$ as:

$$\sigma_R(s) = \frac{\alpha E(t')}{1-\nu} (T_m - T_c) \left(\frac{Nu}{2+Nu} \right) \left[\frac{1}{3} - \left(\frac{s}{W} \right)^2 \right]$$

with:
s   distance from inside wall
α coefficient of thermal expansion
$E(t')$ Young's modulus at time after cooling down
ν Poisson's ratio
Nu   Nusselt number
W   wall thickness

For a HDPE pipe he had determined (compressive) stresses of -3 MPa at the outside and a tensile component of +1 MPa at the inside. Compared with the externally applied values of 15 to 20 MPa the frozen-in tensions are not entirely negligible.

Weld lines. Welding, simple, rapid and straightforward, constitutes the most important technique for assembling thermoplastic construction elements. The final welds mostly distinguish themselves clearly from the main material by their form (upset, beads, camber), thermal history (and ensuing morphological changes), state of orientation and molecular interpenetration. Often less obvious are those weld lines created by the impingement of flowing masses in a mold or around the spider in an extrusion head. There again good molecular interpenetration is essential. This basic element of welding, the build-up of strength through entanglement formation, has been mentioned in an earlier section. At this point some experimental resuls on weld (line) strength will be added.
        Crawford [13] has conceived a mold with typical features such as corners, changes in section, flow welds etc. and has carried out what he calls "pseudo-moldings". In his tests with PP the weld lines turned out to be the most notable defect.
        Moslé et al. [14] molded PC plates with two central holes. By simply changing the stretching direction of these plates in subsequent tensile tests they could either load their specimens parallel **(a)** or **perpendicular (b)** to the weld lines. Their principal observation was **that the yield stresses** were more or less identical within a range of molding temperatures varying between 300 and 340 ^{0}C. The post-yield **strains** , however, were drastically smaller in case **b.** The rapid disintegration of the weld line material during plastic deformation is also underlined by the lower rupture forces (the forces at break in case **b** were only one half of those in **a**). Once again it has been demonstrated that strain hardening, a vital property for engineering thermoplastics, can only occur if the molecules are reliably interconnected so that they do not disentangle at the onset of plastic deformation.

6. Strength of macromolecular single crystals

In oriented and ultra-oriented fibres stress-transfer between microfibrils had been an important strength limiting parameter. Under these circumstances single crystals of macroscopic dimensions should offer an additional advantage. Detailed investigations on high-modulus polydiacetylene crystals (DCHD) have been carried out by Galiotis et al. [15]. Contrary to ultra-highly oriented PE fibres polydiacetylene single crystal fibres show practically no creep (up to 2 % strain and 100 ⁰C). Their strength does depend, however, on fibre diameter (Figure 11).

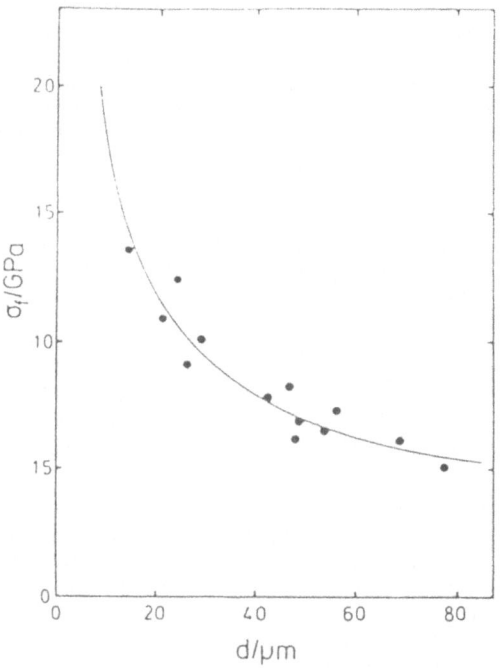

Fig. 11: Dependence of the fracture stress upon effective fiber diameter for polyDCHD.

This is due to the stress concentration caused by steps in the fibre surface from which fracture initiates. Analyzing the geometry of such defects and applying fracture mechanics concepts the authors determine that the theoretical strength of the polydiacetylene is 3 ± 1 GPa which corresponds to about 3 nN per chain molecule. This is in fair agreement with the calculations cited earlier [1].

1. H.H. Kausch, Polymer Fracture, Berlin-Heidelberg-New York: Springer 1978.

2. B. Wunderlich, Macromolecular Physics, Vol. 1-3 , New York-London: Academic Press 1973.

3. Crazing in Polymers, H.H. Kausch Ed., Berlin-Heidelberg-New York: Springer 1983:

   a) M. Dettenmaier, p. 57
   b) K. Friedrich, p. 225.

4. I.M. Ward, Mechanical Properties of Solid Polymers, 2nd ed., Chichester-New York: J. Wiley 1983.

5. J.F. Mandell, D.R. Roberts, F.J. McGarry, Polymer Eng. Sci. 23 /7, 404 (1983).

6. M. Fischer, F. Lohse, R. Schmid, Makromol. Chem. 181 , 1251 (1980).

7. M. Kitagawa, H. Kajiwara, H. Kanzaki, T. Zhang, J. Mater. Sci. 20 - 1945 (1985).

8. J.W. Curtis, J. Phys. D: Appl. Phys. 3 , 1413 (1970).

9. W. Döll, L. Könczöl, Kunststoffe 70 , 563 (1980).

10. G. Sandilands, P. Kalman, J. Bowman, M. Bevis, Polymer Commun. 24- , 273 (1983).

11. P. Stockmayer, S. Wintergerst, 3R international 20 , 274 (1981).

12. J.G. Williams, Plast. Rubber Proc. & Appl. 1 , 369 (1982).

13. R.J. Crawford, 5th Int. Conf. on Deformation, Yield and Fracture of Polymers, Cambridge, 29.3.-1.4.1982, p. 38.1.

14. H.G. Moslé, R.M. Criens, Kunststoffe 72 , 222 (1982).

15. C. Galiotis, R.T. Read, P.H.J. Yeung, R.J. Young, I.F. Chalmers, D. Bloor, J. Polymer Sci., Polym. Phys. Ed. 22 , 1589 (1984).

STRUCTURE OF CONCRETE AND CRACK FORMATION

F.H. Wittmann
Laboratory for Building Materials
Swiss Federal Institute of Technology
Ch. de Bellerive 32
CH-1007 Lausanne (Switzerland)

ABSTRACT. It is shown that three structural levels have to be introduced as to link crack formation and failure of a composite material such as concrete with structural features. On the micro-level properties of the colloidal binding matrix i.e. the xerogel of hardened cement paste are characterized. Special emphasis is placed on the interaction of the xerogel with adsorbed and capillary condensed water. Crack arresting mechanisms and hence the origin of the ductility of the composite material are discussed on the meso-level. Numerical concrete has proved to be a powerful tool to study the behaviour of composite materials with a random structure. Finally material laws are formulated in using all knowledge of the preceding structural levels on the macro-level. The fictitious crack model is probably the most promising approach to describe failure of concrete. Although much progress has been made recently at this moment it is not yet possible to develop systematically constitutive laws based on the structure of the material exclusively. Areas where further research is needed are indicated.

1. INTRODUCTION

There are many papers on specific details of the structure of hardened cement paste and concrete. The wide range of literature begins with microscopical observation and purely phenomenological description and follows through to real mathematical models. In this contribution, we will try to summarize and discuss only those structural aspects which are directly linked with crack formation and failure processes. A more general treatment of modelling the structure and performance of concrete has been published recently /1/.

Until now many structural details of the porous and composite materials were not known well enough. Therefore the existing knowledge has been condensed into several models. This is one possibility to describe the real behaviour in a simplified and approximative way. As new information is available these models can be adjusted continuously.

K. P. Herrmann and L. H. Larsson (eds.), Fracture of Non-Metallic Materials, 309–340.
© *1987 by ECSC, EEC, EAEC, Brussels and Luxembourg.*

It has been proven to be advantageous to subdivide the structure
of concrete into different levels. In Table I a hierarchic system of
three different levels is shown. This means that the models on the
different levels are interrelated in a systematic way, or more precisely
models on a given level are based on the results of the previous level.

At the micro-level the structure of hardened cement paste is
treated. So far, the only model which gives us quantitative results on
the micro-level is the Munich-Model /2/. Therefore we will choose this
model among the different existing materials science models as the basis
for further discussion.

The observed behaviour of concrete cannot be linked directly with
microstructural mechanisms because there are additional factors within
the hierarchic structural system which interfere. In concrete the most
important factors are pores, cracks, and inclusions. These structural
details of a composite material will be introduced on the meso-level.

The final aim of the hierarchic structural system is, of course,
to characterize the macroscopically observed behaviour of a given
material in a realistic and rather general way. On the macro-level the
information which results from the two previous levels will be used to
describe the materials behaviour in such a way that it can be used
directly in structural engineering and design. In this context macroscopic fracture mechanics parameters have to be considered.

TABLE I
Characteristic structural features of hardened cement
paste and concrete subdivided into three different
levels and corresponding types of models

Structural Level	Characteristic Features	Type of Models
Micro-Level	Structure of Hardened Cement Paste, Xerogel	Materials Science Models
Meso-Level	Pores, Cracks, Inclusions, Interfaces	Materials Engineering Models, Mechanical and Numerical Models
Macro-Level	Geometry of Structural Elements	Structural Engineering Models, Material Laws

In the following sections the three different structural levels
will be dealt with consecutively. The interrelationship of results
obtained at different structural levels will be outlined in particular.

2. MICRO-LEVEL : HARDENED CEMENT PASTE

2.1. Introductory Remarks

In concrete, a heterogeneous multi-phase material, different kinds of aggregates may be cemented by hydraulic cement paste to form an artificial stone. The properties of the resulting material depend both on the type of aggregate and the hardening matrix. Most typical aggregates may be assumed to react in reasonable approximation as linearly elastic. Hardened cement paste, however, is known to be a viscoelastic material. Depending on the relative humidity of the surrounding atmosphere the microporous structure of hardened cement paste contains a considerable amount of adsorbed and capillary condensed water. As the water content is changed shrinkage or swelling occurs. The hygral length change of concrete, however, is restrained by the inert aggregates. As a consequence a complex state of internal stresses is built up. The coefficient of thermal dilatation of aggregate and hardened cement paste may vary by a factor of up to 5. It is well known that the elastic moduli also differ to a great extent in most concretes. Therefore a complex state of internal stresses is also created when the temperature is changed or an external load is applied.

Most aggregates used in concrete technology, being approximately linearly elastic within the range of service loads, show no drastic influence of relative humidity of the surrounding air or of temperature on the elastic modulus, on strength and on the coefficient of thermal expansion. The properties of those materials can be realistically described with the help of simple and idealized expressions.

The highly dispersed hydration products of portland cement react to changes of the environment in a far more complicated way. Nearly all characteristic properties of hardened cement paste can be directly linked with the mutual interaction of colloidal particles and/or with the interaction of the total gel structure with water. The presence of aggregates in concrete moderates the actual behaviour of pure hardened cement paste. Therefore on the micro-level we will discuss the structure of hardened cement paste and the interaction of the xerogel with water.

Among the essential reasons why concrete has become the most extensively used building material of our day is the fact that it is comparatively cheap and also that even unskilled personnel under certain conditions may produce concrete with satisfactory serviceability. This situation is not really a challenge for research in this field. As a consequence it is not surprising that so far, very little on the physical basis of the materials properties is really known.

Concrete structures, however, have become increasingly more sophisticated. Cementitious materials are now used under extreme conditions : high temperatures, low temperatures, impact loading conditions, etc. It is essential for a simultaneously economic and safe application of any material that its behaviour while under service can be accurately predicted. This aim, however, can only be achieved via a better understanding of the structure of hardened cement paste and its relation to the macroscopically observed behaviour. As mentioned above this problem will be approached by introducing three different hierarchic levels of structure.

Some investigators tried to develop models of the microstructure of hardened cement paste on the basis of comparatively crude mechanical tests on concrete. These early models had a purely phenomenological character and they can only be used to describe the observed behaviour under given conditions. As fundamental research on the microstructure proceeded, new and more realistic models were created. With the help of these models the actual complex situation within a gel were to be described in a simplified manner. Even these models should not be taken too literally but they should rather be judged by the contribution they provide for a more comprehensive understanding of the total system.

Models will have to be abandoned completely as soon as we have sufficiently detailed and reliable data. So far, however, we definitely need appropriate models. This is the only possibility to accumulate most of the information available. In this way, finally the gap between the research on the microstructure and the engineering properties will be bridged. New results necessarily have to be incorporated in suitable models.

Research on the structure of hardened cement paste can be subdivided into two groups. In one group, mainly a chemical and mineralogical approach is used. In this way the hydration process, the composition, and the crystal structure of hydration products have been studied. An excellent summary of the results obtained has been published by J.F. Young /3/. On the other hand there is a more physics-orientated approach. In the latter mentioned connection the mutual coupling of gel particles and the interaction of gel particles with absorbed vapours are of prime interest. Of course, there is no strict borderline separating these two areas of research.

In this contribution emphasis is placed upon the physical approach because it seems easier to link physical models of the xerogel with crack formation and failure, but some chemical and mineralogical aspects of the structure will be treated first.

2.2. Chemical and Mineralogical Aspects of the Structure

The main compounds in Portland cement are calcium silicates, i.e. $3CaO \cdot SiO_2$ and $2CaO \cdot SiO_2$. The aluminate and ferrite phases are of minor importance with respect to the structure and the mechanical properties of the resulting cementitious material. The hydration of the calcium silicates may be described in a simplified way by the following equations :

$$2(3CaO \cdot SiO_2) + 6H_2O \rightarrow 3CaO \cdot 2SiO_2 \cdot 3H_2O + 3Ca(OH)_2$$
$$2(2CaO \cdot SiO_2) + 4H_2O \rightarrow 3CaO \cdot 2SiO_2 \cdot 3H_2O + Ca(OH)_2$$

These simple expressions, however, are somewhat misleading because the hydration process and the exact determination of the hydration products becomes extremely complex if one goes into detail. Probably the first comprehensive treatise of the chemistry of hydration is provided by the proceedings of the Washington Symposium /4/. In the meantime, a number of additional results on the hydration products have been gathered. Most essential findings have been presented in subsequent meetings in Tokyo /5/, Moscow /6/ and Paris /7/.

It is quite obvious that progress in this field is comparatively slow. This may be explained by the fact that there are few techniques which may be successfully applied to study the structure of hydration products of cements. Only limited information can be deduced from X-ray diffraction. The same applies to the application of electron microscopy. Much effort has been spent to develop appropriate chemical methods. An early break-through has been achieved by sorption methods /8/. As there are no prevailing methods so far, the actual state of knowledge has to be discussed by combining the relevant information of different approaches.

From the early micrographs /9 and 10/ taken from hydrated cement and hydrated clinker components it became clear that a variety of different particles are to be found. Needles, crumbled or rolled foils and tubes have been observed in hardened cement pastes. The variability of layered silicate hydrates is known from natural minerals. If the silicate tetrahedra match the $\{Me(O,OH)_6\}$-octahedra completely plane sheets are formed. This principle of geometrical fitting has been discussed among others by Liebau /11/. If the just mentioned subunits do not fit completely structures like corrugated iron (crumbled) sheets or rolled foils are formed. In the case of calcium silicate hydrates, aluminium and ferric ions among other ions may substitute for both silicon and calcium. In addition sulfur may substitute for silicon. The morphology of the gel particles changes as calcium or silicon are substituted /12/. Richartz and Locher /13/ prepared on this basis a vivid graphic model of different morphologies in hardened pastes.

Kantro, Brunauer and Weise /14/ distinguish three stages of the hydration of cement. According to these authors, immediately after mixing with water a skin adhering to the particles of the anhydrous compounds is formed. This first product has a high C/S ratio. In the second stage splitting-off of particles of the skin is observed. The second product formed in this way has a C/S ratio of 1.0 to 1.5. In the electron microscope one detects foils or platelets at this stage. Finally these particles grow to a thickness of about three layers with a C/S ratio of 1.5 to 2.0. Although modified by several authors this concept basically proved to be correct /15, 16 and 17/.

Depending on the C/S mole ratio calcium silicate hydrates may be subdivided into two groups. In aqueous suspensions at room temperature C-S-H (I) with a C/S ratio of 0.8 to 1.5 is formed. The semi-crystalline calcium silicate hydrate with a C/S ratio of 1.5 or above generally is termed C-S-H (II). In hydrated cement paste semi-crystalline and near amorphous products are found. X-ray powder patterns usually show only three broadened lines. Sometimes this phase was called tobermorite gel. This name was chosen because it was concluded that the hydration products of cement were degenerate varieties of tobermorite. This statement has often been questioned and in fact there is some evidence that some products are degenerate structures of jennite. Therefore it seems reasonable to follow a suggestion made by Taylor /18/ and to use a more general expresion i.e. calcium silicate hydrate or in abbreviated form C-S-H. This general term also takes into consideration the fact that there is not a finite number of phases (each of a definite composition and structure) but instead a large continuous range.

Essentially based on sorption experiments. Powers and Brownyard /8/ determined an average particle radius of 140 $\overset{\circ}{A}$. In this calculation it was assumed that the microstructure is composed of equal spheres. Later Powers /19/ calculated a mean volume of $6.6 \cdot 10^{-6}$ $\overset{\circ}{A}^3$. This value leads to a revised radius of 117 $\overset{\circ}{A}$. There is no doubt that these values should not be taken too literally but they are a strong indication that hydration products are of colloidal dimensions.

More recently Grudemo /20/ suggested a different structure of the hydration products. The fact that basal reflexions are not observed in the region of 9-15 $\overset{\circ}{A}$ may indicate that layered or lamellar structures are rare. Therefore Grudemo concludes that cement gel is a submicrocrystalline mixture of structural elements. Some of them are related to tobermorite and some are related to jennite whereas others are related to CH-portlandite. In this concept, gel pores are formed as silica chains and are left out during the growth of the structure. If this assumption holds true pores with a diameter of 9 $\overset{\circ}{A}$ or multiples thereof must be expected.

If the pore size distribution is determined from sorption data a maximum at a radius of about 18 $\overset{\circ}{A}$ is found /21 and 22/. At much lower radii this method to determine pore size distributions looses its significance. It may be mentioned, however, that often a second maximum is recorded at about 8 to 10 $\overset{\circ}{A}$. This fact would confirm the "Grudemo concept". The observed pore size distribution, however, depends on the preparation of the hydrate sample. Therefore it might be tempting to compare experimentally determined pore size distribution with results of morphology studies of C-S-H.

Another promising approach to the microstructure of hydration products has been put forward by Tamas and Varadi /23/. They studied polymerization of Si O_4^{4-} monomers. In the meantime, this interesting approach to study the microstructure has been considerably extended /24/. Rio and his co-workers /25/ also studied the hydration process on a macromolecular level. They tried to correlate the mechanical behaviour and the morphology with the degree of condensation of the silicate hydrates.

Based on sorption data the internal surface of fully hydrated cement paste is found to be 100-200 cm^2/g. Considerably higher values have been calculated by Winslow /26/ who used small angle X-ray scattering. His findings are discussed in detail with respect to the structure and properties of hardened cement paste by Copeland and Verbeck /27/.

The actual state of knowledge on the structure and composition of hydrates has been reviewed by Taylor and Roy /28/. Structure formation and development in hardened cement pastes have been discussed by Sereda and co-workers /29/. The major components of the microstructure of a xerogel, i.e. solid phase, pores, and water, have been treated separately and discussed in connection with properties of hardened cement paste by Wittmann /30/.

This compilation of information on the microstructure of hardened cement paste is, by no means, complete and it may even seem to be arbitrarily selected in some respect. For the present purpose we may conclude, however, that although there is a wide field of active research, there are not enough well established data available to understand the microstructure in full detail. Lack of knowledge is the main reason why we have to introduce simplifying models on the micro-level. In the following sections therefore we have to deal with the development of appropriate models.

2.3. Earlier Models

Models described in the literature can be subdivided into two groups : Inductive models and deductive models. By taking into consideration all relevant information on the structure such as pore size distribution, mutual interaction of colloidal particles in a xerogel and the characteristic properties of adsorbed water films an inductive model can be developed. The validity of such a model must be checked by a critical comparison of the predictions of the model with the actually observed behaviour of the system. On the other hand one may deduce a model of the microstructure from the experimentally determined macroscopic data. Models of this type have to be tested by comparing them with results of more fundamental work on the structure. Some deductive models are suggested with the aim of covering just one specific point such as influence of moisture content on strength.

A typical example for a deductive model has been given by Ishai /31/. The mechanical behaviour under load has been analyzed and on the basis of the comprehensive results obtained in this way a structural model of hardened cement paste has been established.

All inductive models are based on results of fundamental research on the characteristic properties of the xerogel. There is no doubt that the most detailed information on the structure of hardened cement paste has been gained by sorption methods. The mean particle size and the pore size distribution have been estimated by Powers /19/ as indicated above. Powers summarized his findings in a geometrical model and a physical model /32 and 33/. With the help of the geometrical model the pores of the structure are subdivided into gel pores and capillary pores. The physical model of Powers serves as a basis for a thermodynamic treatment of partially water filled micropores. Within the framework of this inductive model, Powers deals with three basic mechanisms :

a) Change of surface tension of the colloidal particles

b) Change of disjoining pressure in narrow gaps

c) Change of hydrostatic tension

The most important feature of this model probably is the load bearing capacity of water adsorbed in zones of hindered adsorption.

Within the thermodynamic models it is presumed that colloidal particles do not change their structure and/or composition significantly as the moisture content of the system changes. Bernal /34/ has demonstrated, however, that the structure of C-S-H (I) changes during drying. The c-spacing of a unit cell goes down from 14 to 9 Å. It may be concluded that during severe drying, this material loses interlayer water. On this basis Feldman and Sereda proposed another model for hydrated portland cement /35-37/. With the help of this model it is tried to link shrinkage, creep and the influence of moisture content on the elastic modulus with the exchange of interlayer water.

All models described above have been modified by various authors. In some cases major components of a model have been adapted and used for the interpretation of experimental findings. Hope and Brown /38/, for instance, used Feldman and Sereda's model to postulate a possible mechanism of creep.

For many years there has been a lively controversy on the validity of different models. So far, no generally accepted agreement could be reached. But it seems that by now most people concerned have realized that different models may contribute to progress in various ways and that one single model is not able to characterize the complex situation.

Kondo and Daimon /39/ have developed another inductive pore model for C-S-H gel. In this essentially geometrical model clusters of crystallites are separated by inter gel particle pores. In each gel particle there are inter-crystallite pores and finally intra-crystallite pores are found in individual crystallites. This model may partially bridge the gap between the two opposing models of Powers-Brunauer and Feldman-Sereda if it is realistic to compare the inter-crystallite pores with traditional micropores and the intra-crystallite pores with interlayer space. But the more important question about the extent to which water in micropores and in interlayer space influences the mechanical behaviour and in particular crack formation and strength still remains untouched.

2.4. The Munich Model

2.4.1. <u>New data forming the basis of the model</u>. With respect to the xerogel of C-S-H the methods used so far may be subdivided into two groups :

a) Direct observations of characteristic properties
of the gel

b) Investigations into the properties of water adsorbed
   in the colloidal system

Within the first group the determination of van der Waals forces at low
distances, of the surface energy, and of the coupling of individual particles in the gel are of primary interest. The study of the mobility and
of the disjoining pressure of adsorbed films in micropores plays a dominant role in the second group mentioned above. In the following para-
graphs some recent results are briefly summarized.

Classical van der Waals experiments have been restricted to the observation of the interaction at compartively large distances (d>1000 Å).
The properties of a xerogel, however, are only affected by attractive
forces acting between solid surfaces which are separated by micropores.
That means, they are separated by distances of a few Ångströms. If these
micropores will be separated by a crack, van der Waals forces contribute
to the energy consumed.

The theoretical background as well as experimental techniques had
to be extended so that the range of short distances could be investigated
/40/. By evaluating the bending line of a thin quarz plate which was
mounted on a solid quarz support at a certain distance it was possible
to determine the van der Waals attractive force down to about 80 Å /41/.
With the help of theoretical considerations it is possible do extrapo-
late beyond the range of experimental data with a reasonable degree of
reliability to colloidal distances.

The van der Waals attraction is strongly dependent on the dielectric
properties of the medium which is between the interacting surfaces. As a
consequence the van der Waals attraction is diminished as water is adsorbed on two opposing surfaces. The influence of adsorbed films on van der
Waals forces has been studied carefully as a function of film thickness
/41/. Adsorbed water reduces the attractive force approximately by one
order of magnitude.

There is a close correlation between van der Waals attraction and
surface free energy (see e.g. /42/). Therefore the dependence of surface
free energy on the thickness of adsorbed water films may be directly deduced from the observed decrease of van der Waals attractive force. By
comparison of results obtained by independent methods this is shown in
/41/ and /43/. In a more direct way the change of surface free energy
can be calculated with the help of a thermodynamic approach from sorption data /22/. Starting from the dry state the surface energy decreases
sharply as the water vapour pressure is increased. Above $p/p_o=0.5$ the resulting change, however, is compartively small. We will compare this result directly with the influence of moisture content on strength.

Primary bonds and secondary bonds both contribute to the mutual
coupling of gel particles within a xerogel. The mechanical properties of
the system of course depend primarily on these coupling forces. It could
be shown that Mössbauer spectroscopy is a powerful tool to investigate
the coupling of gel particles /44/. If the water vapour pressure is raised above a certain level, the disjoining pressure of water separates

surfaces which are exclusively held together by van der Waals bonds. The disjoining pressure may be subdivided into different components each having a different physical origin /45/. By extending the van der Waals experiments mentioned above to the range of high vapour pressures the action of disjoining pressure could be observed directly /46/. Stockhausen has later discussed the disjoining pressure and its meaning for C-S-H /47/.

The mobility of adsorbed water is a decisive factor in a number of models. If the first adsorbed layers had an ice-like structure the creep mechanism could be traced back to displacements within this modification. Using the Debye theory the viscosity of liquids can be deduced from dielectric measurements. Schlude and Wittmann determined the complex permittivity in the range of microwave frequencies /48/, and Zech and Wittmann in the range of Hertzian spectroscopy /49/. By comparing these results with NMR measurements one can conclude that the viscosity increases with decreasing thickness of the adsorbed layer. In the region of a monolayer, however, the definition of viscosity looses its significance. In this case a high mobility of water molecules along the surface is observed but the molecules are hindered from leaving the surface. Under these conditions the adsorbed films are called a two-dimensional van der Waals gas.

In the next section some general relations will be introduced. With the help of these relations it is possible to describe the xerogel of hydrated portland cement realistically. In this context, of course, drastic simplifications are still inevitable.

2.4.2. <u>Elements of the model</u>. It is well known that a liquid droplet having a radius r is under a hydrostatic pressure P :

$$P = \frac{2\gamma}{r} \tag{1}$$

In equation (1) γ represents the surface tension of the liquid. In a liquid, surface tension and surface energy are numerically equal. In solids these two values are at least in the same order of magnitude. In a colloidal system nonspherical particles can exist. Flood /50/ has shown that the mean pressure in solid particles created by surface tension in such a system can be estimated with the help of the following equation :

$$P = 2\gamma \frac{S}{3} \tag{2}$$

S stands for the specific surface area and has to be expressed as cm^2/cm^3 in this connection. If the specific surface area is introduced instead of the radius the actual particle size distribution is neglected or rather replaced by a mean value. In C-S-H there are particles which are large enough to ensure that no appreciable internal pressure will be created by surface tension. Other particles in the same system will experience comparatively high pressures. The overall response of a system with active and inactive particles has been calculated by Krasilnikov and co-workers /51/. In their paper it is pointed out that expansion of gel particles in a heterogeneous system is not linearely related to the expansion of the total system but a geometrical magnification factor has to be taken into consideration.

Well aware of the implied simplifications we may go back to equation (1). Now r has to be looked at as a characteristic value of a given xerogel and P as a mean internal pressure. The resulting internal pressure changes as the surface energy is changed. The surface energy or rather the interfacial energy of a colloidal system may be changed by adsorption of gases or vapours. If a film of thickness Γ is adsorbed at a given vapour pressure p the interfacial energy measured in vacuum decreases by $\Delta\gamma$ /52/ :

$$\Delta\gamma = \gamma_0 - \gamma = RT \int_o^p \Gamma \, d(\ln p) \tag{3}$$

If γ of equation (3) is inserted into equation (1) the change of internal pressure caused by a changing surface energy can be calculated. Each individual gel particle expands as the internal pressure is reduced. Bangham and co-workers showed that within certain limits, a linear relation exists between the change of interfacial energy and the resulting length change /53/ :

$$\frac{\Delta l}{l} = \lambda \, \Delta\gamma \tag{4}$$

Later Hiller expressed λ in terms of properties of the colloidal system /54/. It is assumed that in the range of low relative humidity (RH) the hygral length change can be described semi-phenomenologically by utilizing equation (4). A more quantitative application of equation (4) is not possible as decisive factors such as the particle size distribution are not sufficiently well known.

As the relative humidity is raised above 50% some surfaces will be separated by disjoining pressure /46/. This leads to additional expansion of the colloidal system. This length change is not caused by a corresponding change of surface energy. Therefore equation (4) cannot be applied in this range. Simultaneously the total structure is weakened by the action of the disjoining pressure.

The Griffith-criterion has proved to be very successful in describing fracture phenomena in hardened cement paste and concrete /55/. According to this concept the square of the related strength is equal to the related interfacial energy :

$$\left(\frac{\sigma}{\sigma_0}\right)^2 = \frac{\gamma}{\gamma_0} = 1 - \frac{\Delta\gamma}{\gamma_0} \tag{5}$$

By inserting equation (3) into equation (5) the relative strength decrease as a function of moisture content can be estimated /56/. By using Bangham's equation (4), equation (5) may be rewritten :

$$\left(\frac{\sigma}{\sigma_0}\right)^2 = 1 - \frac{1}{\lambda\gamma_0} \cdot \frac{\Delta l}{l} \tag{6}$$

Equations (5) and (6) indicate a linear relationship between the square of the related strength and change of interfacial energy and length change respectively. This statement, of course, is only valid in the range of RH in which drying or rewetting changes the interfacial energy only. As mentioned above at high moisture content the action of disjoining pressure cannot be neglected and therefore additional weakening of the structure has to be anticipated.

In conclusion we can say that the Munich Model introduces two terms which can be related to strength and failure of concrete :

  a) Interfacial energy of the xerogel

  b) Disjoining pressure of adsorbed water films

2.4.3. Comparison with experimental results. Theoretical predictions of the Munich Model are compared with experimental results in /2/ and /57/. There the influence of moisture content on hygral length change (swelling and shrinkage), on modulus of elasticity, on damping, and on creep, is discussed. Here we will concentrate on strength exclusively.

Wittmann has measured strength of hardened cement paste as function of moisture content /56/. In this paper /56/ it has been shown that the square of the related strength decreases linearely with increasing swelling. This relation is predicted by equation (6). Above 50% RH, however, there is additional length change observed due to the action of disjoining pressure.

If we replot experimental data /56/ as function of change of interfacial energy we find the relation plotted in Figure 1. For two different values of water/cement ratio, a straight line is obtained in the low humidity region. Above 50% RH the disjoining pressure further weakens the microstructure and hence causes a further decrease of the related strength.

Comparable test series have been carried out under carefully controlled conditions by Pihlajavaara /58/. Experimentally determined values of flexural strength of mortar prisms have been replotted in Figure 2. Within the range of accuracy again a straight line is obtained in the low humidity range and additional weakening is caused by disjoining pressure at humidities above 50%.

The linear relation between the square of the related strength and the change in interfacial energy as documented in Figures 1 and 2 has been predicted by equation (5), a basic relation of the Munich Model.

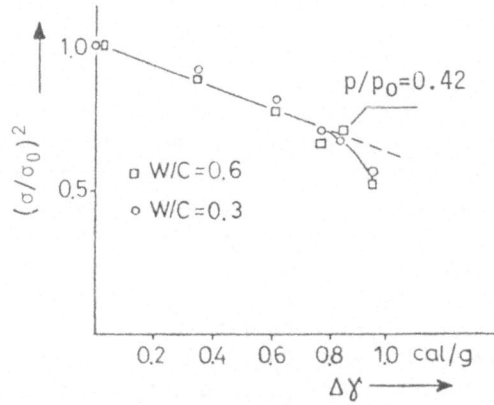

Fig. 1 : Square of the related compressive
strength of hardened cement paste
cylinders as function of change of
interfacial energy. Results of two
test series /56/ with differing
water/cement ratio are shown.

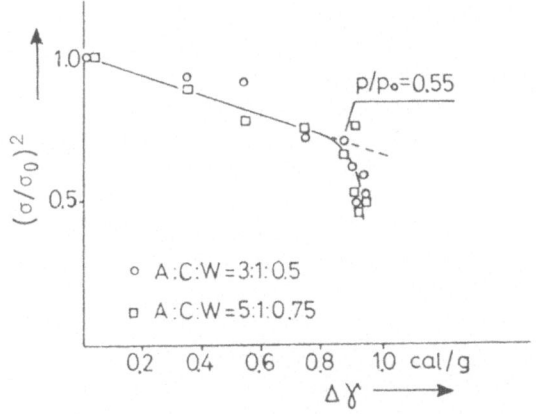

Fig. 2 : Square of the related flexural
strength of mortar prisms as
measured by Pihlajavaara /58/
replotted as function of change
of interfacial energy.

2.5. Porosity

It has often been tried to relate the strength of a porous material directly to the total porosity. In the literature therefore a number of different equations relating strength and porosity can be found. All of them predict an increase of strength as the total porosity decreases. We will not deal with these empirical equations here. By applying these formulae tacitly one assumes that the pore size distribution either is independent of porosity of changes in an analogous way.

This assumption is not fulfilled in many porous building materials. In Figure 3, the pore size distributions of two natural sandstones are shown. In addition the corresponding strengths are indicated in Figure 3. If we based the prediction on the total porosity of the samples only, we would conclude that sample A must have the higher strength. In fact the contrary is observed. The reason for this is that the pore size distribution functions are different. Sample A has more coarse pores than sample B. It is obvious that we have to take into consideration both total porosity and pore size distribution in deriving strength of a porous material.

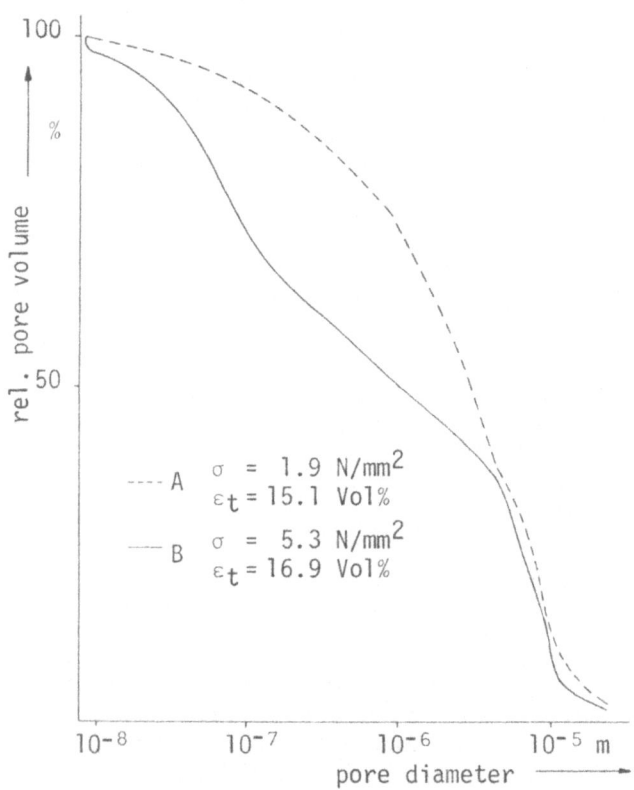

Fig. 3 : Pore size distribution of two different types A and B of porous natural stone (sand stone). In addition the total porosity ε_t and the biaxial flexural strength σ_b of the two materials are indicated.

In concrete technology the water/cement ratio plays a dominant role. In Figure 4 the strength of hardened cement paste is shown as a function of the water/cement ratio. In addition in Figure 4 the elastic modulus is shown as a function of water/cement ratio.

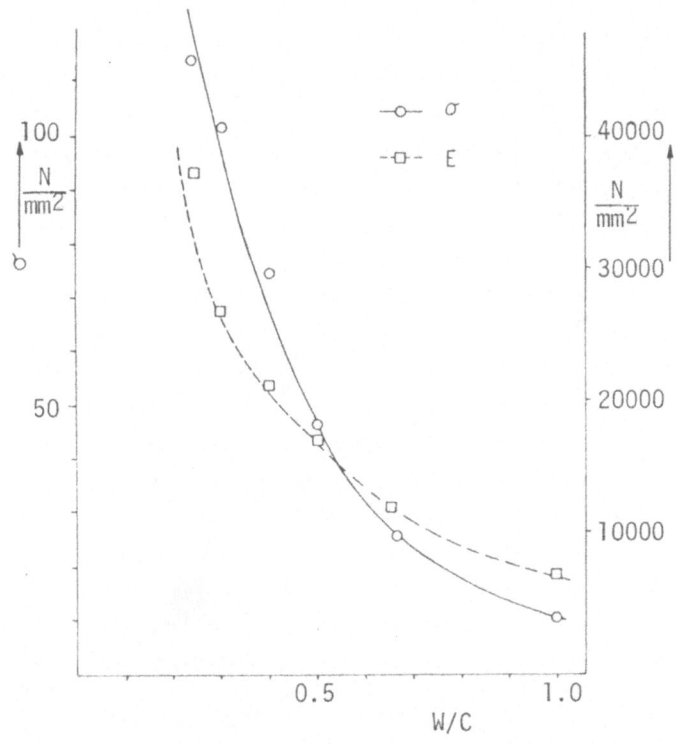

Fig. 4 : Influence of water/cement ratio on strength and modulus of elasticity of hardened cement paste.

In Figures 5 and 6 the pore size distribution of hardened cement paste are given as a function of water/cement ratio and as a function of duration of hydration respectively.

Even though the Griffith criterion is a simplified description of the actual failure process we can still use it to discuss the influence of porosity and pore size distribution on strength of hardened cement paste. If we use as usual, σ for strength, E for elastic modulus, γ for the fracture surface energy and 2c for the crack length, the Griffith equation reads as follows :

$$\sigma = \sqrt{\frac{2E\gamma}{\pi c}} \tag{7}$$

From Figure 4 we know the influence of water/cement ratio on the elastic modulus E. In fact it can be shown that the elastic modulus depends essentially on the total porosity ε :

$$E = E' (\varepsilon) \tag{8}$$

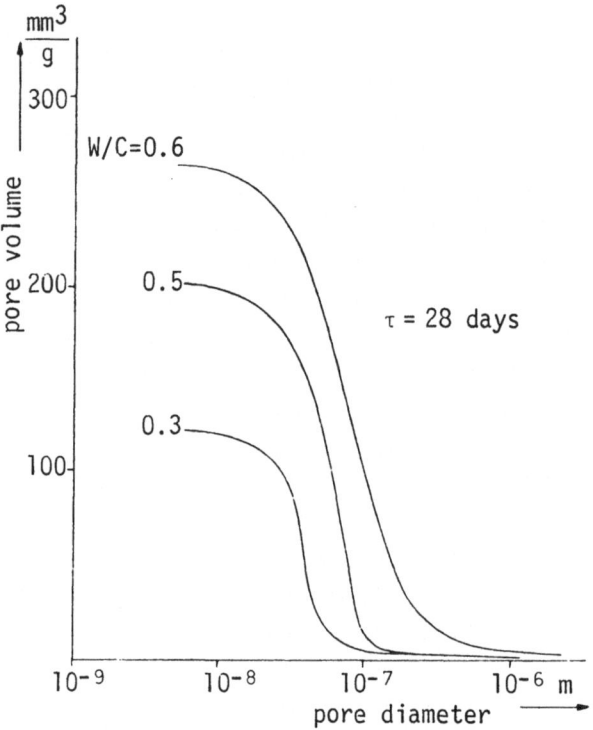

Fig. 5 : Pore size distribution of hardened cement paste after 28 days of hydration for three different water/cement ratios.

In fact, the influence of porosity can be adequately predicted by models of composite materials. The fracture surface energy γ is well defined for a non-porous material. If a crack spreads across a porous structure the energy which is consumed depends on the fracture surface energy of the non-porous material and the number N of particles to be cracked per unit area. In other words, this means that only the part of the fracture surface which goes across solid particles contributes to the energy balance. This is shown schematically in Figure 7. It is obvious that the effective surface energy depends essentially on the total porosity. In material (A) in Figure 7 the crack passes approximately through 55% of solid matter in its way whereas in material (B) only about 25% of the crack length has passed solid particles. In fact this relation holds true even if one considers a tortuous crack path as has been suggested by Higgins and Bailey /59/.

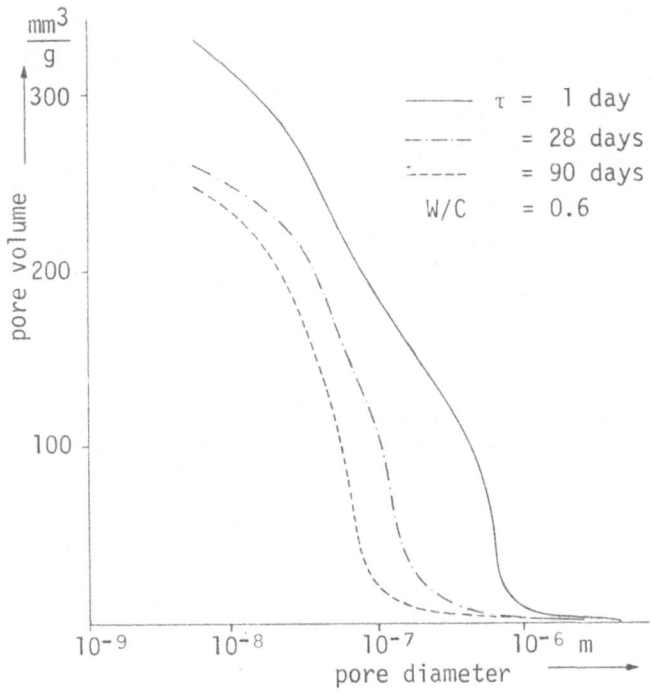

Fig. 6 : Pore size distribution of hardened cement paste having a water/cement ratio of 0.6 after 1, 28, and 90 days of hydration.

Then we can redefine the fracture surface energy of a porous material in the following way :

$$\gamma_p = \gamma \ N = \gamma \ (1-\epsilon) \tag{9}$$

The fracture mechanism in a xerogel such as hardened cement paste as suggested by Higgins and Bailey /59/, is schematically shown in Figure 8. The meaning of crack length 2c in equation (7) will be further discussed in the following section on structural features of the meso-level.

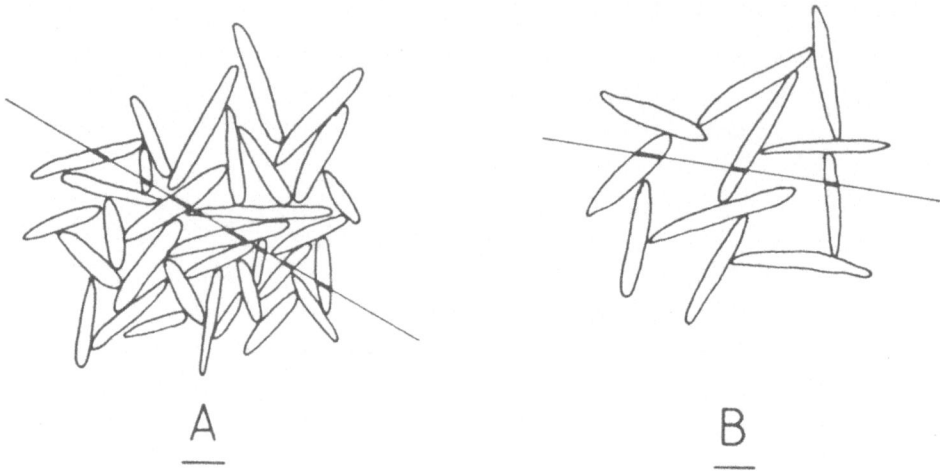

Fig. 7 : Schematic representation of fracture energy consumed
by a crack running through a material (A) with low
porosity and a material (B) with high porosity.

Fig. 8 : Fracture mechanism in a xerogel
such as hardened cement paste
according to Higgins and Bailey
/59/.

3. MESO-LEVEL

3.1. General Remarks

In the preceding section we have described the microstructure of hardened
cement paste and in particular the interaction of the xerogel with adsorbed water. It has been shown that it is possible to predict qualitatively
the influence of moisture content on strength of hardened cement paste on
this basis.

So far we have neglected, however, the heterogeneous structure of the material. Therefore we will introduce big pores, pre-existing cracks, and inclusions as the main characteristic features of the meso-level.

3.2. Big Pores

In Figure 5 the pore size distribution of hardened cement paste with different water/cement ratios are shown. Hardened cement paste prepared under normal conditions always has some big pores which are not covered by the distribution functions shown in Figure 5. These big pores, however, considerably reduce the strength of the material. Recently it has been tried to limit big pores to a minimum, thus increasing the strength.

The crack length 2c in the idealized Griffith criterion (see equation 7) is defined to be the length of one single crack in an otherwise homogeneous infinite plate. In a real porous material it stands for an effective maximum crack length $\bar{C}$ or an equivalent big pore. This value $\bar{C}$ must not necessarily be identical with the existing maximum crack length because there is interaction between large neighboring cracks. In any case it means that C depends sensitively on the pore size distribution (PSD) and in particular on the probability of finding large pores.

$$C = \bar{C} \; (PSD) \tag{10}$$

Then we can rewrite equation (7) by using equations (3) to (10) in the following way :

$$\sigma = \sqrt{\frac{2 \cdot E'(\varepsilon) \cdot \gamma (1-\varepsilon)}{\pi \; \bar{C} \; (PSD)}} \tag{11}$$

In doing so we can link structural details of hardened cement paste directly with strength. In many cases $\bar{C}$ has to be determined from fracture tests. Possibly in the future we will be able to estimate the effective maximum crack length $\bar{C}$ by fitting the pore size distribution to an extreme value distribution function.

3.3. Cracks

It is well known that under usual climatic conditions in hardened cement paste, mortar, and concrete, numerous cracks exist before a load is applied. There are different causes for the development of these cracks, which must be looked upon to be an important feature of the structure of concrete with respect to behaviour under load and with respect to failure.

Considering the total lifetime of a concrete member, the earliest cracks are formed by incomplete compaction. Insufficient compaction can lead to local zones of high porosity which act under load like precracked areas. In Table II different stages of strength development and the corresponding crack formation are compiled.

Immediately after pouring and before hardening, sedimentation (bleeding) takes place. This process causes water filled pockets under coarse aggregates. As a consequence horizontal cracks are formed. This effect obviously causes a certain degree of anisotropy.

TABLE II
Characteristic periods in the lifetime of concrete and corresponding crack formation

Stage within Strength Development	Typical Discontinuity
Pouring and Compaction	Compaction Pores
Fresh Concrete	Bleeding Cavities
Hardening Concrete	Thermal Cracks, Chemical and Capillary Shrinkage Cracks
Drying Concrete	Hygral Shrinkage Cracks
Loaded Concrete	Interfacial Cracks, Crack Growth

As the hydration continues heat of hydration is liberated. Under normal conditions this causes a time-dependent temperature gradient. It can be shown that in many concrete elements thermal cracking takes place in the outer cooler zones. The orientation of these cracks depends on the geometry. There are measures, however, to prevent excessive thermal cracking.

After demoulding, the surface of concrete begins to dry and reaches equilibrium with the environmental humidity very quickly, whereas the centre of a given specimen may remain saturated for many years. This hygral gradient induces shrinkage cracks. Again the orientation of these cracks depends on geometry.

In hardened concrete the interfaces between hardened cement paste and coarse aggregates remain weak for a long time. Therefore comparatively moderate loads, far below the design load, may cause interfacial cracking.

In summary we can conclude that in a concrete specimen some of the causes mentioned in this section and compiled in Table II or combinations of these different causes will introduce cracks into the structure under normal conditions. Some of these cracks are oriented at random and others initiate a certain degree of anisotropy. The observed strength therefore does not only depend on concrete composition but to a large degree on curing conditions.

3.4. Inclusions

According to the Griffith criterion a crack spreads catastrophically once the load becomes critical. There is no stable crack growth and no crack arresting. In a composite material therefore the unmodified Griffith equation is not a realistic approximation.

Consider for a moment the situation in normal concrete, where the aggregates are stronger than the matrix. If a crack starts to spread under a given load from a big pore in the weaker matrix or from an interface there is a chance that it will meet an inclusion and then will be stopped. This simplified approach is schematically shown in Figure 9. $2C_M$ is supposed to be the length of the initial crack in the matrix. As soon as σ_M is reached the crack will spead in an unstable way according to :

$$\sigma_M = \sqrt{\frac{2 \ E \ \gamma_M}{\pi \ C_M}} \qquad\qquad (12)$$

γ_M denotes the fracture surface energy of the matrix. In this simplified approach we consider in fact a few isolated particles in an otherwise homogeneous matrix, i.e. a diluted system.

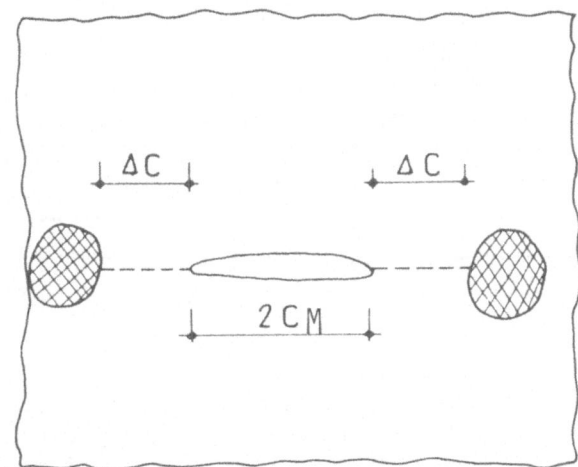

Fig. 9 : Schematic representation of a Griffith crack. Once the crack spreads it meets two inclusions after a crack length increases of ΔC.

Now we suppose that the crack meets an inclusion when it has grown by ΔC. At the crack tip the condition for further crack growth has now changed :

$$\sigma_A = \sqrt{\frac{2 \ E \ \gamma_A}{\pi \ (C_M + \Delta C)}} \qquad\qquad (13)$$

In this equation γ_A stands for the fracture surface energy of the aggregate. In normal concrete this value is higher than the one of the matrix ($\gamma_A > \gamma_M$) and therefore the resulting curve is shifted towards the right as shown in Figure 10. In the example chosen, the crack runs from point P_1 and is arrested at point P_2. Further crack growth is only possible if the load is increased to σ_A. In point P_3 the condition of crack propagation through the inclusion is fulfilled.

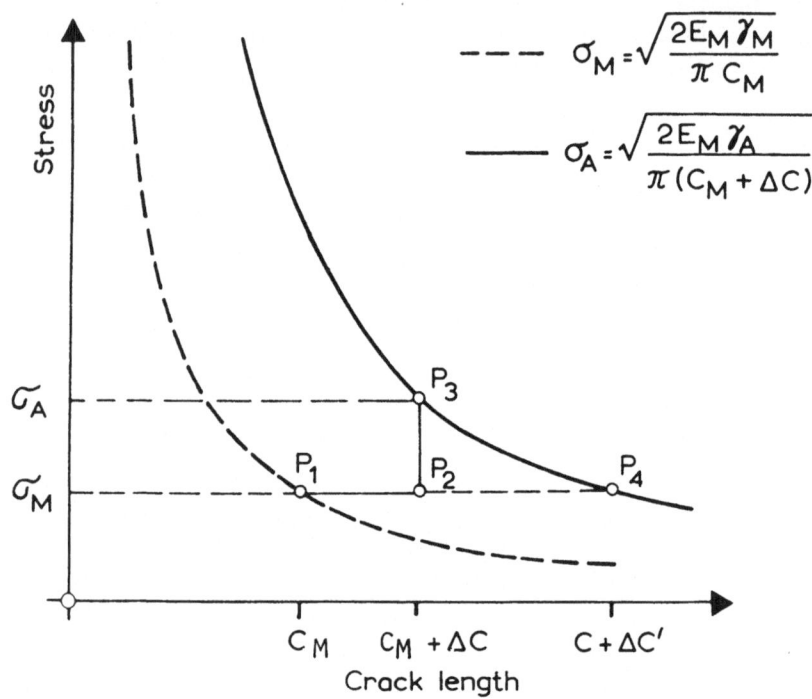

$$\sigma_M = \sqrt{\frac{2E_M\,\gamma_M}{\pi\,C_M}}$$

$$\sigma_A = \sqrt{\frac{2E_M\,\gamma_A}{\pi\,(C_M + \Delta C)}}$$

Fig. 10 : Schematic representation of equations (12) and (13). If the crack in the matrix becomes critical it will spread from P_1 to P_2. At a higher load σ_A the crack can penetrate the inclusion. If ΔC is large enough so that point P_4 is reached there will be no crack arresting.

If, however, the crack has increased at least by $\Delta C'$ before it meets an inclusion, the crack will not be arrested by the aggregate. This means that crack arresting is not only dependent on the mechanical properties of matrix and aggregates but also on the geometrical distribution. If a crack has already reached a critical length before it meets the aggregate, crack arresting has become impossible. This also explains that beyond a certain critical crack length this crack arresting mechanism does not work any more and the composite material fails finally in an unstable way.

3.5. Structural Models of Grudemo

In section 3.4 we have discussed the role of hard and strong aggregates
as crack arresting inclusions. A sharp crack can calso be arrested if it
runs into a wide rounded pore. Grudemo has summarized his observations
with the electron microscope in the form of simplifying structural
sketches /60 and 61/. In this way he designed a possible element of the
composite structure of hardened cement paste in which a crack runs into
the weak zone around an anhydrous cement particle embedded in hydration
products /61/. The sharp crack tip has become rounded and thus the crack
is arrested. This situation is schematically shown in Figure 11a.

With the help of another structural model Grudemo points out that a
crack will pass different zones such as the lamellar CH phase, inner and
outer gel and remaining anhydrous nuclei while it spreads. This simplifying model is shown in Figure 11b and underlines again the fact that
hardened cement paste has to be looked upon to be a heterogeneous composite material. This means that in principle there is no difference between crack formation in hardened cement paste, mortar, and concrete. It
is a difference in scale only.

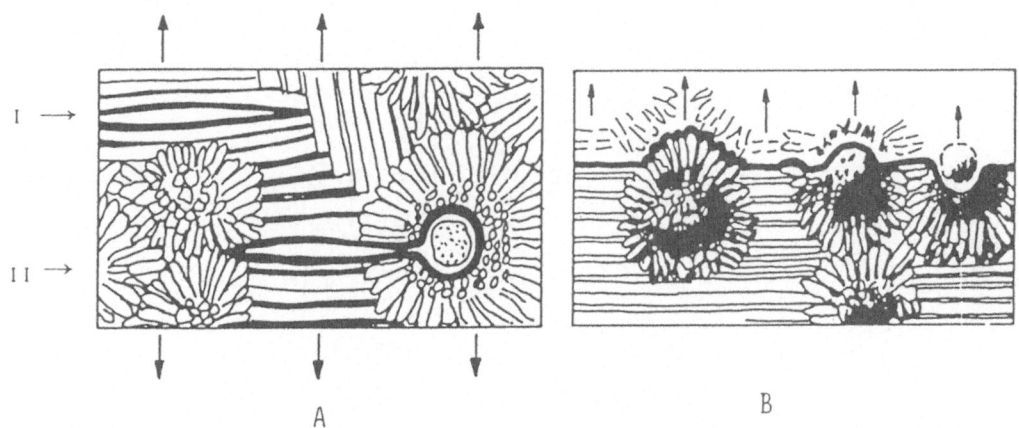

A   B

Fig. 11 : Idealized details of the gel structure according to Grudemo
/61/ :
   a) Cracks are arrested either as they meet stronger zone (I)
      or as they run into rouded-off spaces.
   b) The crack runs along cleavage surfaces in lamellar CH
      phase, zone of contact between gel aggregates, outer and
      inner gel, and remaining anhydrous nuclei.

3.6. Micromechanics and Simulation of Composite Structures

So far we have given a qualitative description of crack propagation and crack arresting in a composite material. The micromechanics of concrete deal with this subject in more detail and in a quantitative way. In fact a crack, if it is arrested, can propagate through the inclusion or along the interface around it. Micromechanics of concrete has been treated elsewhere /62, 63/. Therefore it is not necessary to repeat it here in detail.

We shall, however, mention at this point the possibility of generating composite random structures to simulate the real concrete structure. These computer generated random structures form a basis for the application of micromechanics. In this way many aspects of failure of concrete can be studied systematically. In Figure 12 a computer simulated structure of a matrix with polygonal inclusions is shown. More details are given in /62/. For this type of a composite material, crack growth can be calculated analytically by means of micromechanics. In addition, typical results are shown as different stages of stable crack growth in Figure 12.

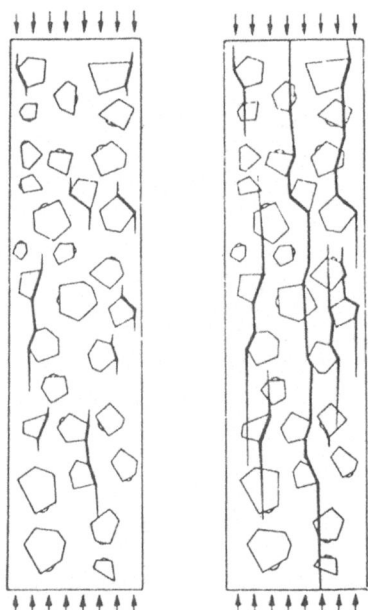

Fig. 12 : Simulated composite structure
and calculated crack pattern
for two different load levels,
according to /62/.

Similar structures can be generated by taking into consideration the random geometry and the size distribution of aggregates /64/. A typical example is shown in Figure 13. In this case, however, crack propagation and crack arresting cannot be treated analytically any more. By means of numerical methods such as finite element analysis it is possible, however, to obtain close approximations of the real fracture process in composite structures. This recent approach to simulate composite structures and to analyse numerically properties such as progressive failure is now called "numerical concrete". Roelfstra's chapter in this volume describes numerical concrete in more detail.

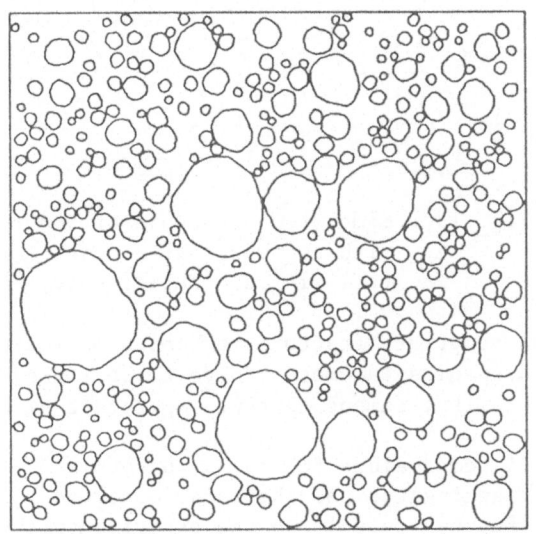

Fig. 13 : Simulated structure of concrete
with aggregates having random
geometry /64/.

As a last example for computer generated composite structures, Figure 14 shows a model of concrete with spherical aggregates of different size where the size distribution of the aggregates follows the Fuller function. It is possible to generate in this way three-dimensional structures with spherical aggregates. These three-dimensional composite structures allow us to study crack formation under uniaxial and under multiaxial states of stress.

4. MACRO-LEVEL

On the macro-level the actual macroscopically observed behaviour of a given material has to be described. For numerous materials this can be done

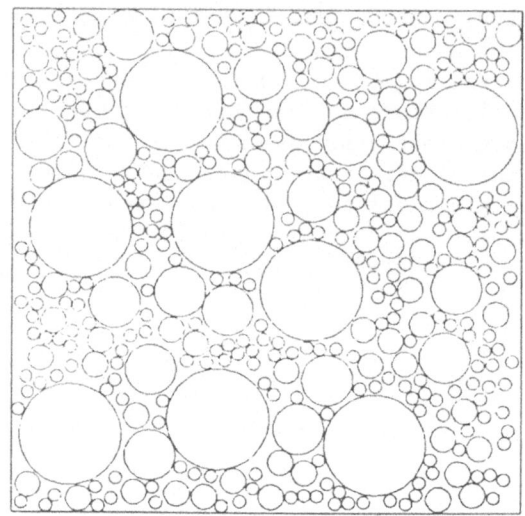

Fig. 14 : Simulated structure of concrete
with aggregate size distribution
following the Fuller curve /64/.

by introducing appropriate fracture mechanics parameters. If possible
these fracture mechanics parameters have to be linked with structural aspects such as porosity, cement content, aggregate size, and aggregate
geometry.

In Table III some values for K_{IC} as determined on hardened cement
paste, different aggregates and the range observed with concrete specimens are given. Further details on the experimental techniques as well
as the results obtained are described in papers mentioned in the list of
references.

There is an obvious influence of water/cement ratio on K_{IC} (see
also /63/). This is related to porosity and pore size distribution. K_{IC}
for different types of aggregates varies within a wide range. The influence of the age of concrete and of mix proportions on K_{IC} has been
studied by various authors /68/. Pak and Trapeznikov have determined experimentally the influence of maximum aggregate size on K_{IC} /69/.

At first glance it seems as if strength and failure processes can
be adequately described by simple fracture mechanics parameters such as
K_{IC} and G_{IC}. Unfortunately this is not the case and these values must be
considered to be rough approximations in the case of concrete in particular.

In an attempt to overcome these difficulties, more complex fracture
mechanics parameters such as crack opening displacement (COD) and J-integral (see f.e. /70/) have been introduced. So far, however, it is not
yet possible to discuss the values in relation with the structure of the
material.

TABLE III
Critical fracture toughness as determined for different
hardened cement pastes, aggregates and concrete

Material		K_{IC} (MN/m$^{3/2}$)	Ref.
Hardened Cement Paste	W/C = 0.30	0.40	
	W/C = 0.36	0.35	66
	W/C = 0.50	0.29	
Aggregates	Limestone	0.70 - 1.00	67
	Marble	1.9	
	Quarz	3.4	68
Concrete		0.70 - 1.50	69

In addition, fracture mechanics parameters of concrete are not in-dependent of the crack history. As a crack extends a damages zone the so-called fracture process zone is created ahead of the crack. As a consequence, fracture mechanics parameters increase with crack length until the damaged zone has fully developed. This situation is described by means of the R-curve approach (/71/ and /72/). The R-curve concept is further discussed in Bui's chapter in this volume. It is obvious that the development of the fracture process zone will be stress and time dependent. At present it is unknown in which way the fracture zone depends on the structure of concrete. Further research is needed to clarify the significance of the fracture process zone in fracture mechanics. The state-of-the-art in this field is well documented in Cedolin's contribution.

Hillerborg and co-workers developed another approach to describe the failure process of concrete, i.e. the fictitious crack model /73-75/. We mentioned already in context with the J-integral and the R-curve that at this moment we cannot discuss these fracture mechanics parameters in terms of structural details mainly because of a lack of suitable data. The same is true for the fictitious crack model. In this concept fracture energy and strain softening of the material are needed in order to predict failure /76/. Extensive studies will be necessary to provide a solid basis for a realistic link between the structure of the composite material and fracture mechanics parameters. It should be mentioned here that recently a RILEM recommendation has been published in which a method to determine fracture energy of concrete is described /77/.

On the macro-level appropriate material laws should be formulated. At present, however, it is not yet possible to predict crack resistance if the composition of concrete is known.

Another approach of the macro-level deals globally with damage, the damage theory. Details are described in the chapter of Bui and Stolz.

5. CONCLUSIONS

The heterogeneous structure of the composite material concrete can be described on three different levels. The physical basis of the xerogel, the binding agent in concrete, is described on the micro-level (Munich Model). This model has proven to be applicable to other cement based xerogels such as aerated concrete /78/. Additional structural details such as pores, cracks and inclusions are introduced on the meso-level. Finally, the macroscopically observed behaviour of concrete is represented on the macro-level.

The influence of porosity, cracks, and inclusions on strength and failure of a composite material can be described in a quasi-quantitative way. K_{IC} and G_{IC} must be looked upon to be rough approximations in the case of concrete. There are different attempts to come to a more realistic description of the failure process such as COD, J-integral, R-curve, and the fictitious crack model (FCM). So far, however, existing experimental data do not allow us a rigorous comparison between these more complex fracture mechanics parameters and structural details.

REFERENCES

1. F.H. Wittmann, 'Modelling of concrete behaviour', *Proc. European Concrete Research,* Swedish Cement and Concrete Research Institute, Stockholm, 1981, pp. 171-189.
2. F.H. Wittmann, 'Grundlagen eines Modells zur Beschreibung charakteristischer Eigenschaften des Betons', *Deutscher Ausschuss für Stahlbeton,* Report Nr. 292, Wilhelm Ernst & Sohn, Berlin (1976).
3. J.F. Young, 'The microstructure of hardened Portland cement paste', in Z.P. Bazant and F.H. Wittmann (Ed.) : *Creep and Shrinkage in Concrete Structures,* Wiley 1982.
4. 'Chemistry of Cement', *Proc. of the 4th International Symposium,* Washington (1960) edited in 1962.
5. 'Chemistry of Cement', *Proc. of the 5th International Symposium,* Tokyo (1968) edited in 1969.
6. 'Chemistry of Cement', *Proc. of the 6th International Symposium,* Moscow (Sept. 1974).
7. 'Chemistry of Cement', *Proc. of the 7th International Symposium,* Paris (August 1980).
8. T.C. Powers and T.L. Brownyard, 'Studies of the physical properties of hardened Portland cement paste', Research Laboratories of the Portland Cement Association, *Bulletin 22* (1948).
9. A. Grudemo, 'The microstructures of cement gel phases', Trans. Royal Inst. Techn., Stockholm, Nr. 242 (1965).
10. A. Grudemo, 'An electronmicroscopic study of the morphology and crystallization properties of calcium silicate hydrates', *Proc. Swedish Cement and Concrete Institute Stockholm,* Nr. 26 (1955).

11. F. Liebau, 'Ein Beitrag zur Kristallchemie der Schichtsilikate', *Acta Cryst.* 824 (1968) pp. 690-699.
12. L.E. Copeland, E. Bodor, T.N. Chang and C.H. Weise, 'Reactions of tobermorite gel with aluminates, ferrites and sulfates', *Res. and Development Laboratories of the Portland Cement Association,* Bulletin 211 (1967).
13 W. Richartz and F.M. Locher, 'Ein Beitrag zur Morphologie und Wasserbindung von Calciumsilikathydraten und zum Gefüge des Zementsteins', *Zement-Kalk-Gips,* 18 (1965) pp. 449-459.
14. D.L. Kantro, S. Brunauer and C.H. Weise, 'Development of surface in the hydration of calcium silicates - II. Extension of investigations to earlier and later stages of hydration', *J. Phys. Chem.,* 65 (1962) p. 1804.
15. R.Sh. Mikhail and S.H. Abo-El-Enein, 'Studies of water and nitrogen adsorption on hardened cement pastes - I. Development of surface in low porosity pastes', *Cem. Concr. Res.,* 2 (1972) pp. 401-414.
16. S. Diamond, 'Identification of hydrated cement constituents using a scanning electron microscope - Energy dispersive X-ray spectrometer combination', *Cem. Concr. Res.,* 2 (1972) pp. 617-632.
17. U. Ludwig, 'Investigation on the hydration mechanism of clinker minerals' (see ref. 6).
18. H.F.W. Taylor, 'Crystal chemistry of Portland cement hydration products' (see ref. 6).
19. T.C. Powers, 'Physical properties of cement paste', *Research and Development Laboratories of the Portland Cement Association,* Res. Dept., Bull. 154 (1960).
20. A. Grudemo, 'On the development of hydrate crystal morphology in silicate cement binders', Liaisons de Contact dans les Matériaux Composites Utilisés en Génie Civil, *RILEM-INSA-Coll.,* Toulouse, France (November 22-24, 1972).
21. F. Wittmann and G. Englert, 'Bestimmung der Mikroporenverteilung im Zementstein', *Mat. Sci. Eng.,* 2 (1967) p. 14.
22. M.J. Setzer and F.H. Wittmann, 'Modified method to calculate pore size distribution using sorption data', *Proc. RILEM-IUPAC International Symposium,* Pore Structure and Properties of Materials, Prague (September 18-21, 1973).
23. F.D. Tamas and T.G. Varadi, 'Role of Poly-reactions in the hydration of cement' (see ref. 6).
24. L.S. Dent-Glasser, E.E. Lachowski, K. Mohan and H.F.W. Taylor, 'A multi-method study of C_3S hydration', *Cem. Concr. Res.,* 8 (1978) p. 733.
25. A. Rio and A. Saini, 'L'industria Italiana del cemento', 39 (1969) p. 857.
26. D.N. Winslow, 'The specific surface of hardened Portland cement paste as measured by low angle X-ray scattering', *Thesis,* Purdue University, La Fayette, Indiana, USA (1973).
27. L.E. Copeland and G.J. Verbeck, 'Structure and properties of hardened cement pastes' (see ref. 6).

28. H.F.W. Taylor and D.M. Roy, 'Structure and compostion of hydrates' (see ref. 7).

29. P.J. Sereda, R.F. Feldman and V.S. Ramachandran, 'Structure formation and development in hardened cement pastes' (see ref. 7).

30. F.H. Wittmann, 'Properties of hardened cement paste' (see ref. 7).

31. O. Ishai, 'The time-dependent deformational behaviour of cement paste, mortar and concrete', *Proc. International Conference, The Structure of Concrete and its Behaviour under Load*, London (September, 1965) p. 345.

32. T.C. Powers, 'Mechanism of shrinkage and reversible creep of hardened cement paste', *Proc. International Conference, The Structure of Concrete and its Behaviour under Load*, London (September 1965) p. 319.

33. T.C. Powers, 'The thermodynamics of volume change and creep', *Materials and Structure*, $\underline{1}$ (1968) pp. 487-507.

34. J.D. Bernal, *Proc. 3th International Symposium on the Chemistry of Cement*, London (1952) p. 216.

35. R.F. Feldman and P.J. Sereda, 'A model for hydrated Portland cement paste as deduced from sorption - Length change and mechanical properties', *Materials and Structures*, $\underline{1}$ (1968) pp. 509-520.

36. R.F. Feldman and P.J. Sereda, 'A new model for hydrated Portland cement and its practical implications', *Engineering Journal*, $\underline{\underline{53}}$ (1970) p. 53.

37. R.F. Feldman, 'Sorption and length change scanning isotherms of methanol and water on hydrated Portland cement', *Proc. 5th International Symposium on the Chemistry of Cement*, Tokyo, Vol. $\underline{\underline{III}}$ (1968) p. 53.

38. B.B. Hope and N.H. Brown, 'A model for the creep of concrete', *Cem. Concr. Res.*, $\underline{5}$ (1975) pp. 577-586.

39. R. Kondo and M. Daimon, 'Phase composition of hardened cement paste' (see ref. 6).

40. F. Wittmann, H. Splittgerber and K. Ebert, *Z. Physik*, $\underline{\underline{245}}$ (1971) p. 354.

41. H. Splittgerber and F. Wittmann, 'Einfluss adsorbierter Wasserfilme auf die van der Waals Kraft zwischen Quarzglasflächen', *Surface Science*, $\underline{\underline{41}}$ (1974) p. 504.

42. H. Krupp, 'Particles adhesion, theory and experiments', *Advan. Colloid Interfaces Sci.*, $\underline{1}$ (1967) p. 116.

43. H.J. Setzer and F.H. Wittmann, 'Surface energy and mechanical behaviour of hardened cement paste', *Appl. Physics*, $\underline{3}$ (1974) pp. 403-409.

44. H. Übelhack and F.H. Wittmann, 'Debye-Waller,factor of colloidal particles in hydro- and xerogels', *Proc. International Conference on Mössbauer Spectroscopy*, Cracow, Vol. $\underline{1}$ (1975) pp. 349-350.

45. B.V. Derjaguin, *J. Colloid Interfaces Sci.*, $\underline{\underline{49}}$ (1974) p. 249.

46. H. Splittgerber, 'Spaltdruck zwischen Festkörpern und Auswirkungen an Probleme in der Technik', *Cem. Concr. Res.*, $\underline{6}$ (1976) pp. 29-36.

47. N. Stockhausen, 'Van der Waals interaction and disjoining pressure between solid surfaces', *Proc. Conference, Hydraulic Cement Pastes; their Structure and Properties*, Sheffield (April, 1976) pp. 219-226.

48. F. Schlude and F.H. Wittmann, 'Uber ein Verfahren zur raschen Bestimmung der komplexen DK im Mikrowellenbereich', *Nachrichtentechn. Zeitung*, 27 (1974) pp. 365-368.

49. B. Zech and F.H. Wittmann, 'Studium des dielektrischen Verhaltens von dünnen adsorbierten Wasserfilmen', *Z. Phys. Chemie*, NF 92 (1974) pp. 45-62.

50. E.A. Flood, 'Adsorption potentials, adsorbent self-potentials and thermodynamic equilibria, solid surfaces and the gas-solid interface', *Advances in Chemistry Series*, Nr. 33 (1961) p. 249.

51. K.G. Krasilnikov, A.M. Podvalny and A.E. Segalov, 'Self-induced deformations in porous bodies', *Kolloidnyi Zhurnal*, 36 (1974) pp. 266-271.

52. J.W. Gibbs, *Collected works*, Yale University Press, New Haven (1957).

53. D.H. Bangham and N. Fakhoury, 'The swelling of charcoal, Part I : Preliminary experiments with water vapour, carbon dioxide, amonia and sulphur dioxide', *Proc. Royal Society*, A130 (1931) pp. 81-89.

54. K.H. Hiller, 'Strength reduction and length changes in porous glass caused by water vapour adsorption', *J. Appl. Phys.*, 35 (1964) pp. 1622-1628.

55. F.H. Wittmann and J. Zaitsev, 'Verformung und Bruchvorgang poröser Baustoffe bei kurzzeitiger Belastung und Dauerlast', *DAfStb Report*, Nr. 232, Wilhelm Ernst & Sohn, Berlin (1972).

56. F. Wittmann, 'Surface tension, shrinkage and strength of hardened cement paste', *Materials and Structures*, 1 (1968) pp. 547-552.

57. F.H. Wittmann, 'The structure of hardened cement paste - A basis for a better understanding of the materials properties', *Proc. Conference, Hydraulic Cement Pastes; their Structure and Properties*, Sheffield (April 1976) pp. 96-117.

58. S.E. Pihlajavaara, 'A review of some of the main results of a research on the ageing phenomena of concrete : effect of moisture conditions on strength, shrinkage and creep of mature concrete', *Cem. Concr. Res.*, 4 (1974) pp. 761-771.

59. D.D. Higgins and J.E. Bailey, 'A microstructural investigation of the failure behaviour of cement paste', *Proc. Conference, Hydraulic Cement Pastes; their Structure and Properties*, Sheffield (April, 1976) pp. 283-296.

60. A. Grudemo, 'Strength-structure relationships of cement paste materials', Part 1 and Part 2 *CBI Research Reports* 6:77 and 8:79, Stockholm (1977 and 1979).

61. A. Grudemo, 'Mircrocracks, fracture mechanism, and strength of the cement paste matrix', *Cem. Concr. Res.*, 9 (1979) pp. 19-34.

62. J.B. Zaitsev and F.H. Wittmann, 'Simulation of crack propagation and failure of concrete', *Mat. and Struct.*, 14 (1981) pp. 357-365.

63. F.H. Wittmann, 'Mechanisms and mechanics of fracture of concrete', *Adv. in Fracture Research*, ICF-5, Vol. 4 (1981) pp. 1467-1487.

64. P.E. Roelfstra and H. Sadouki, 'Simulation des structures composites', *Internal Report,* Laboratory for Building Materials Science, Swiss Federal Institute of Technology, Lausanne (1981).
65. D.D. Higgins and J.E. Bailey, 'Fracture measurements on cement paste', *J. Mat. Sci.,* 11 (1976) pp. 1955-2003.
66. R.A. Schmidt, *Exp. Mech.,* 15 (1976) pp. 161-167.
67. B. Hillemeier and H.K. Hilsdorf, 'Fracture mechanics studies on concrete compounds', *Cem. Concr. Res.,* 7 (1977) pp. 523-536.
68. R.N. Swamy, 'Fracture mechanics applied to concrete', Chapter 6, *in Developments in Concrete Technology,* edited by F.D. Lydon, Applied Science Publishers (1979).
69. A.P. Pak and L.P. Trapeznikov, 'Experimental investigations based on the Griffith-Irwin theory process of the crack development in concrete', Adv. in Fracture Research, *ICF-5,* Vol. 4 (1981) pp. 1531-1539.
70. S. Mindess, F.V. Lawrence and C.E. Kesler, 'The J-integral as a fracture criterion for fiber reinforced concrete', *Cem. Concr. Res.,* 7 (1977) pp. 731-742.
71. C. Sok, J. Baron and D. François, 'Mécanique de la rupture appliquée au béton hydraulique', *Cem. Concr. Res.,* 9 (1979) pp. 641-648.
72. S. Chhuy, M.E. Benkirane, J. Baron and D. François, 'Crack propagation in prestressed concrete, Interaction with reinforcement', Adv. in Fracture Research, *ICF-5,* Vol. 4 (1981) pp. 1507-1514.
73. A. Hillerborg, 'Analysis of fracture by means of the fictitious crack model, particularly for fibre reinforced concrete', *Int. J. of Cement Composites,* 2 (1980) pp. 177-184.
74. A. Hillerborg and P.E. Petersson, 'Fracture mechanical calculations, test methods and results for concrete and similar materials', Adv. in Fracture Research, *ICF-5,* Vol. 4 (1981) pp. 1515-1522.
75. P.E. Petersson, 'Crack growth and development of fracture zone in plain concrete and similar materials', Lund Institute of Technology, *Report TVBM-1006* (1981).
76. F.H. Wittmann, Yiun-Yuan Huang, P.E. Roelfstra, H. Mihashi, N. Nomura and Xin-Hua Zhang, 'Influence of age of loading, water-cement ratio, and rate of loading on fracture energy of concrete', (to be published in *Mat. Struct.,* 19, 1986).
77. 'Determination of the fracture energy of mortar and concrete by means of three-point bend tests on notched beams', RILEM Draft Recommendation, *Mat. Struct.,* 18, 285-290 (1985).
78. Y. Houst, F. Alou and F.H. Wittmann, 'Influence of moisture content on mechanical properties of autoclaved aerated concrete', in F.H. Wittmann (Ed.) *Autoclaved Aerated Concrete, Moisture and Properties,* Elsevier, Amsterdam (1983) pp. 219-234.

FRACTURE MECHANICS PARAMETERS AND FRACTURE PROCESS ZONE OF CONCRETE

L. Cedolin
Department of Structural Engineering
Politecnico
P. Leonardo da Vinci 32
20133 Milano Italy

ABSTRACT. The initiation and propagation of a continuous crack in concrete occurs after the formation at the crack front of a fracture process zone the size of which is not negligible compared to structural dimensions. The results of experimental observations are reported and proposed techniques for the analysis of crack propagation in concrete through nonlinear constitutive relations are illustrated. The concept of R-curve is introduced as a method for the application of linear elastic fracture mechanics to concrete, and the parameters involved in the definition of the R-curve are determined on the basis of the size effect law.

1. INTRODUCTION

The formation of a crack in concrete is a complex phenomenon in which a finite region of material first undergoes severe microcracking and then unloads giving rise to a localization of strain until complete separation occurs. Ahead of the "true" crack there is, then, a process zone composed of a part in which a discontinuity in the displacement field is developing (although forces bridging the crack are still transmitted due to debonding and aggregate interlock) , and a part in which microcracking is diffused in a region of finite width. Several experimental methods can be used for the observation of this phenomenon, but so far only techniques which measure the strain field on the surface of the specimen have been reported in detail and will be illustrated here. It will appear that the length of the process zone is not negligible compared to the structure size, and that the crack tip is not easily definable, so that linear fracture mechanics cannot be directly applied. A nonlinear analysis is then best suited for the prediction of crack propagation, and two strain-softening material models will be illustrated, based one on a discrete crack representation, the other on the smeared crack approach. Frequently, however, the fracture process zone does not alter the remote stress field with respect to that corresponding to a linear fracture mechanics solution for a certain equivalent

K. P. Herrmann and L. H. Larsson (eds.), Fracture of Non-Metallic Materials, 341–358.
© 1987 by ECSC, EEC, EAEC, Brussels and Luxembourg.

crack length. It is possible, then, to analyze fracture in concrete with approximate methods based on linear fracture mechanics introducing the concept of R-curve, i.e. the variation of fracture energy with crack extension. The "raw" experimental data do not allow a direct determination of the R-curve : they must first be "smoothed", and this can be accomplished by using the size effect law.

2. EXPERIMENTAL OBSERVATION OF THE FRACTURE PROCESS ZONE

Since the formation and coalescence of microcracks are responsible for the existence of the fracture process zone, this can be detected by optical microscopy, and a review of its application can be found in [1]. The impregnation of cracks with dyes allows, by sectioning the specimen after the tests, the recognition of the spread of cracking inside the specimen. The limit of optical microscopy is the effective resolution, which can be estimated to be [2] in the order of 20 μm ; it can be augmented through the use of ultraviolet light and fluorescent dye [3]. Also X-rays [4] can be used to detect cracks inside thin slices of concrete, with a resolution similar to the one achieved with microscope and fluorescent dye. Techniques which use the Scanning Electron Microscope have proved [2] the existence on the surface of concrete specimen of tortuous cracks having width of fractions of μm , with branching and multiple fine cracking near the end of each branch. At this scale the phenomenon is essentially three-dimensional, and so complex that it is not possible to discriminate between the advancing true crack and the fracture zone ahead of it. All these techniques for the observation of microcracks are very useful for a phenomenological description of the fracture process, but do not give information expressed in macroscopical terms of strain.

Acoustic methods can also be used in order to give a measure of the extension and level of damage in the material. Methods based on the measure of acoustic emission due to microcracking and other dissipative phenomena are described in [5]. In [6] this technique has been applied to a double cantilever beam specimen, and the cumulative sum of acoustic energy has been correlated to other fracture mechanics parameters. The change in ultrasonic pulse transit time can alternatively be used, and in [7] it has been applied to a variable section beam, obtaining some indication on the extension of the fracture process zone.

A method recently developed is infrared thermography, by which the heat generated by exciting the material beyond its stable reversible limit can, under vibratory loading, show the coalescence of damage in a limited zone of the specimen [8] .

The measure of the displacement (or strain) field in the surroundings of a notch can give a direct indication of the extension of the fracture process zone. Strain gages have been used in [9] and [10] ; however the length of their basis is too large to capture the localization of strain, as it has been proved in [11] and, moreover, they do not allow a continuous measure of the strain field. For this purpose methods based on speckle metrology and holographic interferometry are

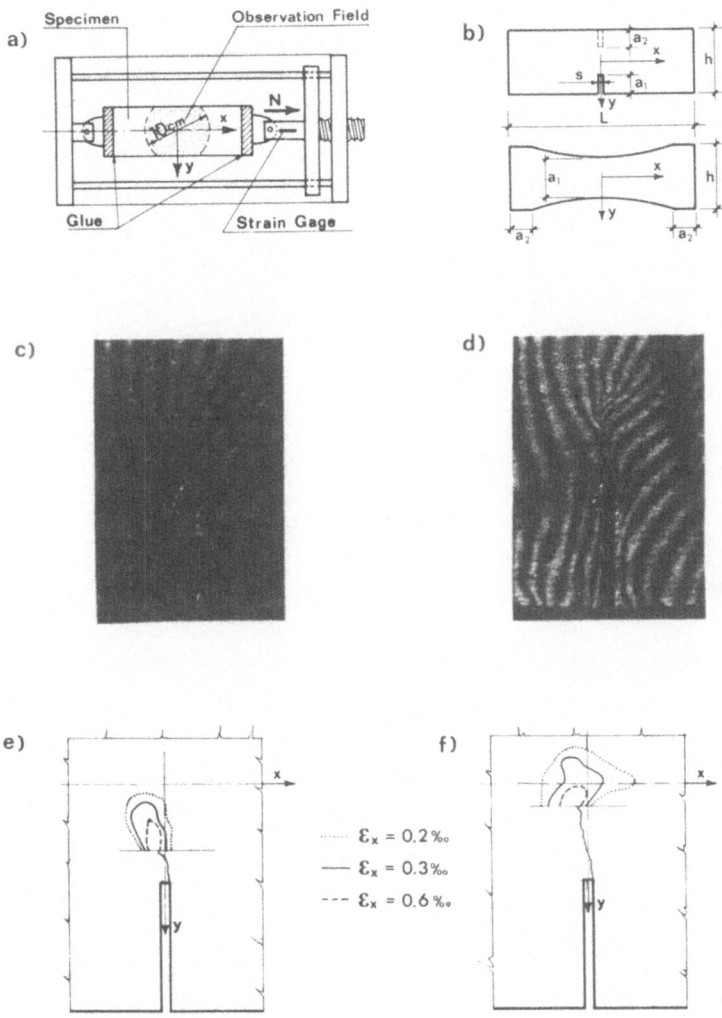

Figure 1 .- (a) Loading Device; (b) Specimen Types; (c),(d) Fringe Patterns for Different Loading Stages; (e),(f) Fracture Zone Extensions.

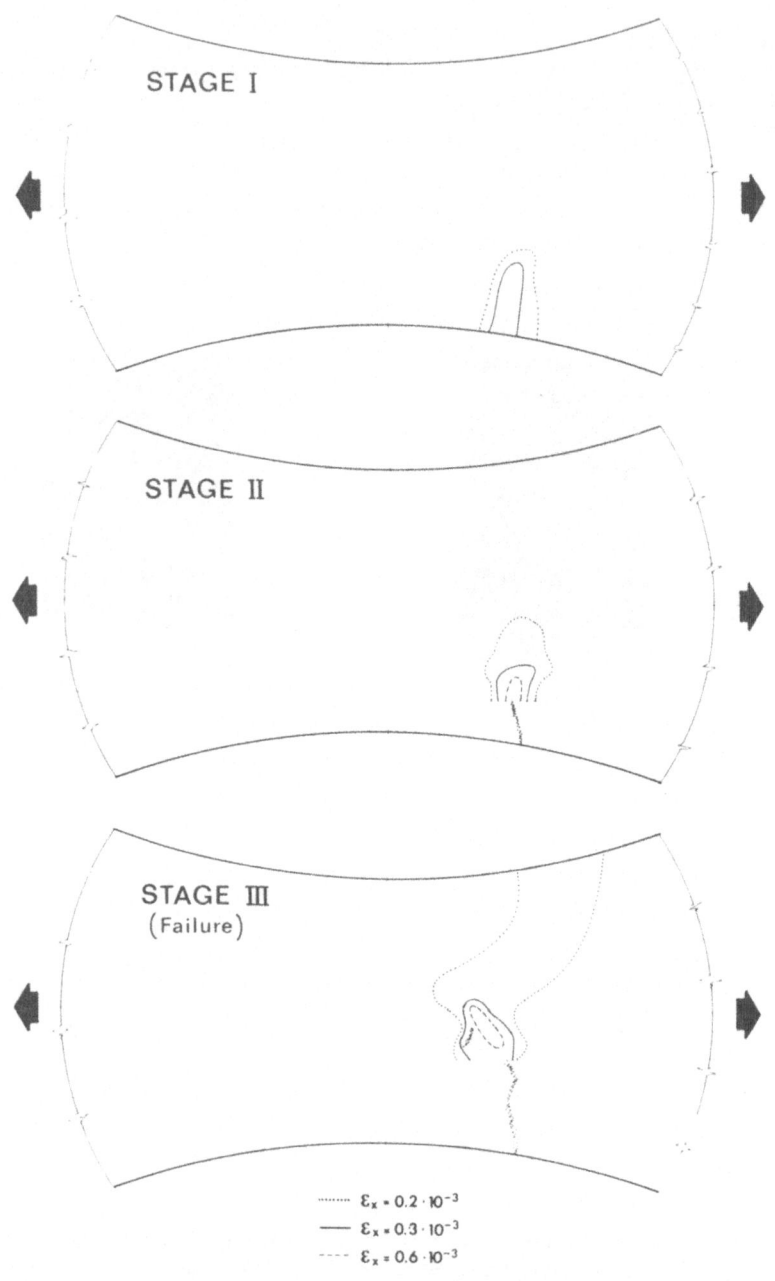

Figure 2 . - Fracture Zone Extensions in Unnotched Specimen

best suited. Their application to fracture mechanics problems is discussed in depth in [12].

A particular type of holographic interferometry has been used with
success for the determination of the fracture zone [11],[13] in concrete. With this method a reference grid of 1000 lines/mm is generated
on the surface of the specimen (Fig. 1a) by interference of two beams
of laser light. This grid is also recorded on a layer of photoresist
placed on the specimen surface, so that it will follow the deformation
of the specimen, while the reference grid remains undeformed. The light
emerging from the two gratings produces a moiré fringe pattern, from
which the strain component in the direction normal to the grid lines can
be determined.

For notched specimen (Fig. 1b) subjected to tensile loading, the
configuration of moiré fringes for two different stages of loading are
reported in Fig. 1c and 1d and the corresponding strain fields are
represented in Fig. 1e and 1f through contour lines of equal deformation. The zone of diffused microcracking which proceeds ahead of the
crack is identified by the contour line corresponding to $\varepsilon_x = 0.2 \times 10^{-3}$.
Below this zone the fringes become so dense that they cannot be distinguished anymore, and indicate a large localized strain. The evidence
presented in Refs. [11] and [13] suggests that the width of the fracture
zone is not affected by the aggregate size, and that the phenomenon is
independent of the notch width or depth. In the same references it is
shown that in the direct tensile tests of unnotched specimens the
strain is not uniformly distributed along the specimen and that the
fracture zone develops from a weak point and advances through the cross-
section (Fig.2).

The high density of the grid used in the above mentioned investigation can be obtained only by isolating the testing apparatus from vibrations through an air cushion. This puts limitations on the weight and
type of loading device. Variations of the same technique are less sensitive to these effects, and may be used for specimen of larger dimensions.

3. STRAIN-SOFTENING FRACTURE MODELS

From a continuum point of view, the effect of localization of the fracture process in a finite zone (Fig.3a) of material can be analyzed with
strain-softening material models.

This kind of approach was first proposed by Hillerborg [14],[15],
who introduced a so called "Fictitious Crack Model" in which the entire
fracture zone is represented with a line crack (Fig.3b) capable of
transferring stresses between the two sides of the crack. The tensile
stress is expressed as a function of crack width δ , and this stress-
displacement relation is considered as a material property. If this
relation is linear

$$\sigma = f_t^1 + C_f \delta \qquad (1)$$

this model involves two material constants, f_t' =tensile strength of concrete, and C_f (<0), slope of the stress-displacement curve. The area enclosed by the stress-displacement relation

$$G_f = \int_{f_t'}^{o} \sigma d\delta = \frac{1}{2} \frac{f_t'^2}{-C_f}$$

(2)

represents the energy adsorbed for the separation of a unit area of material, and is called fracture energy.

A model which is conceptually very similar is the Crack Band Model proposed by Bazant [16]. In this model the entire fracture zone is represented by a finite width of material (w_c in Fig.3c) of uniformly distributed cracking, characterized by a stress-strain relation which assumes that, in the direction z normal to the crack direction, the deformation ε_f due to microcracking is linearly related to σ_z . The uniaxial stress-strain curve is represented in Fig. 1c, where E_c is the Young's modulus of concrete, E_t (<0) indicates the slope of the strain-softening part and ε_o is the strain at which σ_z vanishes. The fracture energy becomes (in the hypothesis of absence of plastic deformation)

$$G_f = w_c \int_0^{\varepsilon_o} \sigma_z \, d\varepsilon_f = \frac{w_c}{2} f_t'^2 \left(\frac{1}{E_c} - \frac{1}{E_t} \right)$$

(3)

In this model w_c appears as an additional material constant, but, as noted at Pag. 164 of the same Ref [16], it does not have a real importance as such, but in connection with ε_z , because it determines the work of the surrounding structure on the fracture process zone. This allows, in finite element calculations, the use of finite elements of arbitrary size, provided that the softening modulus E_t is adjusted so that the fracture energy remains the same.

This second model is particularly suited for finite element calculations, because the use of a smeared crack concept allows the representation of crack propagation through a simple change of material properties. An extensive comparison of the maximum loads obtained with this nonlinear theory with experimental data [16] shows that these can be fitted by a proper calibration of the material parameters. Moreover, in the same Ref.[16] it is shown that the nonlinear theory predicts a dependence of the fracture energy on slow crack growth which is similar to the one found in experiments, i.e. the existence of a so called "R-curve" .

In both models fracture is determined by two independent parameters which have the same meaning and so it is reasonable to expect similar results. The only problem of these models is given by the fact that the determination of the strain-softening modulus with a direct tensile test is not possible because the strain field along the specimen is not uniform, as it was illustrated in Fig. 2 . Alternative ways of dealing with this problem will be shown in the next sections.

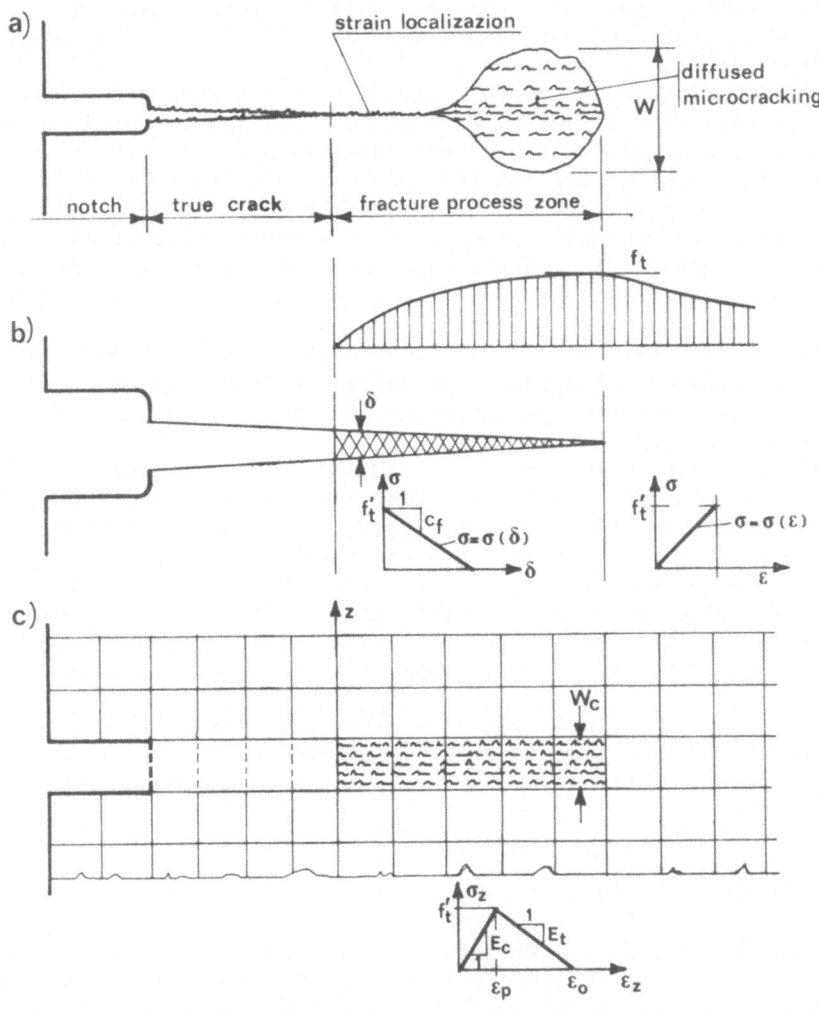

Figure 3 . - (a) Crack Propagation and Fracture Zone ; (b) Fictitious Crack Model ; (c) Crack Band Model

4. R-CURVES

The nonlinear fracture mechanics approach described in the preceding section is rather complicated for engineering applications, and one wonders if, with some approximation, methods based on linear elastic fracture mechanics may still be used. One of these methods is based on the R-curve concept, which was used in [17] and [18] for the interpretation of test results in concrete .

The first ingredient of this approach is the assumption that the fracture process zone is small, although not negligible, with respect to the structural dimensions .The stress and strain fields not in the immediate surroundings of the fracture process zone are then not altered significantly with respect to the ones which would be obtained through a linear fracture mechanics solution for a certain fictitious (equivalent) crack length. This crack length does not obviously correspond to any physical observable crack elongation (which, in any case, would be ambiguous since the existence of the fracture process zone makes hardly definable a crack tip) and its indirect determination through the measure of the specimen's compliance at unloading or reloading is questionable. It can, however, be estimated for the failure condition, as it will be shown further on.

A further ingredient is the assumption that the energy R required for crack growth (also called Fracture Resistance) varies with crack extension: this is shown by tests [17] and it is confirmed by finite element calculations with strain softening models [16] .

A third ingredient is the assumption that the R-curves can be treated as a material property: the same nonlinear models show [19] that this is not rigorously true, but that the variation of the R-curve with specimen geometry is small compared to the statistical scatter of results.

Following the treatment given in Refs. [18] and [19], let $c = a - a_0$ be the crack extension from a notch of length a_0 (Fig. 4a) and $R=R(c)$ the energy required for crack growth. The energy which must be supplied to produce the crack is

$$U(a)= b \int R(c)\ da \quad - W(a) \tag{4}$$

where b is the thickness, $W(a)$ is the total release of strain energy from the structure . Fracture equilibrium occurs when no energy needs to be supplied and none is released in order to change a by δa , i.e. when

$$\delta U = \frac{\partial U}{\partial a}\ \delta a = (bR - \frac{\partial W}{\partial a})\ \delta a \ =0 \tag{5}$$

and, defining the energy release rate of the structure as

$$G = \frac{1}{b} \frac{\partial W}{\partial a} \qquad (6)$$

the equilibrium condition becomes

$$G(a) = R(c) \qquad (7)$$

The fracture equilibrium state is stable if $\delta^2 U$ is positive. Since

$$\delta^2 U = \frac{\partial^2 U}{\partial a^2} \, \delta a^2 = (b \frac{dR}{dc} - b \frac{\partial G}{\partial a}) \, \delta a^2 \qquad (8)$$

the following condition ensures stability

$$R'(c) - G'(a) > 0 \qquad \text{(stable)} \qquad (9)$$

The geometry of a structure may be characterized by a set of dimensions, $d, a, l_1, l_2, \ldots l_n$. Considering geometrically similar structures (except for the crack length a) such that the ratios $\xi_i = l_i / d$ (i=1,n) are the same, the energy release rate can be put in the form [20]

$$G(a) = \frac{W'(a)}{b} = \frac{P^2}{E_c b^2 d} \, g(\alpha, \xi_i) \qquad (10)$$

in which $\alpha = a/d$, P is the loading system and $g(\alpha, \xi_i)$ is a nondimensional function which characterizes the geometry of the structure.

Consider now for a given structure the curves G(a) calculated for increasing values of the applied loads, P_1, P_2, $\ldots\ldots$. According to Eq. (7) the equilibrium states of crack extension are given by the intersection of these curves with the R-curve (Fig. 4b). According to Eq. (9) these equilibrium states are stable if at the point of interception the slope of the R(c) curve is larger than the slope of the G(a) curve. By increasing the load, then, there is a stable crack growth until the load P_{max} is reached (Fig. 4b) for which the slopes become equal and the structure fails.

It is interesting to note that in the case that R(c)=const (Fig. 4c), Eq.(9) becomes G'(a) <0 . Since in general G increases with a, this condition can never be satisfied , so that if a stable crack growth is observed it implies that the R-curve must be increasing. In the case that G decreases with a and R(c) is constant or increasing, Eq. (9) is always satisfied, and the crack is always stable (Fig. 4d).

Considering the case of structures for which G increases with a , one finds that for the same geometry, but for larger values of size d , failure occurs for larger values of c . The complete R-curve, then, could be obtained as an envelope (Fig. 4e) of the energy release curves calculated for the failure load of specimens of the same geometric shape, but of different sizes. This property is very useful, because the only certain information we can have from tests is the failure load, not the crack extension in stable conditions. From the knowledge of the R-curve and of the experimental failure load one can also calculate the equivalent crack extension at failure.

In Ref. [18] it has been tried to obtain the R-curve from the data on maximum loads in fracture tests available in the literature. Chosen a given expression for the R-curve depending on some empirical parameters, these have been determined through the use of an optimization technique. The result was, however, that great fluctuations of the empirical para-meters occurred without appreciable increase of the sum of squares of the deviations between predicted and test results. This indicated that the scatter of experimental data was too great, so that a situation like the one depicted in Fig. 4f was occurring. A smoothing of experimental data is needed, then, before the parameters of the R-curve can be identified, and this may be achieved through the use of the "size effect law" [21] which will be illustrated in the next section. In Ref. [19] this law has been applied to the experimental data reported in [22], obtaining

$$R(c) = G_f \left[1 - (1 - \frac{c}{c_m})^n \right] \qquad \text{for } c < c_m$$

$$\tag{11}$$

$$R(c) = G_f \qquad \text{for } c > c_m$$

with $G_f = 50.8$ N/m , $c_m = 4.81$ cm , $n = 3.6$. The value of G_f is the asymptotic value of R and represents the fracture energy for a very large specimen : defined in this way it becomes a material property (size independent).

5. SIZE EFFECT LAW

The derivation given in [21] is based on the crack band model. Let's consider a crack band of width w_c and length 2a in a panel (Fig. 5a) of width 2d and thickness b. The formation of the crack will relieve strain energy both from the uncracked area (such as 136 and 245 in Fig.5a) and from the crack band (1234 in Fig. 5a). The areas indicated are an approximation which has the only purpose of showing that the total strain energy released W will be a function of a and w_c .

The energy released W must be proportional to the volume $d^2 b$ of the structure and to the characteristic energy density $\sigma_N^2/2E_c$, where

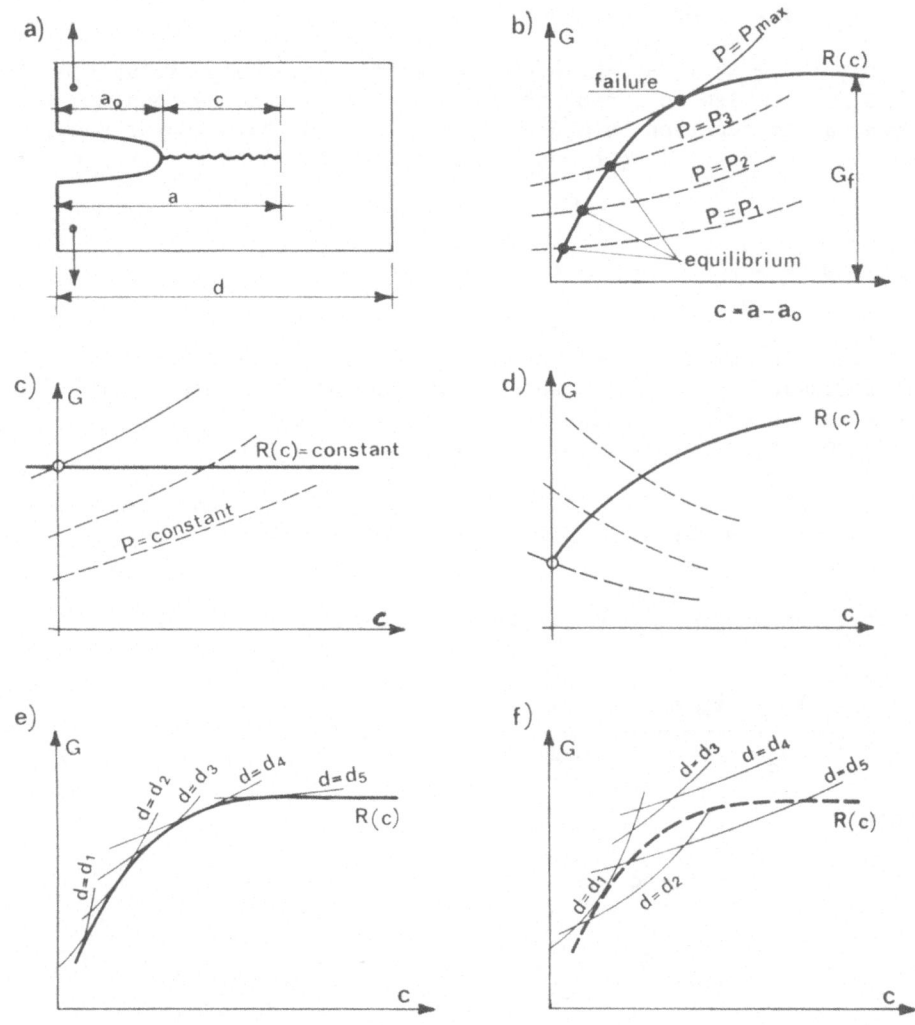

Figure 4 . - (a) Crack Extension from a Notch; (b),(c),(d) R-curve and Energy Release Rate Curve ; (e),(f) R-curve as Envelope of Energy Release Rate Curves Respectively for Smoothed and Unsmoothed Data

$\sigma_N = P/bd$, and for reasons of dimensional analysis, it can be put in the form

$$W = \frac{1}{2E_c} \left(\frac{P}{bd}\right)^2 bd^2 \quad f(\alpha_1, \alpha_2, \xi_i) \tag{12}$$

where $f(\alpha_1, \alpha_2, \xi_i)$ is a continuous and differentiable positive function independent of size d, and, this time, instead of one nondimensional parameter as in function $g(\alpha, \xi_i)$ in Eq.(10), we have two of them, one related to the crack length and the other to the width of the process zone

$$\alpha_1 = \frac{a}{d} \quad ; \quad \alpha_2 = \frac{w_c}{d} \tag{13}$$

This selection of nondimensional variables is different from the approach followed in Ref.[21], and it corresponds to a more general derivation which appears in [23].

The condition for crack propagation is

$$\frac{\partial W}{\partial a} = G_f \, b \tag{14}$$

and, substituting $\sigma_N = P/bd$ we obtain

$$\frac{1}{2 E_c} \frac{\sigma_N^2 b^2 d^2}{b} \frac{\partial f}{\partial \alpha_1} \frac{1}{d} = G_f \, b \tag{15}$$

and then

$$\sigma_N^2 = \frac{2 G_f E_c}{\dfrac{\partial f}{\partial \alpha_1} d} \tag{16}$$

Here the fracture energy is considered to be size independent (asymptotic value of the R-curve). According to the crack band theory

$$G_f = w_c \left(1 - \frac{E_c}{E_t}\right) \frac{f_t'^2}{2 E_c} \tag{17}$$

and substituting in Eq. (16)

$$\sigma_N^2 = \frac{w_c \left(1 - \dfrac{E_c}{E_t}\right) f_t'^2}{\dfrac{\partial f}{\partial \alpha_1} \, d}$$ (18)

We may choose as reference state a structure which is large compared to the crack band width, i.e.

$$\alpha_2 = 0$$ (19)

and introduce a Taylor series expansion about this reference state

$$\frac{\partial f}{\partial \alpha_1} = \left[\frac{\partial f}{\partial \alpha_1}\right]_{\alpha_2=0} + \left[\frac{\partial^2 f}{\partial \alpha_1^2}\right]_{\alpha_2=0} \alpha_2 + \frac{1}{2!}\left[\frac{\partial^3 f}{\partial \alpha_1^3}\right]_{\alpha_2=0} \alpha_2^2 + \dots$$

$$= F_1 + F_2 \alpha_2 + F_3 \alpha_2^2 + \dots$$ (20)

For structures which are not too small, the first two terms in the series expansion will prevail, so that Eq.(18) becomes

$$\sigma_N^2 = \frac{w_c \left(1 - \dfrac{E_c}{E_t}\right) f_t'^2}{(F_1 + F_2 \alpha_2) \, d}$$ (21)

After some manipulations, one obtains

$$\sigma_N \doteq \frac{B \ f_t'}{\sqrt{1 + \dfrac{d}{d_o}}}$$ (22)

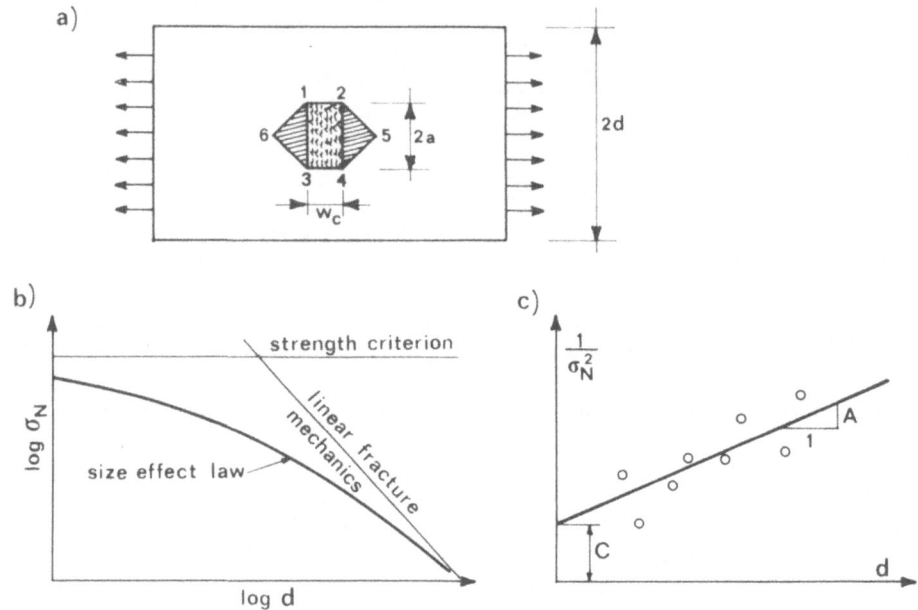

Figure 5 . - (a) Crack Band in Panel Under Uniaxial Tension; (b) Size Effect Law ; (c) Coefficients of Linear Regression Analysis.

where

$$B = \sqrt{\frac{1}{F_2} \left(1 - \frac{E_c}{E_t} \right)} \quad ; \quad d_o = \frac{F_2 \, w_c}{F_1} \tag{23}$$

are parameters which depend on the geometry of the structure .

For a very large structure d/d_o tends to infinity, so that the second term under square root predominates, and we have

$$\sigma_N \cong \frac{B \, f_t'}{\sqrt{\dfrac{d}{d_o}}} \tag{24}$$

which is the well known size dependence according to linear elastic fracture mechanics [24] and corresponds to a straight line of slope 1/2 in a logarithmic scale diagram (Fig. 5b).

For a very small structure

$$\sigma_N \cong B \; f_t' \tag{25}$$

which is the well known criterion of plasticity.

The size effect law of Eq.(22) then corresponds to a gradual transition (Fig.5b) from the strength or yield criterion to the energy criterion of linear elastic fracture mechanics. This law is only approximate because it rests on the hypothesis of a constant size of the fracture process zone at failure.

6. IDENTIFICATION OF THE PARAMETERS OF THE SIZE EFFECT LAW FROM TEST DATA

The values of the parameters B and d_0 (constants for a certain shape of the structure and a certain concrete) can be best determined through a statistical regression of test data. To this purpose, the size effect law can be put in the form

$$\frac{1}{\sigma_N^2} = \frac{1 + \dfrac{d}{d_0}}{B^2 \; f_t'^2} = \frac{1}{B^2 \; f_t'^2} \frac{d}{d_0} + \frac{1}{B^2 \; f_t'^2} = A \, d + C \tag{26}$$

where

$$C = \frac{1}{B^2 \; f_t'^2} \quad ; \quad A = \frac{C}{d_0} \tag{27}$$

and the values of C and A may be obtained by linear regression (Fig. 5c). For example, from the data reported in [22] relevant to three point bent fracture tests of specimen of various depths, the values C = 5.791 MPa^{-2}, A = 29.1 MPa^{-2} m^{-1} have been calculated in Ref. [19], from which one obtains B f_t' = 0.4155 Mpa, d_0 = 19.9 cm .

One can now determine G_f , i.e. the fracture energy for very large sizes $(d/d_0 \gg 1)$, which corresponds to the asymptotic value of R(c). In this case linear elastic fracture mechanics applies, which is represented by the inclined asymptote of the size effect law, given by

$$\frac{1}{\sigma_N^2} = A \ d \tag{28}$$

Now from Eq. (10), introducing for G the value G_f and for α the value $\alpha_0 = a_0/d$ (since for very large specimen R=G =const and no stable crack growth can occur, as we have shown before), one obtains

$$\sigma_N^2 = \frac{P^2}{(b \ d)^2} = \frac{G_f \ E_c}{d \ g(\alpha_0, \xi_i)} \tag{29}$$

From Eqs. (28) and (29) then

$$G_f = \frac{g(\alpha_0)}{E_c \ A} \tag{30}$$

The fracture energy is then inversely proportional to the slope A of the size effect regression line, and, for the same data considered before, one would obtain G_f = 50.8 N/m .
     The same size effect law can be used to smooth the data of the maximum loads for various values of d . As shown before, these values can then be used to obtain the R-curve as an envelope of strain energy release curves (Fig. 4e).

7. REFERENCES

[1]   Slate, F.D., "Microscopic Observation of Cracks in Concrete, with Emphasis on Techniques Developed and Used at Cornell University", in Wittmann F.H. (Ed.) Fracture Mechanics of Concrete, Elsevier, 1983, pp. 75-83.
[2]   Diamond, S. and Bentur, A., "On the Cracking in Concrete and Fiber Reinfoced Cements", in Shah S.P. (Ed.), Application of Fracture Mechanics to Cementitious Composites, Nijhoff, 1985, p. 87-140.
[3]   Knab et al., "Fluorescent Thin Sections to Observe the Fracture Zone in Mortar", Cement and Concrete Research, vol. 14 (1984) pp. 339-344.
[4]   Slate, F.D., "X-Ray Technique for Studying Cracks in Concrete, with Emphasis on Methods Developed and Used at Cornell University", in Wittman F.H. (Ed.) Fracture Mechanics of Concrete, Elsevier, 1983, pp. 85-93.

[5]   Diederichs, U. et al. "Formation and Propagation of Craks and
        Acoustic Emission", in Wittmann F.H. (Ed.) Fracture Mechanics of
        Concrete, Elsevier, 1983, pp. 157-205.
[6]   Izumi et al., "Acoustic Emission Technique to Evaluate Fracture
        Mechanics Parameters of Concrete", International Conference on
        Fracture Mechanics of Concrete (Prepints), Laboratoire Des Mate-
        riaux De Construction, Ecole Polytechnique Federale, Lausanne,
        Oct. 1-3, 1985.
[7]   Alexander, M.G., and Blight, G.E.," The Use of Small and Large
        Beams for Evaluating Concrete Fracture", Ibid.
[8]   Luong, M.P., "Infrared Vibrothermography of Plain Concrete", Ibid.
[9]   Reinhardt, H.W.," Crack Softening Zone in Plain Concrete Under
        Static Loading", Cement and Concrete Research, Vol. 15 (1985), pp.
        42-52.
[10]   Shah, S.P. and Gopalaratnam, V.S., "Softening Response of Plain
        Concrete in Direct Tension", Accepted for Pubblication, ACI
        Journal,1985.
[11]   Cedolin L., Dei Poli S. and Jori,I., "Comportamento a trazione del
        calcestruzzo", Studi e Ricerche, Corso di Perfezionamento per le
        Costruzioni in Cemento Armato, Vol.5, 1983, Politecnico di Milano.
[12]   Jacquot P. and Rastogi P.K.," Spekle Metrology and Holographic
        Interferometry Applied to the Study of Craks in Concrete", in
        Wittmann F.H. (Ed.) Fracture Mechanics of Concrete, Elsevier,
        1983, pp. 113-156.
[13]   Cedolin L. Dei Poli S., and Iori, I.," Experimental Determination
        of the Fracture Process Zone in Concrete", Cement and Concrete
        Research, Vol. 13,1983, pp. 557-567.
[14]   Hillerborg, A. Modéer, M., and Petersson, P.E., "Analysis of Crack
        Formation and Crack Growth in Concrete by Means of Fracture Mecha-
        nics and Finite Elements", Cement and Concrete Research, Vol. 6,
        1976, pp. 773-782.
[15]   Hillerborg, A., "A Model for Fracture Analysis", Report TVBM-3005,
        978, Division of Building Materials, the Lund Institute of Techno-
        logy.
[16]   Bazant, Z.P. and Oh, B. H., "Crack Band Theory for Fracture of
        Concrete", Matériaux et Constructions (Materials and Structures)
        RILEM, Paris, Vol. 16, 1983, pp. 155-177.
[17]   Wecharatana, M. and Shah, S.P.,"Slow crack growth in Cement Compo-
        sites", Journal of Structural Engineering, ASCE, Vol. 108, June
        1982, pp. 1400-1413.
[18]   Bazant, Z.P. and Cedolin, L., "Approximate Linear Analysis of
        Concrete Fracture by R-curves", Journal of Structural Engineering,
        ASCE, Vol. 110, 1984, pp.1336-1355.
[19]   Bazant, Z.P., Kim, J.K., and Pfeiffer, P., "Nonlinear Fracture
        Properties From Size Effect Tests ", Journal of Structural Engi-
        neering, Vol. 112, 1986, pp. 289-307.
[20]   Tada, H., Paris, P.C., and Irwin, G.R.,"The Stress Analysis of
        Cracks Hanbook", DEL Research Corp., K.Hellertown, Pa., 1973.
[21]   Bazant, Z.P., "Size Effect in Blunt Fracture : Concrete, Rock,
        Metal", Journal of Engineering Mechanics, ASCE, Vol. 110, 1984,
        pp.518-535.

[22] Jenq, Y. S. ,and Shah, S.P., "Nonlinear Fracture Parameters for Cement Based Concrete: Theory and Experiments", in S.P. Shah (Ed.), Application of Fracture Mechanics to Cementitiuos Composites, Nijhoff, 1985, p.87-140.

[23] Bazant, Z.P., "Fracture Mechanics and Strain-Softening of Concrete", Seminar on Finite Element Analysis of Reinforced Concrete Structures, Tokyo, May 21-24, 1985.

[24] Carpinteri, A., " Experimental Determination of Fracture Toughness Parameters for Aggregative Materials", Advances in Fracture Research, 5th Int. Conf. on Fracture, Cannes, France, 1981, Vol.4, pp.1491-1498.

NUMERICAL ANALYSIS AND SIMULATION OF CRACK FORMATION IN COMPOSITE MATERIALS SUCH AS CONCRETE

P.E. Roelfstra
Swiss Federal Institute of Technology
Laboratory for Building Materials
Chemin de Bellerive 32, CH-1007 Lausanne
Switzerland

ABSTRACT. A numerical method to simulate crack formation and propagation in computer generated composite structures of concrete is presented. Results obtained by this method can serve as a basis for the development of realistic and comprehensive fracture mechanics models which can be applied in structural engineering analysis. The versatile possibilities of this method are demonstrated by some examples. Three other relevant models are discussed and compared with the presented method.

1. INTRODUCTION

In the contribution titled "Structure of Concrete and Crack Formation" by F.H. Wittmann a hierarchic system of three different levels for the structure of concrete has been introduced. In addition models which describe the physical and mechanical properties of the material on each level have been classified as follows:

micro-level: materials science models

meso -level: materials engineering models

macro-level: structural engineering models

The structural aspects of the micro-level which are linked with crack formation and failure processes together with appropriate materials science models, such as e.g. the "Munich Model" /1/, are explained and discussed extensively in the contribution of F.H. Wittmann.

Structural engineering models on the macro-level, such as the "Fictitious Crack Model" of Hillerborg /2-3/ and the "Blunt Crack Model" of Bazant /4/ as well as a proposed simple R-curve method, are described by L. Cedolin in the contribution titled "Fracture Mechanics Parameters and Fracture Process Zone of Concrete".

In this contribution materials engineering models on the meso-level are treated. The salient aspect of these models is that they take the heterogeneous character of the material into consideration. Some of these materials engineering models are based on assumed statistical distributions of heterogeneities and their properties, while

K. P. Herrmann and L. H. Larsson (eds.), Fracture of Non-Metallic Materials, 359–384.
© *1987 by ECSC, EEC, EAEC, Brussels and Luxembourg.*

other models are based on numerical simulation methods of the "real" composite structure of concrete. Allthough the assumptions and methods used to develop these models on the meso-level differ, the final aim is the same: development of constitutive laws or methods which can describe failure of concrete structures in a realistic way.

In the next section three typical materials engineering models are outlined. In the following sections a more general method which has been developed recently in our laboratory is explained in detail.

2. REVIEW OF MATERIALS ENGINEERING MODELS

2.1. Stochastic Model

Mihashi has developed a stochastic model to describe failure behaviour of concrete /5-7/. In this model failure behaviour of concrete under variable and sustained load is described by mathematical expressions which formulate the survival (non-failure) probability of a chain of linked elements. These elements are considered to be characteristic volume elements of the material and should not be confused with finite elements. The mechanical properties of each element are described by a state parameter.

For example, state 0 means that there are only some initial microcracks in the structure of the element. State 1 means that some microcracks have grown and are collapsed to larger cracks which are still stable. In this way the gradual fracturing of an element up to failure (the last state) can be described. Mihashi showed that 3 states (0, 1, and 2) are sufficient for concrete elements under tensile loads. In that case state 2 means that a crack runs completely through the element and that the chain is broken. In Fig. 1 a chain of linked elements is shown and the different states of an element are indicated.

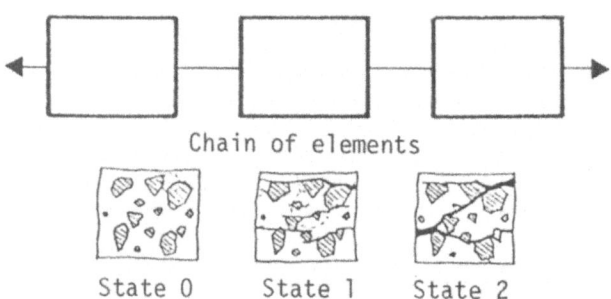

Chain of elements

State 0   State 1   State 2

Fig. 1 : Chain of linked elements and different states of an element.

It is assumed that the rate of change of state of an element (in other words the rate of crack formation) can be described as a Markoff process. The survival probability of state i of an element before a given time t can then be formulated as:

$$p_i^e(t) = e^{-\int_o^t \bar{p}_{ij}(t)\, dt} \tag{1}$$

where $\bar{p}_{ij}(t)$ is the mean value of the transition probability from state i to state j.

The non-failure probability of an element with 3 state possibilities (0, 1, and 2) can be expressed as follows:

$$p^e(t) = p_0^e(t) + p_1^e(t) \tag{2}$$

and finally the non-failure probability of a chain with m elements as:

$$p(t) = \{p^e(t)\}^m \tag{3}$$

In this model the random variable of the survival probability is not allways time t but other quantities can be chosen, in general depending on loading conditions as is shown in Fig. 2.

In that way different kinds of survival probabilities can be determined. The nucleation process theory of solids has been used to formulate the mean transition survival probability of states $\bar{p}_{ij}(t)$ of Eq. (1). This has led to the following expression:

$$\bar{p}_{ij}(t) = L_i\,\{\sigma(t)\}^\beta \tag{4}$$

where β is a temperature dependent material constant and L_i is material parameter which depends on Young's Modulus, surface energy, temperature, and heterogeneity. The heterogeneity of the material is characterized by the expected value of the largest microcrack length. The lower case subscript of L refers to state i.

As an example the survival probability of a specimen under a monotonically increasing load will be determined. In that case stress σ is chosen as random variable and $\sigma\{t\}$ is substituted by $\dot{\sigma}t$ in Eq. (4). After some mathematical manipulations one will find for Eq. (3):

$$p(\sigma) = e^{-\frac{m\,L}{(\beta+1)\,\dot{\sigma}}\,\sigma^{\beta+1}} \tag{5}$$

where m is the number of elements in the chain. m is therefore a volume parameter (L and β are already explained in the text).

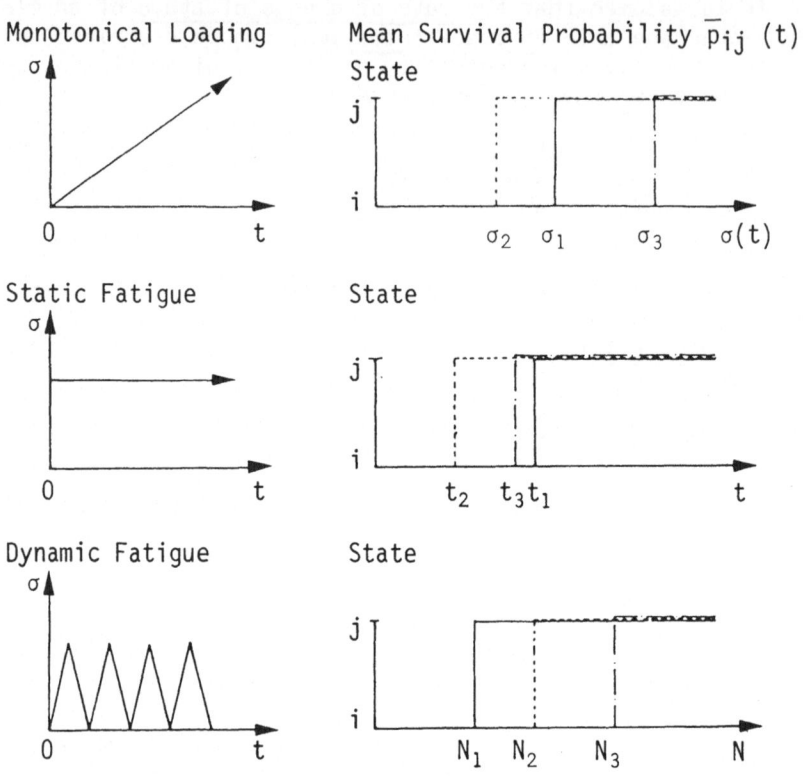

<u>Fig. 2</u> : Loading conditions, random variable and transition
of elements.

Strength of the material is supposed to be the stress level on
wich failure occurs and therefore Eq. (5) can be used to determine the
mean strength as follows:

$$\bar{f} = \int_0^\infty \sigma \frac{d(1-P(\sigma))}{d\sigma} \, d\sigma = \{\frac{(\beta+1) \; \dot{\sigma}}{mL}\}^{\frac{1}{\beta+1}} \; \Gamma \; (\frac{\beta+2}{\beta+1}) \tag{6}$$

where Γ is the gamma function. Eq. 6 indicates that the mean strength
of the material decreases as the heterogeneity (L) and volume (m) of
the specimen increases. From Eq. (6) a simple expression can be
deduced which describes the influence of the <u>relative</u> rate of loading
on the <u>relative</u> mean strength:

$$\frac{\bar{f}}{\bar{f}_0} = (\frac{\dot{\sigma}}{\dot{\sigma}_0})^{\frac{1}{\lambda+1}} \tag{7}$$

where $\bar{f}$ and $\bar{f}_0$ stand e.g. for mean strength under a high rate of
loading (dynamic) and a low rate of loading (static) respectively.

Eq. (7) is in good agreement with many experimental results. An example is shown in Fig. 3.

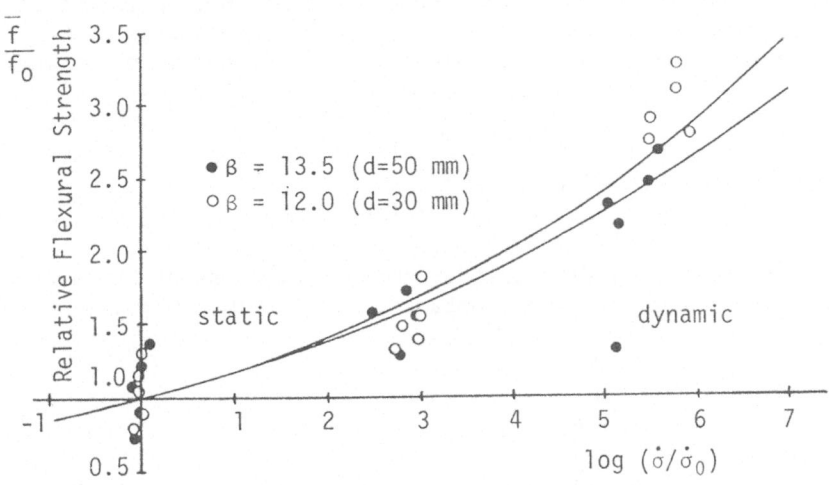

: Related flexural strength at a concrete slab as function of related rate of loading /8/.

2.2. Microplane Model

The microplane model has been developed by Bazant and Oh /9-10/ to describe progressive fracturing of concrete and rock under general loading conditions. This model is based on the assumption that the nonlinear behaviour (e.g. microcracking) only occurs in the thin mortar layers which separate the coarse aggregate grains. These thin mortar layers have been called "microplanes". Fig. 4 shows schematically the randomly oriented microplanes in the composite structure of concrete.

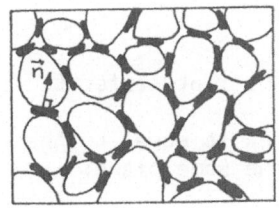

Fig. 4 : Microplanes in the structure of concrete.

The nonlinear behaviour of each microplane is described for simplicity by only a uniaxial stress-strain relation:

$$\sigma_n = F(e_n) \, e_n \qquad\qquad (8)$$

where σ_n is the stress normal to the microplane; e_n is the strain to the microplane; $F(e_n)$ is the relation function.

A proposed relation is given by Eq. (8) and is shown in Fig. 5.

$$\sigma_n = E_n \, e^{-(ke_n^p)} \, e_n \quad : \text{ for } e_n > 0$$
$$\sigma_n = E_n \, e_n \qquad\qquad : \text{ for } e_n < 0 \qquad (9)$$

In which E_n is the initial normal stiffness of the microplane and k and p are positive constants; $k = 1.8 \times 10^7$, $p = 2$.

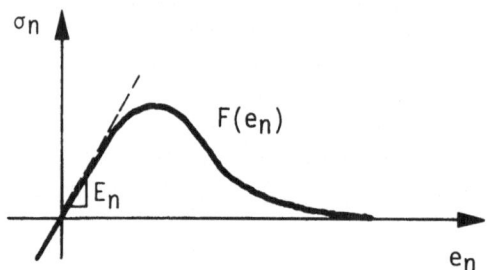

Fig. 5 : Uniaxial stress-strain relation of a microplane.

In order to describe the mechanical behaviour of the material (macro-level) the nonlinear responses of the random oriented micro-planes and the linear responses of the inclusions have to be super-imposed. Therefore the following assumptions have been made:
• The macroscopic strain tensor, ε_{ij}, is the sum of a purely elastic macrostrain ε_{ij}^a and an inelastic macrostrain e_{ij}:

$$\varepsilon_{ij} = \varepsilon_{ij}^a + e_{ij} \qquad\qquad (10)$$

The Latin lower case subscripts refer to Cartesian coordinates x_i (i = 1, 2, 3).
• The normal microstrain e_n is equal to the resolved macroscopic strain tensor e_{ij} for the same plane:

$$e_n = n_i \, n_j \, e_{ij} \qquad\qquad (11)$$

in which n_i = direction cosines of the unit normal n of the microplane and the repeated Latin lower case subscripts indicate a summation over 1, 2, 3.

Based on these assumptions and equilibrium conditions the following relation was obtained:

$$d\sigma_{ij} = D_{ijkm} \, d\varepsilon_{km} \qquad (12)$$

with $\quad [D_{ijkm}] = [C^a_{ijkm} + C^c_{ijkm}]^{-1}$

in which C^a_{ijkm} is tangent compliances tensor of the linear part and C^c_{ijkm} is the tangent compliances tensor of the nonlinear part.

The tangent compliances C^c_{ijkm} is related to the microplane system as follows:

$$C^c_{ijkm}{}^{-1} = D^c_{ijkm} = \int_S n_i \, n_j \, n_k \, n_m \, \frac{dF(e_n)}{de_n} \, f(n) \, dS \qquad (13)$$

where D^c_{ijkm} is the tangent stiffnesses tensor; n_i, n_j, n_k, n_m are direction cosines of the normal to the microplane; $F(e_n)$ is the stress-strain function of the microplane; $f(n)$ is the relative frequency function of planes of various orientations n (for isotropic solids $f(n) = 1$); S is the relative total surface, which is in this model the surface of a unit hemisphere.

The integral of Eq. (13) has to be evaluated numerically. Bazant and Oh /11/ developed numerical integration formulas which give very consistent results.

It turned out, that the obtained value for Poisson's ratio ν was too high if Eq. (13) was applied for purely elastic deformations. This is caused by the neglection of the microplane shear stiffness. Bazant and Oh /9/ showed that this inconvenience can be overcome by adopting the bulk compliance, in which the coefficients of the compliance tensor C^a_{ijkm} are expressed.

The microplane model is governed by the uniaxial stress-strain relation $F(e_n)$ of the microplanes. If this relation includes unloading and reloading in tension and compression, then very complex load paths in 3D on the macro-level can be analyzed. This has been demonstrated, for example, by Bazant and Gambarova /12/ who applied the microplane model to describe shear crack within the crack band theory. The assumed stress-strain relation of the microplane for this analysis is shown in Fig. 6 and the comparison with experimental results is shown in Fig. 7.

The microplane model has proven to be a powerful method to describe macroscopic mechanical behaviour of concrete in 3D. Whether the simple uniaxial stress-strain relation for the microplanes suffices to handle all situations is still questionable. The number of applications which has been realized, however, is still in progress.

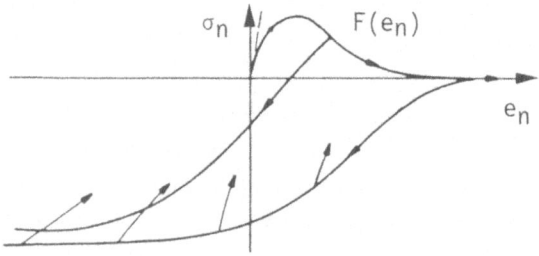

Fig. 6 : Stress-strain relation of the microplane with
unloading curves.

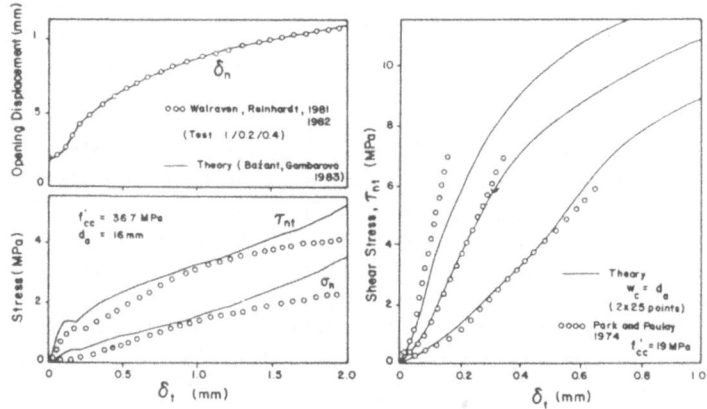

Fig. 7 : Comparison of crack band model with measurements by
Walraven and Reinhardt (1981, 1982) and Park and
Paulay (1974) (after Bazant and Gambarova, 1983).

2.3. LEFM model

Zaitsev and Wittmann /13-15/ have put forward a 2D model to analyse
crack formation and crack propagation in computer simulated composite
structures of concrete. In this model concrete is considered to be a
two-phase material: aggregate grains embedded in a matrix of hardened
cement paste. The composite structure of concrete is generated randomly in a given cross-section. In addition pre-existing cracks in the
interface (contact zone between inclusions and matrix) due to e.g.
bleeding can be simulated. A typical example of such a generated
composite structure is shown in Fig. 8.

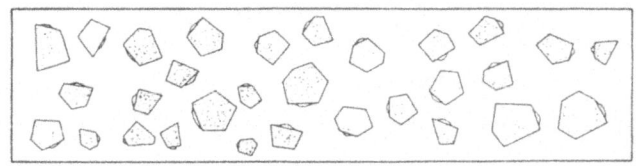

Fig. 8 : A typical computer simulation of the random
structure.

The mechanical behaviour of these structures under uni- or multi-
axial monotically increasing load is simulated as follows:
. For each crack in the structure the external load is determined
  needed to propagate that crack. These loads are calculated with
  mathematical expressions which are deduced with the help of the LEFM
  method.
. From the calculated loads the minimum load is chosen and the corresponding crack is propagated.
  This sequence is repeated until a crack runs completely through
the material, which corresponds with failure.
  To perform realistic simulations, different kinds of crack
propagation are included in this model. A crack can run through the
matrix, along interfaces, through inclusions and collapse with other
cracks.
  An example of a complex situation is shown in Fig. 9.

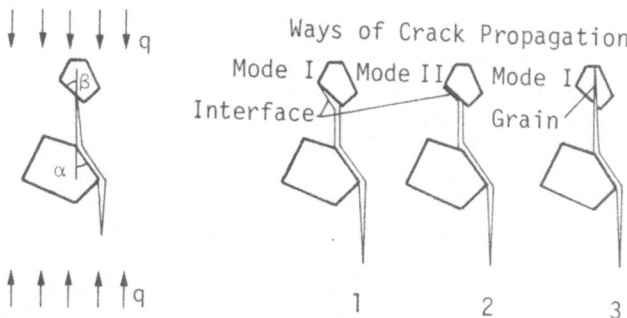

Fig. 9 : Situation of a crack and possible ways to
propagate.

A crack has encountered on inclusion and there are 3 possible
ways to propagate. The external load needed to propagate the crack in
one of the ways can be calculated with Eq. (14a-c):

$$q_I^{INT} = \frac{2\ K_{IC}^{INT}\ \sqrt{\pi}\ \ell_2/L_1}{A(\alpha,\rho)\ f_1(\beta)\ -\ 3\ C(\alpha,\rho)\ g(\beta)} \qquad (14a)$$

$$q_{II}^{INT} = \frac{2\ K_{IIC}^{INT}\ \sqrt{\pi}\ \ell_2/L_1}{A(\alpha,\rho)\ g(\beta)\ +\ C(\alpha,\rho)\ f_2(\beta)} \qquad (14b)$$

$$q_I^{INCL} = \frac{K_{IC}^{INCL}\ \sqrt{\pi}\ \ell_2}{A(\alpha,\rho)\ 2\ L_1} \qquad (14c)$$

where q is the external load (upper case subscripts refer to type of material and lower case subscripts to crack mode), K_{IC} is the fracture toughness in mode I, K_{IIC} is the fracture toughness in mode II, ℓ_2/L_1 is the relative extension of the crack. $A(\alpha,k)$, $C(\alpha,\rho)$, $f_1(\beta)$, $F_2(\beta)$ and $g(\beta)$ are given by:

$$A(\alpha,\rho) = \sin^2\alpha\ \cos\ \alpha\ -\ \rho\ \sin^2\alpha \qquad (15a)$$

$$C(\alpha,\rho) = \sin\ \alpha\ \cos^2\alpha\ -\ \rho\ \sin^2\alpha\ \cos\ \alpha \qquad (15b)$$

$$f_1(\beta) = 3\ \cos\ \beta/2\ +\ \cos\ 3\ \beta/2 \qquad (15c)$$

$$f_2(\beta) = \cos\ \beta/2\ +\ 3\ \cos\ 3\ \beta/2 \qquad (15d)$$

$$g(\beta) = \sin\ \beta/2\ +\ \sin\ 3\ \beta/2 \qquad (15e)$$

where ρ a friction coefficient. Angles α and β are defined in Fig. 9.

The way the crack propagates can be found by comparing the calculated values of q_I^{INT}, q_{II}^{INT} and q_I^{INCL}. The fracture thoughness of aggregate grains of normal concrete is relatively high with respect to the fracture thoughness of the mortar matrix and of the interfaces. Cracks propagate therefore predominantly in the interface along the aggregate grains and through the mortar matrix. This in contrast with lightweight concrete, where the fracture thoughness of the inclusions is relatively low and cracks will propagate through the inclusions. Both simulations can be performed with this model as shown in Fig. 10.

The described model can be used to study the influence of inclusions, porosity, and pre-existing cracks on the mechanical properties of composite materials. Many simulations have been performed and results agreed well with experimental findings /13-15/. At the beginning behaviour of a composite material under compression has been analyzed /14/. More recently Hu, Cotterell and Mai /16/ have applied the same approach to simulate the behaviour under tensile load. Refinements of this approach are possible and the complex behaviour such as strain softening can in principle be linked with processes in the composite structure. Inelastic deformations and shrinkage, however, cannot be taken into account with this approach in a realistic way.

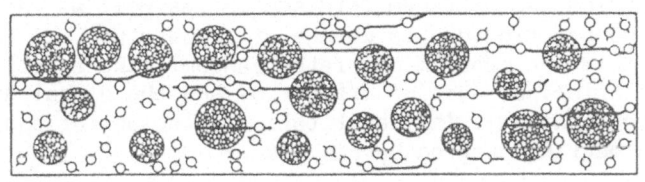

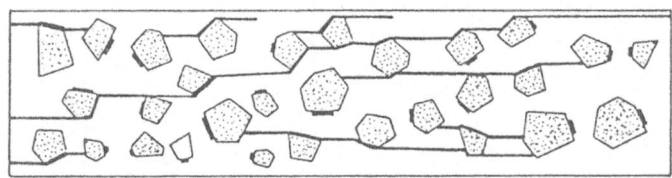

<u>Fig. 10</u> : Calculated crack patterns of a lightweight and a
normal concrete.

3. NUMERICAL CONCRETE

3.1. Introduction

Most concrete structures are in general submitted to variable loads
and are exposed to variable climatic conditions. So far, no general
model exists to describe the physical and mechanical behaviour of
concrete under these conditions. The chief cause is the rather complex
structure of concrete. In order to find a solution the TL-approach has
been developed, which is described in detail in the contribution of
F.H. Wittmann to this book.

One of the basic concepts of this approach is to use advanced
numerical methods to analyse physical and mechanical behaviour of
concrete on the meso-level. In order to realize these kinds of analyses, realistic composite structures of concrete are simulated with
the computer. These structures have been called "Numerical Concrete".
Numerical Concrete was first used to determine <u>effective</u> physical and
mechanical properties /17, 18/. Actually, Numerical Concrete is used
to study time-dependent behaviour (such as creep), the nonlinear
drying process, frost resistance and durability. In this contribution
the latest application of Numerical Concrete with respect to failure
analysis will be explained.

3.2. Constitutive relations of the Components

Aggregate grains

For the present only normal concrete is considered in the analysis and
therefore it is assumed that aggregate grains behave in a perfectly
linear elastic manner without any failure. Some types of aggregate
grains have orthotropical properties due to well defined layered micro
structures (e.g. Gneiss). The orthotropical properties of these

aggregate grains can be taken into account as well as thermal expansion or contraction due to change of temperature.

The simulation program generates randomly the rotation angle ϕ between the principal axes of orthotropy and the global axes x-y, as indicated in figure 11. The constitutive relation for aggregate grains is given in matrix notation by Eq. (16).

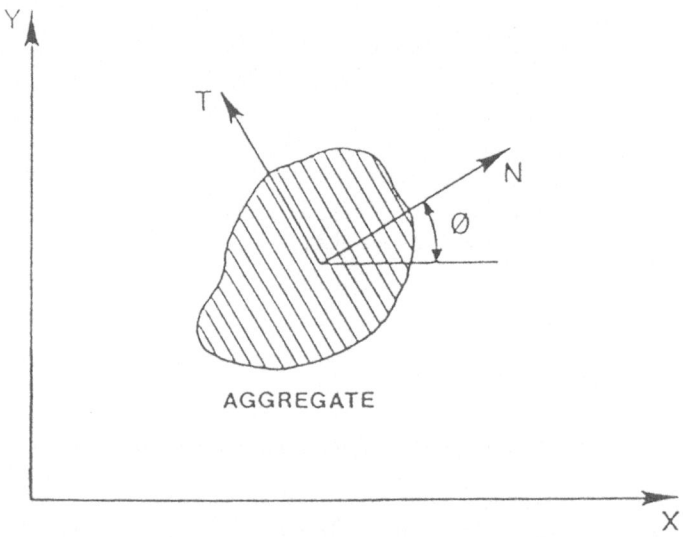

Fig. 11 : Rotation between principle axes of orthotropy and principles axes of the 2D space.

$$\{\sigma\}_x = [T]^T [D]_n [T] (\{\varepsilon\}_x - [T]^{-1} \{\varepsilon\}_n^i) \qquad (16)$$

where: $\{\sigma\}_x^T = \{\sigma_{xx} \; \sigma_{yy} \; \sigma_{xy}\}$; stress vector (in xy)

$\{\varepsilon\}_x^T = \{\varepsilon_{xx} \; \varepsilon_{yy} \; \varepsilon_{xy}\}$; strain vector (in xy)

$\{\varepsilon\}_n^{iT} = \{\alpha_n \Delta T \; \alpha_t \Delta T \; o\}$; initial strain vector (in nt)

α_n and α_t are thermal dilatancy coefficients in directions n and t; ΔT is a temperature increment;

$$T = \begin{bmatrix} C^2 & S^2 & CS \\ S^2 & C^2 & CS \\ -2CS & 2CS & C^2-S^2 \end{bmatrix} \qquad \text{transformation matrix;}$$

$C = \cos \phi$ and $S = \sin \phi$;

$$[D] = \frac{E_t}{(1-n\nu^2)} \begin{bmatrix} n & n\nu & 0 \\ n\nu & 1 & 0 \\ 0 & 0 & m(1-n\nu^2) \end{bmatrix}$$

elasticity matrix

where: $n = \dfrac{E_n}{E_t}$ and $m = \dfrac{G}{E_t}$

E_n and E_t are Young's Moduli in the directions of the axes of orthotropy, ν is Poisson's ratio, and G is the shear modulus.

Interface

The interface between aggregate grains and hardened cement paste (also called "auréole de transition") plays an important role in the complex fracture behaviour of concrete. In normal concrete crack propagation occurs predominantly in the interface along the aggregate grains, because the mechanical properties (such as strength) of the interface are inferior to those of aggregate grains and of the matrix of hardened cement paste.

The average thickness of the interface is 20 μm, which is relatively small as compared with the dimensions of the aggregate grains. Therefore, only a framework of spring and friction elements has been defined between the intersection points of the polygonial lines, wich describe the contour of the aggregate grains and the mortar matrix, as shown in Fig. 12.

The force-displacement relation for the linear elastic situation is given by:

$$\frac{EL}{2d} \begin{bmatrix} 1 & 0 & -1 & 0 \\ \cdot & 0.5 & 0 & -0.5 \\ \cdot & \cdot & 1 & 0 \\ \cdot & \cdot & \cdot & 0.5 \end{bmatrix} \begin{bmatrix} U_i \\ V_i \\ U_j \\ V_j \end{bmatrix} = \begin{bmatrix} F_{ni} \\ F_{ti} \\ F_{nj} \\ F_{tj} \end{bmatrix} \tag{17}$$

where E = Young's modulus; L = considered length of the interface; d = thickness of the interface (average 20 μm).

It should be noted that Eq. (17) only holds for d ≪ L.

The normal stress σ_{nn} and the shear stress σ_{nt} can be calculated according to Eq. (18)

$$\sigma_{nn} = E(U_i - U_j)/d$$
$$\sigma_{nt} = E(V_i - V_j)/2d \tag{18}$$

Failure occurs if $\sigma_{nn} > f_t$ and/or $|\sigma_{nt}| > c - \sigma_{nn}\,tg\phi$, where f_t is the tensile strength, c the cohesion and ϕ the friction angle. c and ϕ depend to a large extend on the roughness of the aggregate surfáce.

In the case where shear failure takes place and where stress is negative (compression) friction forces are introduced. These friction forces are a function of the normal stress.

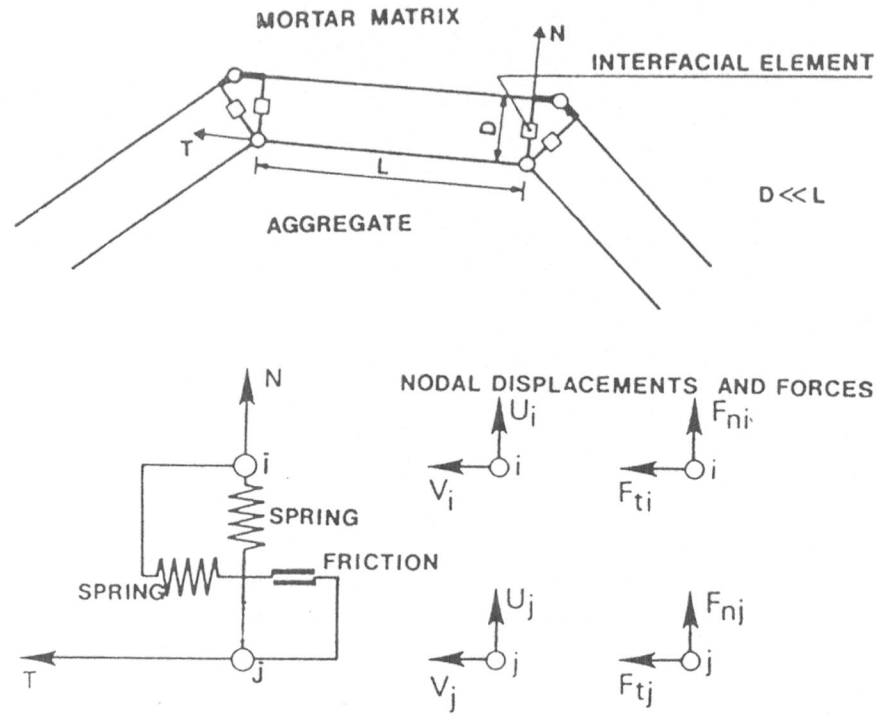

Fig. 12 : Interfacial finite elements between the boundary nodes of aggregates and those of the mortar matrix.

Mortar Matrix

It is assumed that the mortar matrix behaves in a linear elastic manner until tensile strength f_t is reached. Then, the nonlinear behaviour is localized in a so-called fictitious crack, where stress transfer perpendicular to the crack still can take place. The stress transfer capacity in the fictitious crack decreases as the width of the crack increases. A typical stress-crack-width relation (softening diagram) is shown in Fig. 13. The area under the softening diagram is equal to fracture energy G_f.

In the finite element analysis of Numerical Concrete the softening behaviour is smeared out over the "projected" length $\underline{L}$ of the applied triangular elements, as shown in Fig. 14.

A combined (linear+softening) stress-strain relation of an element with a typical projected length of 2 mm is shown in Fig. 15.

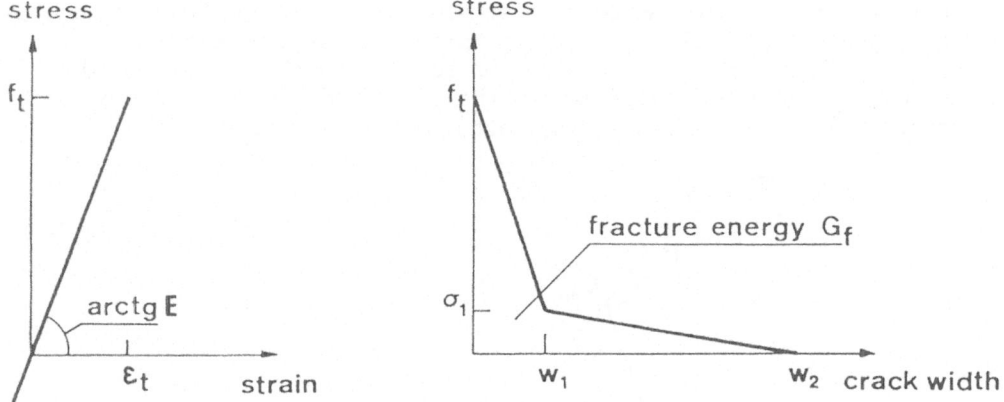

Fig. 13 : Stress-Strain and Stress-Crack width relation.

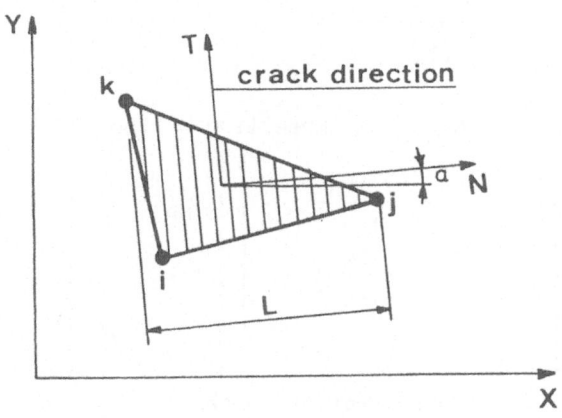

Fig. 14 : Definition of projected length L.

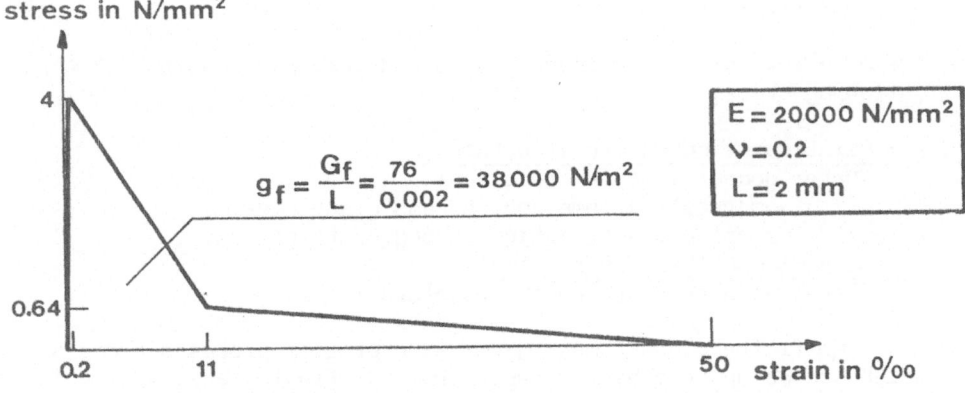

$$g_f = \frac{G_f}{L} = \frac{76}{0.002} = 38000 \text{ N/m}^2$$

E = 20000 N/mm²
ν = 0.2
L = 2 mm

Fig. 15 : A typical stress-strain relation for mortar matrix element.

The numerical modelling of the combined stress-strain relation is indicated in Fig. 16. As tensile strain ε_{nn} increases, Young's modulus E_n is gradually "softened" and the maximum tensile stress σ_{nn} is updated. The model accounts also for some residual deformation ε_{nn}^p in oder to describe realistically the unloading situations. The general constitutive law of the mortar matrix is given by:

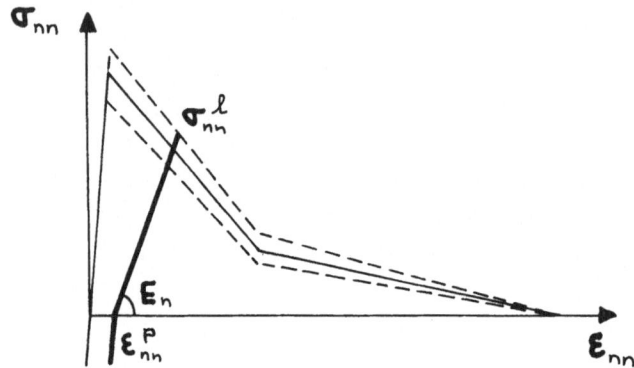

Fig. 16 : Numerical Modelling of the stress-strain reaction.

$$
\begin{bmatrix} \sigma_{nn} \\ \sigma_{tt} \\ \sigma_{nt} \end{bmatrix} = \begin{bmatrix} \dfrac{E_n}{1-\alpha v^2} & \dfrac{v E_n}{1-\alpha v^2} & \cdot \\ \cdot & \dfrac{E_t}{1-\alpha v^2} & \cdot \\ \cdot & \cdot & \beta G \end{bmatrix} \begin{bmatrix} \varepsilon_{nn} - \varepsilon_{nn}^i \\ \varepsilon_{tt} - \varepsilon_{tt}^i \\ \gamma_{nt} - \gamma_{nt}^i \end{bmatrix} \qquad (19)
$$

where $\alpha = E_n/E_t$, β = shear retention factor. By means of the initial strains ε_{nn}^i, ε_{tt}^i, γ_{nt}^i thermal dilatancy, residual deformation, creep and shrinkage can be taken into account.

3.3. Numerical Procedure

The numerical procedure which has been developped consists of the following steps:

- Simulation of composite structures -
    Data: geometry of the specimen;
          aggregate volume content and granulometry;
          shape characteristics of aggregate grains.
- Finite Element Mesh Generation -
    Data: simulated composite structures;

The finite element mesh generation program proposes a coarse which can be updated by the user. From the final coarse mesh a finer finite element mesh is generated. A typical example is shown in Fig. 17.

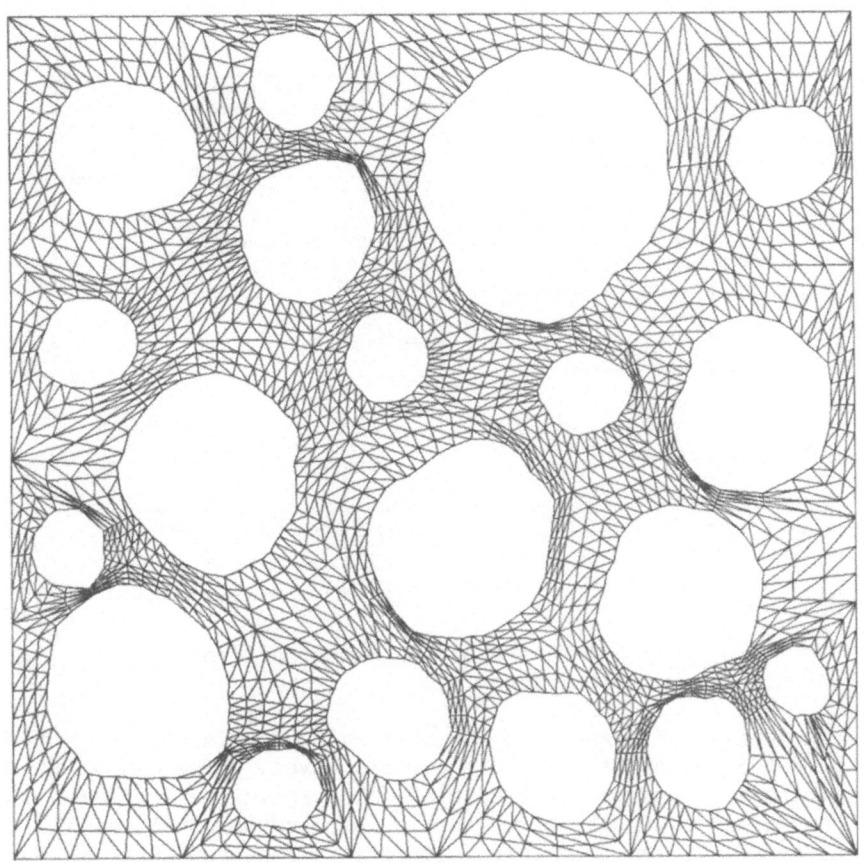

Fig. 17 : Computer generated finite element idealization of
a composite structure.

In order to reduce the equation solving effort the substructure
technique has been used. Around each inclusion a substructure is
defined. The Reverse Cuthill-McKee (RCM) method has been used to
obtain an efficient numbering scheme of nodal variables.

- Simulation of Crack Formation and Propagation -
      Data: materials properties
            loading conditions

The flow-chart of the simulation is shown in Fig. 18.
The initial state of the composite structure is calculated on the
basis of temperature and moisture distributions. It should be noted
that even in the initial state crack formation can take place. The
module detects first that substructure in which failure criteria have
been exceeded. The calculated displacements of the boundary of this

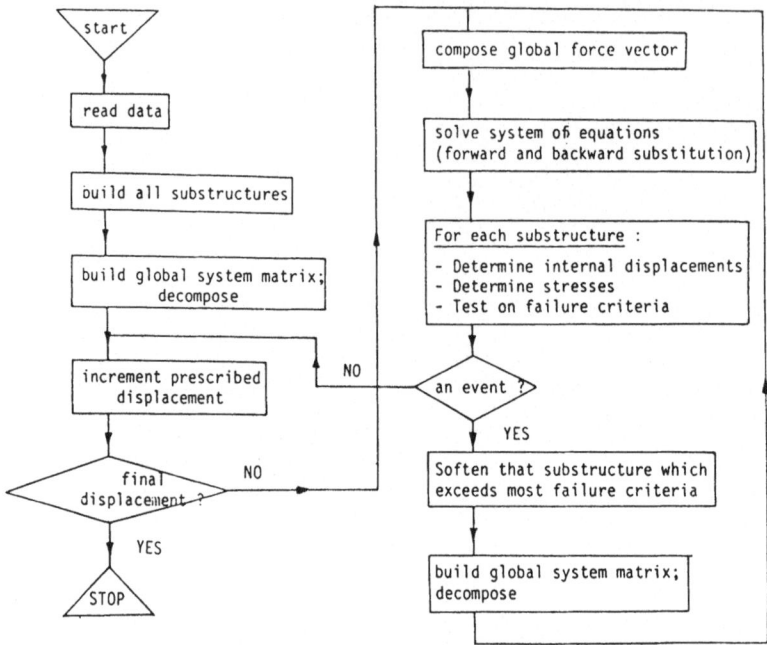

Fig. 18 : Flow-Chart of the Numerical Procedure for the
Simulation.

substructure are "frozen" and the stiffness of this substructure is
"softened" internatively up to a certain level, which can be adjusted
by setting a so-called event-counter. Once again the stiffness matrix
of the overall composite structure is composed and analyses of
stresses are carried out. This procedure is repeated until all conditions have been satisfied.

    After the initial situation has been treated, the response to an
increasing load or displacement on a boundary of the structure is
calculated. Almost the same procedure as for the calculations of the
initial state is followed.

    In general several thousand finite elements are needed for a
realistic analysis of Numerical Concrete. It is evident that only very
high speed computers can solve the large system of equations within a
reasonable time (some hours). Up to now, only crack formation in very
simple composite structures has been simulated. More realistic composite structures can be analyzed and parameter studies can be undertaken at the beginning of 1986, when our institute will have a vec-
torial computer (CRAY-1).

Example 1

In Fig. 19 the results of the analysis of the crack formation and
propagation of a simple composite structure (only 1 inclusion) are
shown. This structure was submitted to a monotonically increasing

prescribed displacement of one of the boundaries, and the reaction force was calculated. At loading stage 1 the interface collapsed in a brittle manner and the reaction force dropped. At stage 2 the maximum load bearing capacity of the matrix was reached and the softening behaviour started. At stage 3 an unloading cycle was simulated to determine the hysteresis caused by the friction forces between the matrix and the aggregate grain. In this case the hysteresis caused by the energy consumption was small, as can be seen in Fig. 19.

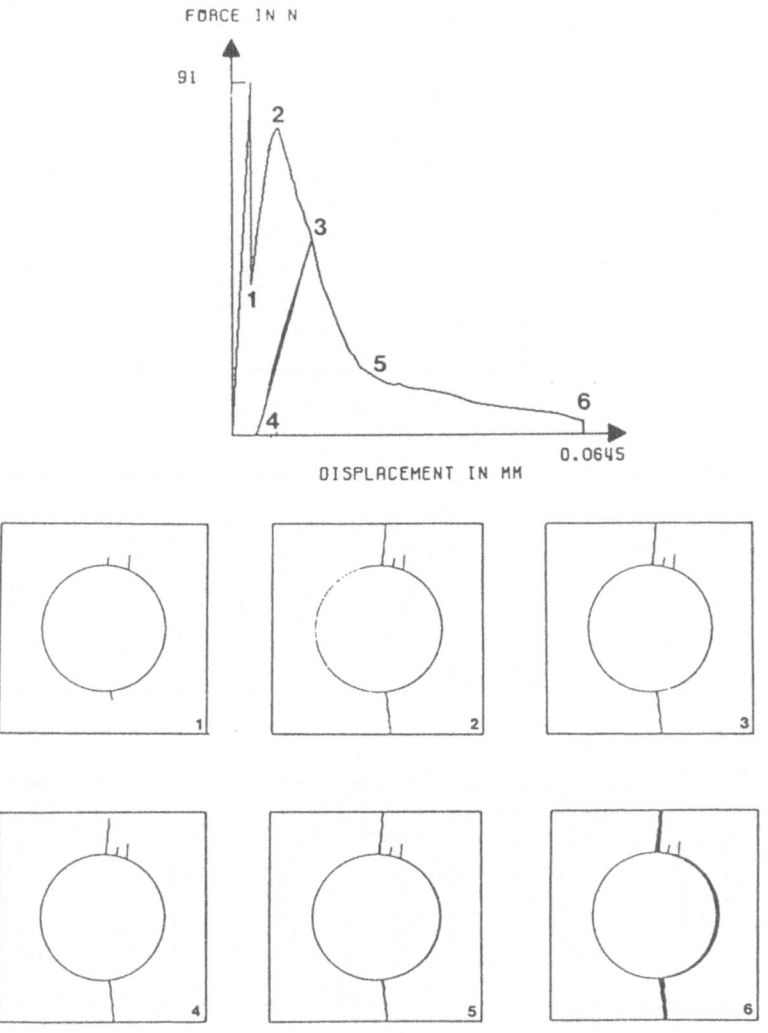

Fig. 19 : Calculated force-displacement diagram and crack patterns of different load stages.

The same structure was used to study the influence of an initial strain (e.g. caused by shrinkage or temperature) on the macroscopic failure behaviour. As can be seen in Fig. 20 the maximum load increased from 91 to 114 N. This can be explained by the prestressing effect of the interface (weakest link of the system) caused by the shrinkage strain. This effect is also observed experimentally.

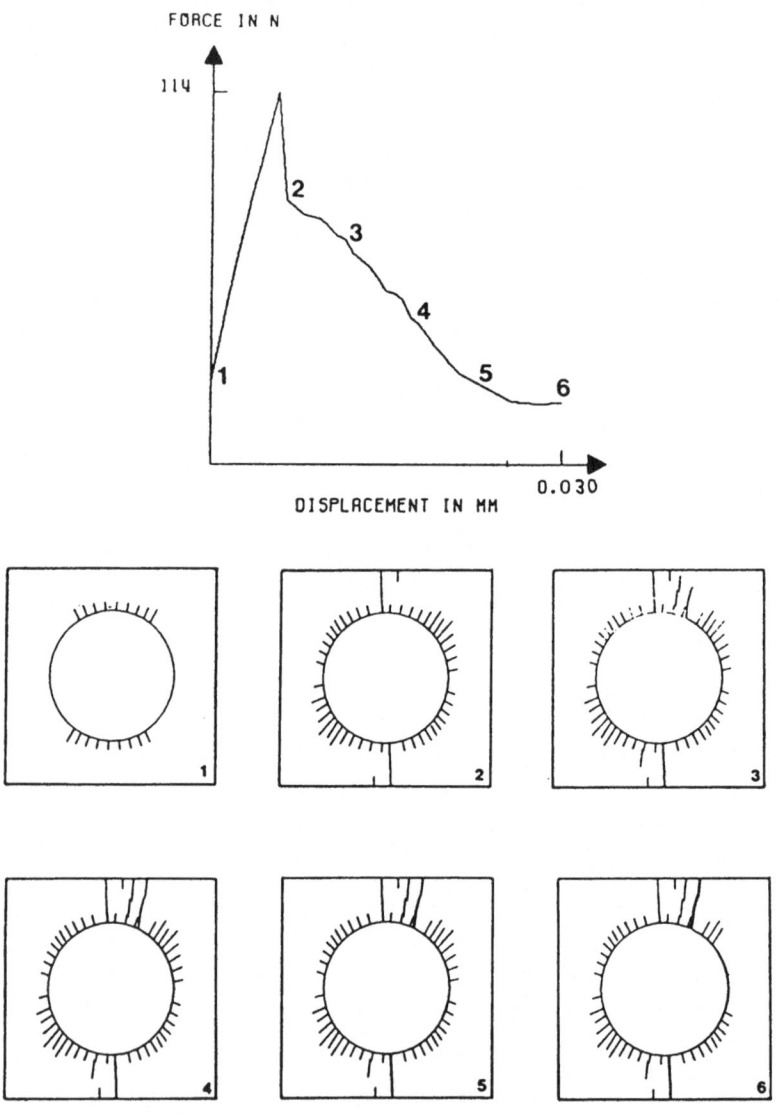

Fig. 20 : Calculated force-displacement diagram and crack patterns of different load stages, taking the effect of initial strain into consideration.

Example 2

Sadouki /19/ applied Numerical Concrete to study the influence of the shape of aggregate grains on failure behaviour. For that purpose two composite structures with each 19 aggregate grains were generated with the computer. The shapes of the aggregate grains in one of the structure were determined with a morphological law. In the other structure aggregate grains with spherical shapes were applied. The granulometry and aggregate content of each structure was the same. The structures were, like in the first example, submitted to an increasing prescribed displacement of one of the boundaries and the reaction forces were calculated.

Fig. 21 shows the distribution of areas with softening behaviour (indicated with small lines) and with microcracks (indicated with small lines marked with a point), both structures determined at an intermediate state of the analysis.

In Fig. 22 the calculated corresponding deformations of the structure with non-spherical aggregate grains is shown. The calculated load-displacement diagrams are compared in Fig. 23.

The comparison between the two diagrams shows that, up to a certain prescribed displacement, less energy (which is equal to the area under the curves) is consumed for crack formation in the structure with spherical aggregate grains than in the structure with non-spherical aggregate grains. This effect has also been observed experimentally.

Some improvements, however, must be made before Numerical Concrete can be used to develop realistic constitutive relations on the macro-level. For example, as can be seen in Fig. 23 the calculated reaction force does not gradually decrease to zero as the prescribed displacement increases. This can be explained as follows. In the analysis the major stress (which is the maximum of the principal stresses) and its direction are determined in each finite element. If the major stress exceeds the tensile strength, the corresponding direction is fixed and tensile strain softening takes place in that direction. The direction of the major stress does, however, not always coincide precisely with the fixed direction in the next loading steps. This is mainly due to redistributions of stresses and strains in the structure. The applied constitutive relation for the mortar matrix does not account for these situations. This means that some finite elements keep a small rigidity in the major stress direction. Hence, the global stiffness of the structure will not decrease to zero in the loading direction. This behaviour can be improved by using a simple isotropc strain softening relation, which can be justified on grounds of the relatively small size of the finite elements, or by a more realistic sophisticated multi-axial relation, which will consequently increase computer time.

Another improvement must be made with respect to the failure behaviour of the interfaces. The brittle failure behaviour of the interfaces causes sudden drops of the external force in the force-displacement diagrams, as can be seen in Fig. 19, 20, and 23. This behaviour resembles not completely the real failure behaviour of concrete, even if more aggregate grains are simulated in the struc-

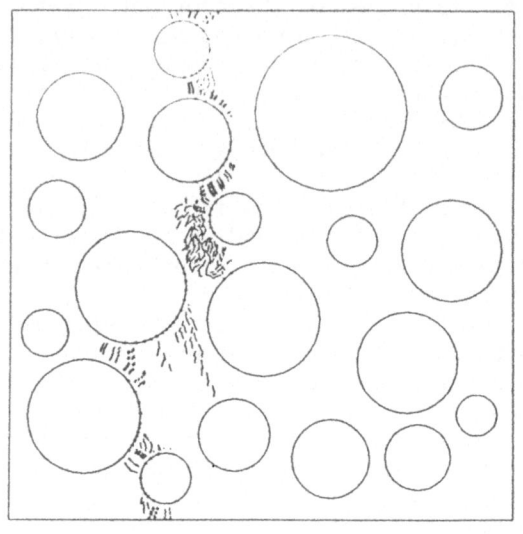

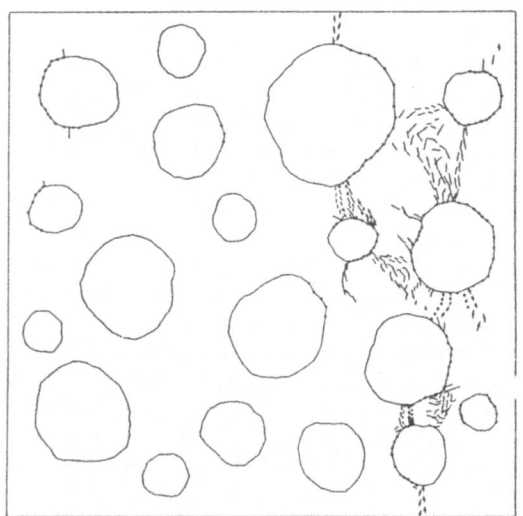

Fig. 21 : Distribution of areas with softening behaviour and
formed microcrack at an intermediate state of the
failure analysis of two different structures.

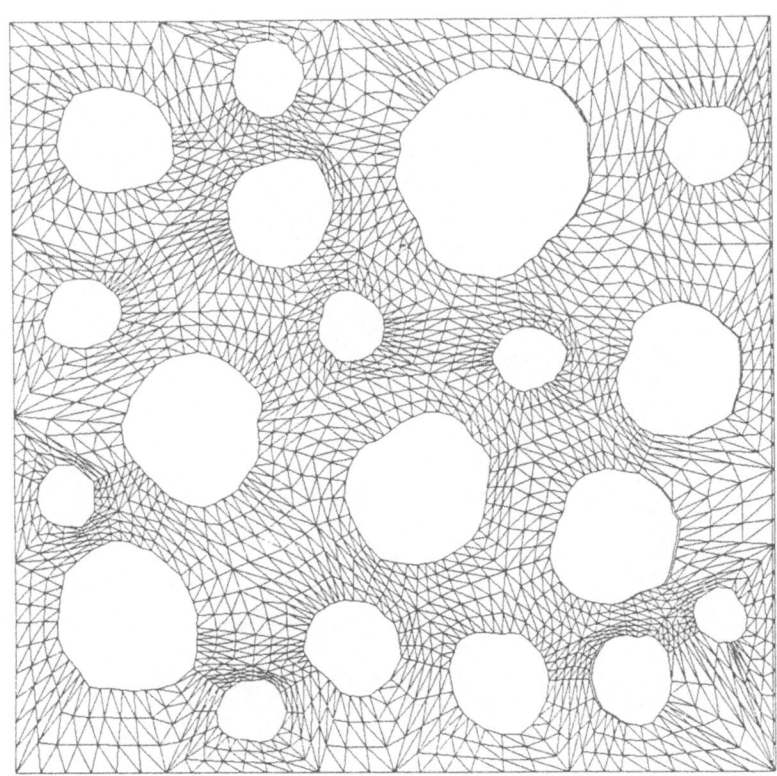

Fig. 22 : Calculated deformations of the structure with
non-spherical aggregate grains.

ture. The reason is that only a 2D analysis is carried out. A crack in
the 2D cylindrical interface runs necessarely across the entire thickness of the specimen. In a realistic 3D simulated structure crack
arresting in the third dimension is taken into consideration. This
inconvenience in the 2D analysis can be possibly overcome by assuming
an "effective thickness" of the interface as function of its direction
with respect to the direction of the applied load. Actually, studies
are undertaken to these improvements in the software modules of
Numerical Concrete. In future, Numerical Concrete will be extended to
3D. Comparison of the 2D and 3D versions of the model can demonstrate
whether the 2D analysis suffices to describe the complex real situation.

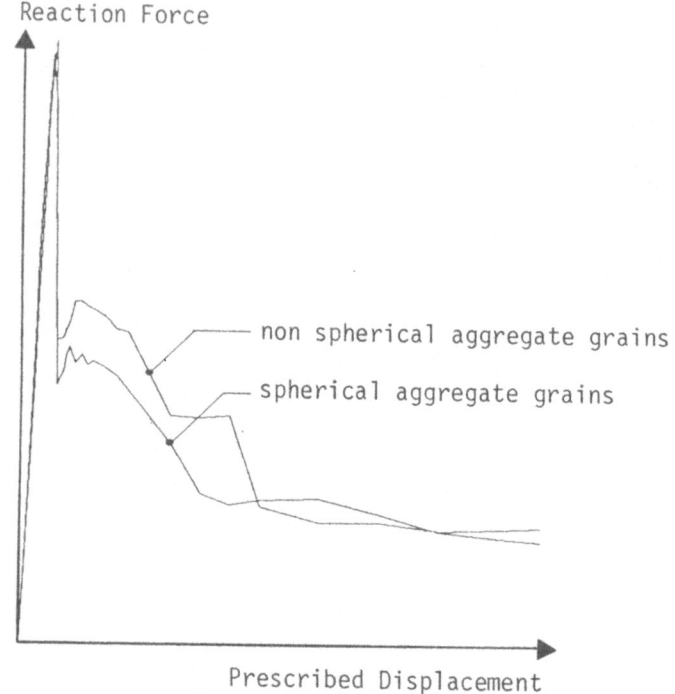

Reaction Force

— non spherical aggregate grains

— spherical aggregate grains

Prescribed Displacement

Fig. 23 : Comparison of the calculated load-displacement
diagrams.

4. CONCLUSIONS

In this contribution the basic concepts of four materials engineering
models are outlined. The methods which have been used to take the
heterogeneous character of the material into consideration are
explained in detail. The specific features of these models are briefly
summarized here:
- With the stochastic model a set of mathematical expressions is
  deduced which formulate the influence of heterogeneity, temperature,
  volume and rate of loading on mechanical properties.
- With the microplane model constitutive relations in 3D on the
  macro-level can be obtained which can handle complex loading paths.
- The simulation model based on LEFM can be used to study the influence of inclusions, pre-existing cracks, flaws, and porosity on
  crack propagation in and failure of composite structure.
- With Numerical Concrete the gradual failure of composite structures
  can be simulated. Initial states of stress and strain due to temperature and moisture gradients can be taken into consideration.

REFERENCES

1. WITTMANN, F.H., Grundlagen eines Modells zur Beschreibung charac-
   teristischer Eigenschften des Betons, Deutscher Ausschuss für
   Stahlbeton, Report Nr. 292, Wilhelm Ernst & Sohn, Berlin, 1976.
2. HILLERBORG, A., A Model for Fracture Analysis, Report TVBM-3005,
   Lund Institute of Technology, Division of Building Materials,
   Sweden, 1978.
3. HILLERBORG, A., MODEER, M., and PETERSSON, P.E., "Analisis of
   Crack Formation and Crack Growth in Concrete by Means of Fracture
   Mechanics and Finite Elements", Cement and Concrete Research,
   Vol. 6, 1976, pp. 773-782.
4. BAZANT, U.P., and OH, B.H., "Crack Band Theory for Fracture of
   Concrete", Matériaux et Constructions (Materials and Structures)
   RILEM, Paris, Vol. 16, 1983, pp. 155-177.
5. MIHASHI, H., and IZUMI, M., A Stochastic Theory for Concrete
   Fracture, Cement and Concrete Research, 7, pp. 411-422, 1977.
6. MIHASHI, H., and WITTMANN, F.H., Stochastic Approach to Study the
   Influence of Rate of Loading on Strength of Concrete, HERON, 25,
   The Netherlands, 1980.
7. MIHASHI, H., A Stochastic Theory for Fracture of Concrete, in
   Fracture Mechanics of Concrete, ed. by F.H. Wittmann, Elsevier
   Science Publishers B.V., Amsterdam, The Netherlands, 1983.
8. TAKEDA, J., TACHIKAWA, H., and FUJIMOTO, K., Influence of
   Straining Rate and Propagation Stress Wave on Deformation and
   Fracture of Concrete, Proc. of the Second Int. Conf. on Mechanical
   Behaviour of Materials, Boston, pp. 1468-1472, 1976.
9. BAZANT, Z.P., and OH, B.H., Model of Weak Planes for Progressive
   Fracture of Concrete and Rock, Report No. 83-2/448 m, Center for
   Concrete and Geomaterials, Northwestern University, Evanstone,
   Illinois, 1983.
10. BAZANT, Z.P., and OH, B.H., Microplane Model for Fracture Analysis
    of Concrete Structures, Proc. of the Symposium on the "Interaction
    of Non-Nuclear Munitions with Structures", U.S. Air Force Academy,
    Colorado Springs, pp. 49-55, 1983.
11. BAZANT, Z.P., and OH, B.H., Efficient Numerical Integration on the
    Surface of a Sphere, Center for Concrete and Geomaterials, North-
    western University, Evanstone, Illinois, 1982.
12. BAZANT, Z.P., and GAMBAROVA, P.G., Crack Shear in Concrete: Crack
    Band Microplane Model, Journal of Structural Engineering, ASCE,
    Vol. 110, No. 9, pp. 2015-2035, 1984.
13. ZAITSEV, Y.U., and WITTMANN, F.H., Crack Propagation in a Two-
    Phase Material such as Concrete, Proc. 4th Int. Conf. on Fracture
    (ICF-4), Waterloo, Canada, Vol. 3, pp. 1197-1204, 1977.
14. ZAITSEV, Y.U., and WITTMANN, F.H., Simulation of Crack Propagation
    and Failure of Concrete, Matériaux et Constructions, 14, No. 83,
    pp. 357-365, 1981.
15. ZAITSEV, Y.U., Fracture Mechanism and Strength of Concrete under
    Triaxial Compression, Proc. 5th Int. Conf. on Fracture (ICF-5),
    Cannes, France, pp. 2281-2282, 1981.

16. HU, X.Z., COTTERELL, B., and MAI, Y.W., Computer Simulation Models of Fracture in Concrete, Proceedings of the International Conference on Fracture Mechanics of Concrete, Lausanne, Switzerland, pp. 73-82, 1985.
17. WITTMANN, F.H., ROELFSTRA, P.E., and SADOUKI, H., Simulation and Analysis of Composite Structures, Materials Science and Engineering, 68, pp. 239-248, 1984-1985.
18. ROELFSTRA, P.E., SADOUKI, H., and WITTMANN, F.H., Le Béton Numérique, Matériaux et Constructions, 107, 1985.
19. SADOUKI, H., Simulation et Analyse Numérique du Comportement Mécanique des Structures Composites, Ph.D. Thesis, Swiss Federal Institute of Technology, Laboratory for Building Materials, Lausanne, to be published, 1986.

INDEX

Craze, craze zone, crazing 228, 232, 257-259, 263-267, 271-274, 276-283, 294-296, 298-301, 303, 304
Creep 87, 90, 93, 113, 114, 270, 273, 274, 296, 374
Creep crazes 304
Criteria for fracture initiation and propagation 1, 125, 182, 216, 352, 355. See also Directional criteria
Critical crack length 9, 173, 270
Critical crack opening displacement 232
Critical strain energy release rate 145, 211, 212
Critical stress 9, 89
Critical stress intensity factor 81, 157, 160, 214-219, 235, 243. See also Fracture toughness
Cross-link, cross-linking 291, 293, 300-302
Cross-link density 300, 301
Crystal lamellae 291, 292, 300
Crystalline 77, 78, 118
Crystallinity 291, 292, 298, 299
Cutting (wood) 213-216
Cutting loads 214, 215
Cutting tools 82
Damage 21, 26-31, 37, 38, 44, 58, 293, 335. Anisotropic d. 38. Continuous d. 26. Sudden d. 26
Damage criterion 22, 35
Damage fracture model 30
Damage front 27
Damage theory 21, 33, 36, 37, 335. Continuum d.t. 33, 58. Scalar d.t. 36
Damaged zone 26, 27, 29, 31, 36, 38, 42, 44
Deductive models 315
Defect 291, 292, 294, 296, 303, 304
Delayed fracture 84, 100, 173-175, 177
Diffusion 169-171, 173
Directional criteria 181, 193
Discretization 53
Disjoining pressure 317-320
Dispersion toughened ceramics 137, 150, 154
Displacement rate 164
Dissipation rate 22, 28, 34, 36
Drying process (concrete) 369
Drying stresses (wood) 213, 214
Ductile failure, ductile fracture 2, 23, 26, 28, 300, 304
Dugdale model 24, 231, 258, 265-267, 272, 273, 279, 283
Durability 369,
Dynamic effects 82, 255

Plane strain 7, 8, 24, 48, 51, 56, 162, 187, 199, 214
Plane stress 7, 8, 48, 51, 56, 161, 162, 199, 211, 214
Planing 213
Plastic deformation 2, 3, 293, 297, 298, 300
Plastic strain rate 22
Plastic work 25
Plastic zone 22, 24, 25, 27, 29, 235, 257, 265, 279
Plasticity 22, 79, 137, 235. Deformation p. 51. Incremental p. 51, 52.
    Perfect p. 29
PMMA 66, 235, 255, 257, 260-264, 266, 267, 269-280, 282, 283, 287, 288,
    295, 297, 301, 303
Poisson's ratio 7, 88, 130, 131, 184, 210, 365
Polycarbonate (PC) 294, 295
Polycristal 75, 79, 81, 88. Polycristalline 42, 79, 82, 118
Polymers 227-307. Amorphous p. 301. Brittle p. 257-290. Crystalline p.
    301. Glassy p. 294, 295, 297, 301. Semi-crystalline p. 295, 298,
    300. Thermoplastic p. 293, 294
Polymer strength 291, 292, 294, 295, 301
Polymer structure 292
Polymerization 275, 314
Pore 41
Pore size distribution 322-326
Porosity 77, 100, 132, 321-324, 334, 368, 382
Porous materials 322
Prestressing effect 378
Principal stress trajectories 181, 184, 190-192
Probabilistic fracture mechanics 84, 102, 103, 207, 220
Probability 223
Probability of failure (fracture) 84, 103, 107, 174, 175, 178, 221, 224
Probability of survival 86, 103, 225, 360, 361
Process zone, fracture p.c. 21, 23, 27, 28, 30, 89, 90, 138, 142-144,
    335, 341, 348, 352
Progressive fracturing 363
Proof test, proof testing 85, 87, 105, 107, 108, 110, 112, 157, 178
PVC 235, 238, 243, 255, 257, 266, 267, 275, 280-283
Rate of crack formation 361
Rate of loading 177, 293, 297, 362, 382
R-curve 21, 22, 24, 25, 30, 56, 60, 63, 90, 92, 131, 143, 335, 336, 341,
    342, 346, 348-350, 352
Reflection optics 257, 259
Relative humidity 319
Reliability 70, 84
Residual deformation 374

Residual stresses (internal tensions, self-stresses) 88, 89, 122, 138, 146, 149, 150, 181, 184, 304

Rheological mutation (glue) 220

Rocks 38, 43, 75, 76

Rubber 228

Running cracks 232

Scanning electron microscope 342

Scattering of strength values 82

Self-stresses. See Residual stresses

Shadow optical method of caustics 181, 267-269, 287

Shape memory effect 138, 151, 152

Shear crack 365

Shear failure 371

Shear modulus 210, 371

Shear retention factor 374

Shear yielding 228, 258

Shorts cracks 57, 129

Shrinkage 78, 368, 374, 378. Shrinkage cracks 87, 328

Silicon nitride 113

Simulation of Composite structures 332, 374

Singularity 50, 51, 54, 55, 211. $R^{-\frac{1}{2}}$ s. 54, 55. R^{-1} s. 54, 55

Sintering 76, 78

Size effect law 341, 342, 350, 355

Slicing 213

Slow crack growth: see subcritical c.g.

Softening behaviour 37, 39, 372

Softening diagram 372, 377, 379

Specimen: Bimaterial s. 193. Compact s. 48, 58, 60. Double-cantilever beam (DCB) s. 63-65, 90, 157, 159, 212, 217, 228. Double torsion (DT) s. 84, 157, 160-165, 229. Notched s. 345. Three-point bend s. 31, 49, 83. Unnotched s. 345

Speckle metrology 342

Spherulites 292, 299, 300

Splitting 213

SPT diagram: see Strength-Probability-Time diagram

Stable crack growth. See Subcritical c.g.

State parameter 360

Static fatigue 84, 157, 173-175

Statistical distribution of heterogeneities 359

Statistical effects 102, 174

Stochastic model 360, 382

Strain concentration 292

Strain energy 8, 50, 198-203, 211-213, 273, 356